Interfaces Between Nanomaterials and Microbes

Editors

Munishwar Nath Gupta
Former Emeritus Professor
Department of Biochemical Engineering and Biotechnology
Indian Institute of Technology Delhi
New Delhi, India

Sunil Kumar Khare
Institute Chair Professor of Biochemistry
Enzyme and Microbial Biochemistry Laboratory
Department of Chemistry
Indian Institute of Technology Delhi
New Delhi, India

Rajeshwari Sinha
Independent Researcher
New Delhi, India

CRC Press is an imprint of the
Taylor & Francis Group, an **informa** business
A SCIENCE PUBLISHERS BOOK

Cover credit: Image used on the cover has been taken from Chapter 10. Reproduced with kind permission of the authors.

First edition published 2021
by CRC Press
6000 Broken Sound Parkway NW, Suite 300, Boca Raton, FL 33487-2742

and by CRC Press
2 Park Square, Milton Park, Abingdon, Oxon, OX14 4RN

CRC Press is an imprint of Taylor & Francis Group, LLC

Library of Congress Cataloging-in-Publication Data

Names: Gupta, Munishwar Nath, editor. | Khare, Sunil Kumar, 1960- editor. | Sinha, Rajeshwari, 1986- editor.
Title: Interfaces between nanomaterials and microbes / editors, Munishwar Nath Gupta, Sunil Kumar Khare, Rajeshwari Sinha.
Description: First edition. | Boca Raton : CRC Press, Taylor & Francis Group, 2021. | "A science publishers book." | Includes bibliographical references and index.
Identifiers: LCCN 2020057666 | ISBN 9780367271824 (hardcover)
Subjects: MESH: Microbiological Phenomena | Nanostructures--microbiology | Nanostructures--toxicity | Anti-Infective Agents | Drug Carriers
Classification: LCC QR46 | NLM QW 4 | DDC 579--dc23
LC record available at https://lccn.loc.gov/2020057666

ISBN: 978-0-367-27182-4 (hbk)
ISBN: 978-0-367-70349-3 (pbk)
ISBN: 978-0-429-32126-9 (ebk)

Typeset in Times New Roman
by Shubham Creation

Preface

The fascination with nanomaterials started with material scientists. The reports of their early applications slowly spread the curiosity about these materials to workers belonging to other disciplines. Soon, biologists became interested. The molecular biologists among them were already working at the nano-level. Isolation, purification, and structural characterization of proteins, lipids, nucleic acids, and carbohydrates involved working with nanosized species. So, in a way, they were the first nanotechnologists! It is not surprising that many techniques, such as electron microscopy, ultracentrifugation, and light scattering techniques used by biochemists were adopted by nanoscientists.

Some realizations/concerns started trending early and since many of these continue to be important [and actually are the subject of various chapters in the present book], these are listed below.

- The high surface to volume ratio of nanomaterials was noticed and exploited in many applications.
- Their possible toxicity was a cause for concern.
- Use of organisms and other green methods for synthesis began to be explored.
- The nano-level materials are associated with some unique scientific properties; the simplest of this which attracted early attention is surface plasmon resonance.

A common thread running through some of the above was their interaction with microbes. This became very important as nanomaterials became part of the drug delivery and drug design. Today, theranostics combines both diagnosis/detection and therapeutics. This area testifies to the power of biologists, medical scientists, and nanoscientists working together. Much of all that has happened or is happening in the nanoscience/nanotechnology have the interactions of nanomaterials with microbes at its core. Hence, it was thought that it is a ripe time to bring out this book which can be used as a knowledge base by scientists aiming to join these multidisciplinary efforts, or even those who are already engaged in such efforts.

To make the book a standalone source for such people, the first two chapters provide a quick introduction to nanomaterials and microbes. The third chapter describes how various microorganisms [bacteria, algae, and protozoa] interact with metallic nanoparticles. The fourth chapter comprehensively looks at the toxicity aspects. The fifth and sixth chapters focus on nanomaterials based upon cellulose and titanium dioxide. The former has started drawing serious attention and the latter continues to throw up more and more interesting applications. Synthesis [with a focus on green routes] and applications in diverse areas are covered in the next few chapters. The last chapter deals with an old but certainly not yet outdated nanomaterial, that is, liposomes.

We thank all the authors who agreed to contribute to this book and graciously agreed to our suggestions regarding their chapters. We thank the CRC Press, Taylor and Francis Group, particularly, Mr Raju Primlani, for his help and outstanding patience. We hope the book will be found useful by all who work or intend to work in this area.

Editors

Contents

1

Introduction to the Microbial World

Munishwar Nath Gupta[1], Sunil Kumar Khare[2*] and Rajeshwari Sinha[3]

[1]Former Emeritus Professor
Department of Biochemical Engineering and Biotechnology
Indian Institute of Technology Delhi, Hauz Khas, New Delhi 110016, India
Email: mn7gupta@gmail.com

[2] Institute Chair Professor of Biochemistry,
Enzyme and Microbial Biochemistry Laboratory, Department of Chemistry
Indian Institute of Technology Delhi, Hauz Khas, New Delhi 110016, India
Tel: +91 11 2659 6533, Fax: +91 11 2658 1102
Email: skhare@rocketmail.com, skkhare@chemistry.iitd.ac.in

[3]Independent Researcher, M25-Kalkaji, New Delhi 110019, India
Email: rajeshwari.sinha@yahoo.com

INTRODUCTION: A PRIMER FOR MICROBIOLOGY

The first sentence in the excellent book by Frankel and Whitesides (2009) reads "Small is different". Very small is very different. Microorganisms are one such class which is very small indeed. The term micro comes from "micros", which means small. These organisms were so named because most of these cannot be seen with naked eyes. Various microscopic techniques have been critical to studies on microorganisms.

It was John Ray who coined the term "species" in the 17th century for living organisms. The current binomial system of nomenclature (the name has two words) was given to us by Carolus Linnaeus in the 18th century. While early scientists

*For Correspondence: skhare@rocketmail.com, skkhare@chemistry.iitd.ac.in

thought that there were two kinds of species (two kingdoms), Whittaker (1959), on the basis of the updated knowledge, proposed 5 kingdoms. These kingdoms are called: Monera, Protista, Plantae, Fungi, and Animalia. His classification used three parameters: cellular structure, structure of organism, and how an organism derived its nutrition (essential molecules for the metabolic activity) and energy. Before we further discuss the five kingdoms, let us first discuss these three criteria which define these kingdoms for the sake of clarity. Our discussion, of course, is based on the updated information since the times of Whittaker.

Microorganisms include unicellular organisms (archae, bacteria, some protista, some fungi, and some chlorophyta), few multicellular organisms (large fungi and some chlorophyta), and viruses. It is helpful to remember that classifying any organism as a microorganism had no biological criterion; it was purely a question of size. However, cell size is not an incidental property. Directly or indirectly, there are biological requirements which tailor the cell sizes.

The availability of the light microscope had revealed that at the cellular level, organisms can be called **prokaryotes** or **eukaryotes**. In 1977, Carl Woese proposed that archaebacteria are different from bacteria. Hence, many scientists accept that the primordial cell evolved into archaea, bacteria (including cyanobacteria), and eukarya (eukaryotes). The eukaryotes include protists, fungi, plants, and animals.

The primordial cell, the ancestor of all cells, is believed to have come into existence about 3 billion years ago. There are some conflicting views about the evolutionary relationship among archaea, present day bacteria, and eukaryotes. However, there is general agreement on archaea being the most ancient and these and bacteria evolved somewhere around 1.3–1.5 billion years ago. Initially, archaea were found in harsh environments, but have been later found in diverse habitats, and hence include both mesophiles and extremophiles. The major 3 groups are methanogens (who produce methane by reduction of carbon dioxide), halophiles (which occur and even require high saline conditions), and thermoacidophiles. It should be added that halophilic organisms are not restricted to archaea; there are bacterial halophiles, and even some eukaryotes such as algae *Dunaliella salina* or fungus *Wallemia ichthyophaga* are known. The niches for thermoacidophiles are hot springs and hydrothermal vents. Some archaea are also alkalophiles, but these are a subclass of halophiles, and are called haloacidophiles.

So, prokaryotes are of two kinds: archaea and bacteria. Prokaryotes, except planctomycetes, have free deoxyribonucleic acid (DNA) in the cytoplasm, which means that DNA is not enclosed in any membrane. Chromosomes are this DNA associated with some "histone-like" proteins, and may contain extrachromosomal DNA as part of the bacterial genome. Cell division is by binary fission; the exchange of genetic material takes place by conjugation, transduction, or transformation. Eukaryotes have 70S ribosomes. The specialized pathways like for ammonia oxidation or photosynthesis (in cases these pathways exist) are enclosed by internal membranes. Some prokaryotes have flagella (used for motion); these have a single protein called flagellin.

In eukaryotes, there are always more than one chromosome, and these have DNA bound with histones and are inside a nucleus enclosed in a membrane. In fact, there are many cellular organelles enclosed in membranes, such as golgi bodies, chloroplasts (in photosynthetic eukaryotes), and endoplasmic reticulum, the membranous structures present throughout the cytoplasm. Extrachromosomal DNA is present only inside organelles (e.g., mitochondria). Flagella (in those eukaryotes which have these) consist of multi-protein fibrils. The eukaryotic cells have 80S ribosomes.

There are a couple of aspects in which not only the two prokaryotes differ even between themselves, but even bacteria are not all similarly constructed.

CELL WALL AND CELL MEMBRANES

One of the earliest ways of classifying bacteria is by staining technique invented by Hans Christian Gram in 1894. Treated with the dye crystal violet and iodine solutions, and then followed by an alcohol wash, if cells retain the deep purple color, these are called gram-positive. Those which don't are called gram-negative bacteria. Treatment with another dye safranin turns these into pink, whereas this counterstain is hidden by the deep purple color of the stained gram-positive bacterial population. The distinct result is due to the outermost cellular component called cell wall. The gram-positive bacteria have a 20–80 nm thick cell wall, which is responsible for giving rigidity/shape as well as retaining the color of iodine. This is made of peptidoglycan (also called murein as its backbone chain has alternating groups of N-acetylglucosamine and N-acetylmuramic acid; the enzyme lysozyme hydrolyzes murein and hence is also called muramidase!) in which the carbohydrate backbone is crosslinked with amino acids and diaaminopimelic acid. Teichoic acid and teichuronic acids are also physically linked to this peptidoglycan layer.

The diverse behavior of bacteria is underlined by some gram-positive bacteria becoming gram variable. This may be the case with old cultures; in some cases, such as *Bacillus*, *Butylvibrio*, and *Clostridium*, growth results in thinning out of the peptidoglycan wall and cells not being able to retain the stain. In yet another kind called gram-indeterminate bacteria (notable examples being *M. tuberculosis* and *M. leprae*), gram stain gives unpredictable response. Both Mycobacteria and Nocardia do not get stained (by gram stain) as their cell wall has polymers of arabinogalactan esters (of mycolic acid) attached to the peptidoglycan backbone. Coryneform also have a similar kind of cell wall. Planctomycetes, again are an exception and do not have a peptidoglycan cell wall. Their shape instead is maintained by S-layers.

In gram-negative bacteria, the outermost part is a bilayer of lipopolysaccharides (LPS). The composition of LPS varies among species. Both kinds of bacteria have the cytoplasm enclosed in the bilayer of phospholipids (the well-known fluid mosaic model), in which proteins are interspersed (integral proteins) or bound to the outsides of the phospholipid bilayer (peripheral proteins). In gram-negative bacteria, in between the LPS bilayer and the inner biomembrane, there is the periplasmic space, and there is a thin (5–10 nm) murein layer present in that space. Some scientists prefer to call this outermost component cell envelope and reserve the term cell wall for the outer component of gram-positive bacteria.

Another exceptional case with respect to cell wall presence is that of the mycoplasma. These smallest (some as small as 0.3 μm in diameter) and simplest kind of bacteria have no cell wall, but only cell membrane.

Gram-positive bacteria without cell walls are called protoplasts; gram-negative bacteria without cell walls (but retaining the outer membrane) are called spheroplasts.

Many gram-negative bacteria (e.g., *N. meningitis*, *K. pneumoniae*, *H. influenzae*, *P. aeruginosa*, *Salmonella* sp.) and even some strains of *E. coli* have a capsule in addition to the cell envelope. These capsules are made up of polysaccharides (usually) or glycoproteins; *B. anthracis* has a capsule of poly (D-glutamic acid).

Most of the eukaryotes do not have cell walls, but their exterior is strengthened by microtubular cycloskeletons. Plants have cellulosic cell wall, and fungal cell walls are made of chitin (a polymer of N-acetylglucosamine which is partially deacylated; the degree of deacetylation varies with the fungus).

Some gram-positive bacteria also have capsules, but their composition varies with the bacteria. *B. megaterium* capsule has both protein and polysaccharide; *S. pyogenes* has a hyaluronic acid capsule.

Alternatively, a less dense slime layer is loosely associated with the bacterial cell (please also see the later discussion on biofilms). The term glycocalyx is used for this, but also often includes capsular structures. The composition/thickness of the slime layer depends upon the metabolic stage of the bacteria.

Archae exist under diverse hostile conditions, hence their cell walls also evolved into multiple kinds for survival. Thermoacidophiles in fact do not have cell walls, except *Sulfolabus* spp., which has cell wall made up of protein. Cell walls of methanogens have pseudomurein, in which N-acetylmuramic acid of peptidoglycan is changed to N-acetyltalosaminuronic acid and L-amino acids are linked to the polysaccharide backbone. Halobacteria have cell walls made up of proteins; an exception is *Halococcus* which has cell walls constructed out of heterpolysacharides. Consequently, while other halobacteria lyse at low salt concentration, this one does not!

Archaea especially differ from bacteria and eukaryotes with respect to the structure of their cell membranes (also called cytoplasmic membranes). In archaea, the phospholipids:

- Have sn-glycerol-1-phosphate as the backbone instead of sn-glycerol-3-phosphate (in bacteria and eukaryotes).
- Have fatty acids with long isoprenoid chains with branches which sometimes contain cyclopropane or cyclohexane rings. Bacterial (and those in eukaryotes) membrane phospholipids have straight chain fatty acids.
- Have fatty acids forming ether bonds (rather than esters) with two alcoholic groups of glycerol. In some cases, these two fatty acid chains are linked to another glycerol structure at other end (forming a closed structure). So, instead of diesters, archaea have either diethers or tetraethers.
- Are fused at the tails mostly to form monolayers of phospholipids, which may be interrupted by small stretches of bilayer.

The net result is that this membrane is lot more stable under harsh conditions, in which most archaea exist.

In general, archaea membranes have a lot less fluidity. While eukaryotic membranes have sterol (e.g., cholesterol) and bacterial cells have hopanoids (pentacyclic compounds) to disrupt regular bilayer structures, no corresponding molecules are there to modulate membrane fluidity in archaea.

Eukaryotes have many structures enclosed by the membrane: mitochondria, chloroplasts, golgi bodies, vacuoles, lysosomes, peroxisomes, glyoxysomes, and (as already mentioned) nucleus (which also has nucleolus), and arrays of internal membranes called endoplasmic reticulum. Eukaryotes, unlike prokaryotes, also have microtubular structures which are part of flagella, cilia, basal bodies, mitotic spindle, and centrioles.

NUTRITION AND ENERGY METABOLISM

All organisms require nutrition to survive and grow. In most cases (especially higher organisms), the nutrient molecules undergo fermentation, or are oxidized to produce the required energy. Most of this energy is used up for biosynthetic purposes and these anabolic processes involve reduction, and nutrient via catabolism is indirectly also the source of reduced form of Nicotinamide Adenine Dinucleotide (NAD) coenzymes. Even nutrition and energy metabolism vary over a wide range to reflect microbial diversity. The best way to understand various terms used in this context is to consider the following:

a) All terms have a suffix-troph (of Greek origin, it means ‘to feed’)
b) Source of energy
c) Source of carbon
d) Dependence on oxygen.

Source of Energy

Based upon source of energy, there are **Phototrophs** (from light), **Lithotrophs** (inorganic chemicals as energy source), and **Heterotrophs/Organotrophs** (organic compounds as source of energy). Energy from chemicals is derived by a redox process in which there is an electron donor (reducing agent which gets oxidized) and an electron acceptor (oxidizing agent which gets reduced). In lithotrophs, inorganic chemical is the electron donor. Often, the electron donor (the chemical being oxidized) becomes part of the description of lithotrophs. Thus, there are two kinds of ammonia-oxidizing bacteria (nitrosifyers and nitrite-oxidizing bacteria), and sulfur-oxidizing bacteria (*Thiobacillus* sp., *Sulfolobus* sp., *Beggiatoa* sp., etc). Those of the latter, which grow at acidic pH (*Sulfolobus* sp., *T. ferrodoxans*, for example), can also use ferrous ions as electron donors. The various species of sulfur (sulfur element, hydrogen sulfide, other sulfides or thiosulfates) are used by this class. Lithotrophs also include hydrogen oxidizing bacteria; these can be either gram-positive or gram-negative, but all contain hydrogenase enzyme. In these lithotrophs, oxygen is the ultimate electron acceptor. Organotrophs include **methanotrophs** (which oxidize methane) and the broader category of **methylotrophs**, which use 1-C compounds (methanol, methylamine, formate, dimethyl ether, carbon monoxide, dimethyl carbonate, etc.) to grow. These have oxygen as the ultimate electron acceptor. Many organotrophs have sulfate ion as the electron acceptor, but use organic acids, fatty acids, alcohols, and even hydrogen as the electron donor; these are all called sulfate reducing bacteria. These are to be distinguished from sulfur-reducing bacteria, which simultaneously oxidize compounds, such as acetate or ethanol. Homoacetogenic bacteria use carbon dioxide as the electron acceptor and utilize a variety of electron donors to produce acetate as the single product of anaerobic process.

Source of Carbon

As is obvious from the terms associated with the energy source, the issue of the source of carbon cannot be often decoupled with the source of energy. Thus, different terms are also used to describe organisms. For the sake of clarity, these are also explained at the risk of some overlap in the discussion.

Autotrophs are able to reduce carbon oxide to obtain the organic compounds needed as the starting building blocks. These may obtain energy from light (**Photoautotrophs)** or by oxidation of the available inorganic elements (**Chemoautotrophs).** Examples of photoautotrophs are purple sulfur bacteria, green sulfur bacteria, and cyanobacteria. These all are able to carry out photosynthesis. Chemoautotrophs include few bacteria involved in nitrogen cycle and nitromonas. Chemoautotrophs are also sometimes called chemolithotrophs ('lithos' in Greek means stone and the reference is to them getting energy from oxidation of inorganic compounds). **Heterotrophs** cannot produce all needed organic compounds themselves and use organic compounds produced by autotrophs for sourcing their carbon. Few organisms (such as purple non-sulfur bacteria) get energy from light but cannot fix carbon oxide; these use acetate or formate or methanol, etc. (simple organic compounds) as their carbon source and are called **Photoheterotrophs. Chemoheterotrophs** require oxidation of the metabolites to obtain energy, and also require organic compounds as a source of carbon. Majority of living organisms come under photoautotrophic and chemoautotrophic classes; most of the microorganisms are chemoautotrophs.

Dependence on Oxygen

Tied to nutrition and energy metabolism is the issue of the various ways the microbial growth is affected by the presence of oxygen. **Obligate aerobes** require the presence of oxygen for their growth. **Obligate anaerobes** cannot tolerate the presence of oxygen. Spores of those obligate anaerobes which are capable of forming them can survive, however, even in the presence of oxygen. Also, these can often be isolated from sites which are aerobic, but also have enough facultative anaerobes (see below for their definition) to use up available oxygen. An abscess with a good blood supply is an example of such a site (Heritage et al. 1999). **Aerotolerant anaerobes** are microbes whose growth is not affected by the oxygen presence. **Facultative anaerobes** can grow even in the absence of oxygen, but their growth is better if oxygen is present. These have both fermentative and respiratory pathways for energy generation. **Microaerophiles** are aerobes which require small level of oxygen (2–10.5%) to grow, but cannot tolerate normal atmospheric level of oxygen. *Neisseria gonorrhoeae* exemplifies another peculiar behavior of requiring oxygen, but also increased level of carbon oxide (5–10%). Oxygen involvement produces superoxide radical during metabolism. If present, superoxide dismutase converts this highly reactive species to hydrogen peroxide, which in turn is decomposed by catalase to water and oxygen. Obligate anaerobes lack both dismutase and catalase, while aerotolerant anaerobes have dismutase but not catalase.

EXTREMOPHILES

Microbial life survives under extreme conditions of various kinds. These extremophiles survive under extreme temperatures and extreme pH (**Thermophiles, Psychrophiles, Acidophiles, Alkaliphiles**), dessicated conditions (**Xerophiles**), high weight or pressure (**Barophiles, Piezophiles**), radiation (*Dienococcus radiodurans*), and high salinity (**Halophiles**) (Rothschild and Mancinelli 2001). Many archaea, cyanobacteria, and green algae such as *Dunaliella salina* can thrive in saturated solutions of sodium chloride. *Ferroplasma acidarmanus* survives not only pH around 0, it does so when simultaneously many metal ions are present. It has no cell wall, only cell membrane. These extremophiles represent exquisite design at both cellular and molecular level and testify to what adaption can achieve. In that respect, considering Homo sapiens as the highest form of evolution overlooks that evolution is an exercise in trade-off between various traits.

MICROBIAL KINGDOMS

With above background in place, we can revert to our discussion of the three kingdoms which have microorganisms, the other two plantae and animalia are not discussed, as these are not relevant to our subject.

Monera

This kingdom includes all prokaryotic organisms: archaea, bacteria (including filamentous bacteria *Actinomycetes* spp. and photosynthetic bacteria *Cyanobacteria* spp.). Some bacteria also have flagellum (e.g., *Rhizobium* sp.) or flagella (e.g., *Azetobacter* sp.); slime layer/capsule; pigments associated with membranes (e.g., *Chromatium* sp.); plasmids; filamentous structures: one or many pili or shorter fimbriae (e.g., *E. coli*, *Salmonella* sp.), and folded invaginations of cell surface membranes called mesosomes. Most of the bacteria have the diameter in the range of 0.5–1.0 µm, and all are in the range of 0.2–100 µm. Shape-wise, these are cocci (spherical), bacilli (straight rods), spirilla (curved helically), and vibrios (comma-shaped). In some cases, these are found to occur in association in different ways. Notably, diplococci (in pair), streptococci (chain-like formation), staphylococci (bunches of cocci). Bacteria reproduce/multiply asexually by a process called binary fission. Several (including the well-known *E. coli*) display bacterial conjugation in which plasmids are transferred, exhibiting genetic recombination, which is one of the ways of diversification of organisms in nature.

Protista

This kingdom includes fungi (molds and yeasts), algae, and protozoa. Many protista are microscopic in size and hence are included among microorganisms (Caron et al. 2009). Unicellular eukaryotes can be classified into autotroph algae, decomposers (slime molds), and predator protozoans. In adverse conditions, some protists (e.g., euglena when water around them dries up) form cysts (metabolically inert with a wall formed from their secretions), which revert to the normal form once the stress condition disappears.

Fungi are heterotrophic. Those decomposing dead plant or animal matter for deriving their nutrition are called saprophytes. Others are parasites (living in or on another living organism); many of these cause diseases. Most of the fungi are obligate aerobes, though few yeasts are facultative anaerobes. Fungi reproduce both asexually (multiple modes are known) or sexually. Fungi have thread-like structures called hyphae which entangle to form mycelium. Fungi have 5 categories.

- Mixomycetes: Also called slime molds
- Phycomycetes: Their habitat is water; both saprophytes and parasites are there. Examples are *Mucor* and *Albugo*
- Ascomycetes: Can be unicellular or multicellular, characterized by a cup-shaped fruiting body called ascocarp. Examples are yeasts, *Penicillium*, and *Aspergillus*
- Basidomycetes: Characterized by basidium (club shaped end of mycelium); common examples are mushrooms (*Agaricus*), smut (*Ustalago*), and rust (*Puccinia*)
- Deuteromycetes: These have only asexual reproduction and are also sometimes called Fungi Imperfecti, as their hyphae remain separate and don't form mycelium. Examples are *Alternaria, Helminthosporium*, and *Colletotrichum.*

Algae

Another diverse group, they have six major divisions, but three (Rhodophyta{red algae}, Phaeophyta {brown algae}, and Chlorophyta {green algae}) actually belong to the plant kingdom and are not discussed here. All protistan algae are photosynthetic, and are estimated to be responsible for 80% of carbon oxide fixed in the biosphere.

- Euglenophyta (euglinas): Unicellular, few form colonies; have flagellate but no cell wall. Only asexual reproduction and contain only chlorophyll as the pigment. Normally found in freshwater ponds, ditches, and even in moist soils
- Bacillariophyta (diatoms): Again, unicellular with few forming colonies, have a cellulose impregnated with silica as walls. It is the fossilized form of their wall that is called diatomaceous earth. Diatoms have both chlorophyll and brown pigments. Both asexual and sexual reproduction is known. Diatoms constitute an important part of phytoplanktons in oceans
- Dinophyta (dinoflagellates): These are either unicellular or filamentous; have cellulosic cell walls; mostly asexual reproduction, but sexual reproduction is also known. Some are heterotrophic, others have chlorophyll. They have two flagella and mostly marine organisms. *Gonyaulax*, when its number at sea surface increases, suddenly give rise to "red tide". Some dinoflagellates are phosphorescent and are responsible for the glowing sea surface in the dark.

Protozoa

These are important in the food chain of aquatic communities. There, and in wetlands, these are either saprophytic or ingest bacteria. The four groups are-Zooflagellates, Amoeba (sarcodines), Sporazoans, and Ciliates.

- Zooflagellates do not have walls, and reproduce asexually. Some important parasites in this group are trypanosomes (causes African sleeping sickness), leishmania (causes kala-azar). Others which show mutualism are exemplified by those which are present in the guts of termites and wood cockroaches
- Amoeba and related sarcodines: These organisms have diverse kinds of extensions from the main body (like pseudopodia in amoeba, which uses them for locomotion as well), which trap and engulf food particles. Their frequent habitats are bottom of pools of water or on objects submerged in water. They reproduce asexually. Some like amoeba have only cell surface membrane, others like heliozoans (which occur in fresh water) have silica shells through the pores of which filopodia project to get their prey. The genus entamoeba occur in intestines of vertebrates
- Sporazoans are all parasites for one or more animals. These often have complex life cycles with different stages in different hosts and switching between asexual and sexual forms. They feed on the cells/body fluids of the hosts. The most well-known example is *Plasmodium*, which causes malaria.

Ciliates

These are heterotrophic and generally feed on bacteria, single cell algae, and detritus swept into their funnel shaped oral grooves. Generally free, but are known to be parasitic as well. Occur in diverse habitats with some water in it: oceans, rivers, lakes, ponds, and soils. Their size ranges from 10 μm to 4 mm in length. Because of their cilia, these can move fast and change directions freely. Their outer cover is a thin flexible pellicle made up of proteins which surrounds the surface membranes. Common examples are *Paramecium* and *Colapada* (feeds even on paramecia!). Both asexual reproduction and sexual reproduction (by conjugation) are known; sporulation is very rare. Most ciliates have two nuclei, a macronucleus with many sets of chromosomes, and a diploid micronucleus involved in reproduction.

Viruses

Species belonging to all five kingdoms are hosts for viruses which are "obligate intracellular parasites'' (Baker et al. 2011). Viruses can replicate only inside host cells. Nucleoproteins which insert their DNA or ribonucleic acid (RNA) genome during infection and exploit the host machinery for multiplying, are considered the border between living and nonliving. Size-wise too, they are at the border of nano-world; generally, they are 20–400 nm but a few filoviruses (RNA viruses which cause hemorrhagic fevers in humans and primates and include the dreaded Ebola virus) can be as long as 1000 nm. The more complex viruses have envelope (the glycoprotein membrane surrounding capsid, the protein coat with nucleic acid in its interior). Herpes simplex virus has a loose-fitting envelope; rabies virus and HIV have a tight-fitting envelope. In general, the matrix; the space between capsid and envelope has virally encoded proteins.

Among the viruses infecting microorganisms, bacterial viruses, more commonly called bacteriophages or simply phages, are most important. Bacteriophages are either lytic (virulent) or temperate (avirulent). In the former, host cells are destroyed, and the multiplied population of the virus is released at the end of the incubation period. In temperate viruses, replication in the host continues in generations of the host bacteria (lysogeny). In lysogeny, the viral DNA gets integrated with DNA (to become a prophase) or extrachromosomal part of the host. Sometimes, viral DNA dissociates, and lysis takes place. In fact, lysogeny has many variants, especially with some filamentous phages. In chronic cases, the virus particles slowly leak (without host lysis). In pseudolysogeny, replication rate of virus is very high. In polylysogeny, more than one kind of virus infect the bacterial cell (Warwick–Dugdale et al. 2019).

While bacterial viruses have a wide distribution; they are the "most abundant biological entities in the ocean" (Warwick–Dugdale et al. 2019). Working with *Pseudoalteromonas* infected with a filamentous phage, it was found that the infection "increased the fitness of the infected cell phenotype" by restricting its growth under the arctic conditions of limited nutrition (Warwick–Dugdale et al. 2019).

This work throws up two indications which may have general consequences. Firstly, considering that microbiome (or microbiota as it is called frequently) has been found to have profound influences on human health (e.g., via gut-brain and gut-immune system axes), the role of viral infection in affecting it throws up interesting opportunities in engineering the gut biome. Secondly, to the old debate about viruses being "living" or potential living systems, or nonliving, this adds another conundrum: do viruses switch more frequently from being a parasite to aiding the organism in surviving? For all we know, the latter role of viruses may not be limited to "high seas"!

There is another important aspect in which prokaryotes, eukaryotes, and viruses differ. The old belief that proteins are structured molecules has undergone revision in the last decade. Proteomes of all organisms have some proteins which have significant level of disorder. In general, viruses and eukaryotes have considerably larger content of these intrinsically disordered proteins (IDPs) or proteins with intrinsically disordered regions. In the case of eukaryotes, many regulatory proteins or those involved in signal transduction are IDPs. Not only is this disorder necessary for the biological function, it is believed that this disorder is necessary in proteins for them to diversify/evolve. Considering that viruses have a much higher rate of mutations and can evolve rapidly, it is not surprising that high disorder in their proteins is a part of their survival mechanism. Finally, protein disorder is an extreme case of protein flexibility. The trade-off between stability and activity in proteins is also part of the survival mechanism of thermophiles. The enzymes from thermophiles have low catalytic efficiency as their flexibility is poor, and this in turn contributes to the higher thermostability of these proteins.

Prions

Disordered structure in proteins is associated with the proneness to form abnormal protein deposits called amyloids (which are associated with several neurodegenerative diseases). Amyloids are also associated with transmissible spongiform encephalopathies (TSE), which are caused by prions which have just proteins and not even nucleic acid. Prions use the genetic machinery of the host to replicate. We still do not have clarity on what switches a normal prion to become infectious and an amyloid forming form. Again, these share the grey zone of living/nonliving like viruses, but are the last frontier of microbiology.

MICROBIAL COMMUNITIES

It is widely believed that while evolutionary transition from single cell organisms to multicellular organisms lost the "minimalistic lifestyle" coupled with swift adaptability, it brought in more sophisticated regulation via collaboration among cells. Perhaps, unicellular organisms forming colonies was a step in that direction. Alberts et al. (2002) cite the example of staying together of Myxobacteria in loose colonies, in which the hydrolases secreted by individual organisms are pooled for degrading the available organic matter in the soil; a cooperation described as "wolf-pack effect". Along with other examples, the book also mentions division of labor within colonies of *Volvox* cells.

Another good example of this cooperation between individual members of an organism is the formation of biofilms. Planktons are collections of organisms drifting together in large water bodies, but planktonic form now also includes unattached (to a surface) organisms in suspension. It has been found that when a bacteria in planktonic form becomes a part of biofilm, some genes (e.g., expressing flagella molecules) are downregulated, while others (related to slime product formation) are upregulated. Let us look at biofilms in little more detail.

Biofilms

Archaea, bacteria, protozoa, fungi, and algae alone or together can form this cross-feeding and functional community, which in many ways behaves like a single multicellular organism. These aggregates can be like a free mat or (often) anchored to a surface. There are water channels running through the biofilm, which allows flow of nutrients and waste products. An essential part is extracellular polymeric substances (EPS), in which cells are embedded. EPS consists of proteins, carbohydrates, lipids, nucleic acids, and almost any other kind of material which gets picked up. This is produced by microbial cells, but can also be a product of host (anchoring surface), such as a plant root or animal epithelium.

The main and essential constituents of EPS in case of most of the prokaryotes (except yeast and fungi where these are less common) are exopolysachharides. These form part of the biofilm as capsular structure around microbes, but have been observed as loose slime associated with outer surface of microbes. The exopolysachharides include glucans (scleroglucan, curdlan, cellulose, dextran), alginates, pullulan, emulsan, gellan, hyaluronic acid and heparin, xanthan, and many others with more complex structures having organic and inorganic substituents (Sutherland 1990).

Biofilms seem to have evolved at least 3.5 billion years ago (Hall–Stoodley et al. 2004), as these have been found on old fossils. They are also ubiquitous, as these have been found on land, underground, deep sea vents, and hot springs. Their presence on medical devices and implants is fairly common, and has been causes of deep concern to the medical community. The reason is that by becoming part of biofilms, microbes become more resistant to numerous stress conditions, such as lack of water, extreme pH, and to heavy metals and even antibiotics. EPS is largely responsible for most of such attributes. It even protects bacteria from damage due to ultraviolet radiations.

Biofilms are multilayered structures. The secretion of larger amounts of EPS by the community members, adequate supply of oxygen (in case of aerobics), less shear stress (like still water in a pond rather than heavy water currents over the anchored biofilm) facilitate build-up of layers. A phenomenon called "seeding dispersal" involves breaking away a part of biofilm, which can become a starting point of another biofilm elsewhere (Kaplan 2010). Seeding dispersal is the last step in the life cycle of biofilms. This is an important way microbes move from environmental niche to the humans, and spread within the body and among the population. The seeding dispersal has several mechanisms, including ones based upon signal transduction. While it is responsible for the spread of the infection, it also destroys the existing biofilm, and hence inducing it is of interest from the viewpoint of controlling infection. Enzymes such as Dispersin B and DNAase cause dispersal. Nitric oxide does that in several cases, and *cis*-2-decenoic acid induces dispersion of biofilms containing *P. aeruginosa* and *Candida albicans*. About

17 million new infections involving biofilms occur per year in USA alone, and are reported to result in nearly 550,000 fatalities per year (Wolcott and Dowd 2011). Bacterial and fungal biofilms can form on prosthetics, catheters/pacemakers. Some well-known examples of infections involving biofilms include formation of dental plaques (*S. mutans* and *Candida albicans* are prominent infective agents), vaginitis, oropharyngeal candidiasis, pneumonia in cystic fibrosis, and infective endocarditis.

Bacterial populations, both in suspensions and frequently inside the biofilms, have small amounts of metabolically inactive cells (resting cells) called persistor cells. Exposure to antibiotics is known to increase their percentage from 0.01–80 (Wood 2017). This is yet another dimension to the challenge of treating antibiotic resistance infections which involve biofilms.

Cells, upon becoming a part of a biofilm, undergo a phenotypic shift. A biofilm is a regulated consortium with physiological heterogeneity (cells in stationary phase to active phase) due to mass transfer limitation (of nutrients) over a very short distance within the biofilm. There is an intracellular communication including quorum sensing. If both autotrophic and heterotrophic organisms are present, cross-feeding takes place. Baty et al. (2000) have discussed an example of division of labor chitin substratum degraded by one set of microbes to provide nutrient to a "detached planktonic subset".

Biofilms have found numerous biotechnological applications. The early example of immobilized biocatalyst was microbes encased in biofilms (Tsoligkas et al. 2011, Todhanakasem 2017). They have been used in bioremediation, nitrification, and in design of microbial fuel cells (by forming them on electrodes) (Dinamarca et al. 2018, Bayat et al. 2015, Schramm et al. 1996, Gatti and Milocco 2017).

A NEW TAXONOMICAL CLASSIFICATION?

Moving even further away from evolutionary taxonomy by using the molecular biology-based techniques, a monophyletic (based upon a single ancestor) taxonomy is being developed (Baker et al. 2011). For example, doing away with the five kingdoms, the three supergroups of eukaryotes being proposed are: Archaeplastida, Excavata, and Chromalveolata. These developments are likely to impact interfacial areas, like how microorganisms interact with nanomaterials in the coming years; meanwhile, the above primer should be more than adequate for nanotechnologists to navigate through the microbiological space.

A READY RECKONER FOR SIZES

Table 1.1 Size estimates

Size estimates	*Size of particles*
1 µm = 10^{-6} m = 10^{-4} cm = 10^{3} nm = 10^{4} Å	Solutions: about 1 nm
1 nm = 10^{-9} m = 10^{-3} µm = 10 Å	Suspensions: >1000 nm
1 Å = 10^{-10} m = 10^{-8} cm	Colloids: 10–1000 Å or 1–100 nm
1 pm = 10^{-12} m	

Table 1.2 Sizes of nanostructures associated with the human body

Nanostructure	*Size* (nm)
Glucose	1.0
Deoxyribonucleic acid	2.2–2.6
Average size of protein (rubisco monomer)	3.0–6.0
Haemoglobin	6.5
Micelle	13.0
Ribosomes	25.0
Enzymes and Antibodies	2.0–200.0

Source: Jeevanandam et al. 2018.

Table 1.3 Sizes of various microorganisms

Microbe	*Size* (μm)
Gram-negative bacteria	0.5–5.0
Gram-positive bacteria	0.5–5.0
Fungal spores	2.0–10.0
Pollen	6.0–100.0
Microphytes	1.0–500.0

Source: Westmeier et al. 2018.

Table 1.4 Sizes of various microorganisms

Microbe	*Size*
Bacteria	0.2–100.0 μm
Fungi (yeasts)	5.0–10.0 μm
Fungi (molds)	2.0–10.0 μm by several mm
Protozoa	2.0–200.0 μm
Algae	1.0 μm to many feet
Viruses	0.015–0.20 μm

Source: Pelczar et al. 2010.

Some Interesting Facts

- Human eyes can see objects which are ~0.1 mm long; for seeing smaller objects microscopes are needed.
- Light microscopes can see objects which are 500 nm or bigger in size; for anything smaller, electron microscopes are required.
- Human hair has a thickness of about 100 μm
- A haemoglobin molecule has the width of about 64 Å, whereas haem itself has a diameter of just 12 Å.
- Benzene molecule has a diameter of about 6 Å.

Note: micrometer (μm); meter (m); centimeter (cm); nanometer (nm); Angstrom (Å); picometer (pm).

References

Alberts, B., A. Johnson, J. Lewis, M. Raff, K. Roberts and P. Walter. 2002. Molecular Biology of the Cell, (Fourth edition). Garland Science, New York.

Baker, S., J. Nicklin and C. Griffiths. 2011. Microbiology (Fourth edition). Garland Science: Taylor and Sciences Group, New York.

Baty, A.M., C.C. Eastburn, S. Techkarnjanaruk, A.E. Goodman and G.G. Geesey. 2000. Spatial and temporal variations in chitinolytic gene expression and bacterial biomass production during chitin degradation. Appl. Environ. Microbiol. 66: 3574–3585.

Bayat, Z., M. Hassanshahian and S. Cappello. 2015. Immobilization of microbes for bioremediation of crude oil polluted environments: A mini review. Open Microbiol. J. 9: 48–54.

Caron, D.A., A.Z. Worden, P.D. Countway, E. Demir and K.B. Heidelberg. 2009. Protists are microbes too: A perspective. ISME J. 3: 4–12.

Dinamarca, M.A., J. Eyzaguirre, P. Baeza, P. Aballay, C. Canales and J. Ojeda. 2018. A new functional biofilm biocatalyst for the simultaneous removal of dibenzothiophene and quinoline using *Rhodococcus rhodochrous* and curli amyloid overproducer mutants derived from *Cobetia* sp. strain MM1IDA2H-1. Biotechnol. Rep. 20: e00286.

Frankel, F.C. and G.M. Whitesides. 2009. No Small Matter. Harvard University Press, Cambridge.

Gatti, M.N. and R.H. Milocco. 2017. A biofilm model of microbial fuel cells for engineering applications. Int. J. Energy Environ. Eng. 8: 303–315.

Hall–Stoodley, L., J.W. Costerton and P. Stoodley. 2004. Bacterial biofilms: From the natural environment to infectious diseases. Nat. Rev. Microbiol. 2: 95–108.

Heritage, J., E.G.V. Evans and R.A. Killington. 1999. Microbiology in Action. Cambridge University Press, Cambridge.

Jeevanandam, J., A. Barhoum, Y.S. Chan, A. Dufresne and M.K. Danquah. 2018. Review on nanoparticles and nanostructured materials: History, sources, toxicity and regulations. Beilstein. J. Nanotechnol. 9: 1050–1074.

Kaplan, J.B. 2010. Biofilm dispersal: Mechanisms, clinical implications, and potential therapeutic uses. J. Dent. Res. 89: 205–218.

Pelczar, M.J., E.C.S. Chan and N.R. Krieg. 2010. Microbiology (Fifth edition). Tata McGraw-Hill, New Delhi.

Rothschild, L.J. and R.L. Mancinelli. 2001. Life in extreme environments. Nature 409: 1092–1101.

Schramm, A., L.H. Larsen, N.P. Revsbech, N.B. Ramsing, R. Amann and K.H. Schleifer. 1996. Structure and function of a nitrifying biofilm as determined by in situ hybridization and the use of microelectrodes. Appl. Environ. Microbiol. 62: 4641–4647.

Sutherland, I.W. 1990. Biotechnology of Microbial Exopolysachharides. Cambridge University Press, New York.

Todhanakasem, T. 2017. Developing microbial biofilm as a robust biocatalyst and its challenges. Biocatal. Biotransformation. 35: 86–95.

Tsoligkas, A.N., M. Winn, J. Bowen, T.W. Overton, M.J.H. Simmons and R.J.M. Goss. 2011. Engineering biofilms for biocatalysis. ChemBioChem. 12: 1391–1395.

Warwick–Dugdale, J., H.H. Buchholz, M.J. Allen and B. Temperton. 2019. Host-hijacking and planktonic piracy: How phages command the microbial high seas. Virol. J. 16: 15.

Westmeier, D., A. Hahlbrock, C. Reinhardt, J. Fröhlich-Nowoisky, S. Wessler, C. Vallet, U. Pöschl, S.K. Knauer and R.H. Stauber. 2018. Nanomaterial–microbe cross-talk: physicochemical principles and (patho) biological consequences. Chem. Soc. Rev. 47: 5312–5337.

Whittaker, R.H. 1959. On the broad classification of organisms. Q. Rev. Biol. 34: 210–226.

Wolcott, R. and S. Dowd. 2011. The role of biofilms: Are we hitting the right target? Plast. Reconstr. Surg. 127: 28S–35S.

Wood, T.K. 2017. Strategies for combating persister cell and biofilm infections. Microb. Biotechnol. 10: 1054–1056.

2

An Overview of Interactions between Microorganisms and Nanomaterials

Munishwar Nath Gupta[1], Sunil Kumar Khare[2*] and Rajeshwari Sinha[3]

[1]Former Emeritus Professor
Department of Biochemical Engineering and Biotechnology
Indian Institute of Technology Delhi, Hauz Khas, New Delhi 110016, India
Email: mn7gupta@gmail.com

[2]Institute Chair Professor of Biochemistry,
Enzyme and Microbial Biochemistry Laboratory
Department of Chemistry
Indian Institute of Technology Delhi, Hauz Khas, New Delhi 110016, India
Tel: +91 11 2659 6533, Fax: +91 11 2658 1102
Email: skhare@rocketmail.com, skkhare@chemistry.iitd.ac.in

[3]Independent Researcher, M25–Kalkaji, New Delhi 110019, India
Email: rajeshwari.sinha@yahoo.com

INTRODUCTION

Broadly, nanoparticles can be divided into organic and inorganic classes. Organic nanoparticles include dendrimers, liposomes, carbon-based nanoparticles (such as carbon nanotubes, graphene), and polymeric micelles. Inorganic nanoparticles include those which are made of just metals/metal oxides and quantum dots. Besides, magnetic nanoparticles have more complex compositions, but are mostly

*For Correspondence: skhare@rocketmail.com, skkhare@chemistry.iitd.ac.in

inorganic in their chemical nature. Both synthetic and natural polymers have been used for preparing nanoparticles. The natural polymers include polyamino acids, proteins (e.g., albumin, gelatin, etc.), chitosan, dextran, and form biodegradable nanomaterials. Synthetic polymers such as poly(D,L-lactide-co-glycolide), polycaprolactam, polycyanoacrylates, polyvinyl alcohol, and polyethylene glycol are especially useful as non-immunogenic materials for some applications. The applications of nanomaterials include their use in diagnostics and as therapeutics. Gold particles have been extensively studied for their use as parts of biosensor devices. Superparamagnetic nanoparticles have been used in protein separation, protein immobilization, and in imaging for enhancing the sensitivity of magnetic resonance methods. Polymeric nanoparticles, metallic nanoparticles, and quantum dots are all useful as fluorescent nanoprobes. Hybrid structures which involve these have been around for quite some time now. Metal-dye conjugates, apart from their use via fluorescence, are also used in electron microscopy. Dye-doped silica shells make dyes more photostable. Microspheres containing quantum dots are envisaged for bar-coding biological molecules. Engineering the nanoparticles by functionalization and conjugation allows the in-building of properties, such as targeting and controlled release, and even evading the immune surveillance in applications related to therapeutics.

THE NANO WORLD

There are three kinds of nanomaterials with respect to their origin (Jeevanandam et al. 2018). The first kind has always existed even before we became conscious of the special inherent properties of these materials. Most of our biological macromolecules (proteins, nucleic acids, polysaccharides, lipids etc.) are of nano dimensions. The nanoparticles occur in dust storms, volcanic eruptions, cosmic dust. These are also produced in forest fires, photochemical reactions taking place in our ecosystem, and when plants and animals shed skin or hair. The second source is incidental; the anthropogenic activities such as transportation burns fuels (e.g., the exhausts of vehicle engines contain nanoparticles). The burning of diesel is estimated to account for 90% of carbon nanoparticles in our atmosphere. Our industrial activities, such as welding, ore smelting/refining, mechanical grinding, and few other chemical manufacturing practices are known to produce nanoparticles. The cigarette smoke is reported to contain about 100,000 different kinds of nanoparticles (Jeevanandam et al. 2018)! The third is of more recent origin; when we started synthesizing/fabricating nanomaterials intentionally with specific design and application in mind. These are termed as "engineered nanoparticles" by some authors (Westmeier et al. 2018). Their level of complexity ranges from silver/gold nanoparticles to microbots (conjugates of nanoparticles) (Westmeier et al. 2018).

The synthesis of nanomaterials can be carried out by diverse methods (Roy et al. 2017); lately there has been more emphasis on developing green routes for their synthesis (Shukla and Iravani 2019). Of special interest are the approaches wherein proteins have been used as templates for growth of nanoparticles (Yang et al. 2004), or wherein ultrasonic irradiation during the preparation can influence the shape of the nanomaterial (Malhotra et al. 2013). There has been some confusion in the literature about the definition of nanomaterials. Let us briefly look at the current views (Jeevanandam et al. 2018).

The British Standards Institution considers nanoscale as about 1–1000 nm. The more commonly held view restricts it to about 100 nm only. The United States Food and Drug Administration (USFDA) recognizes any materials that have "at least one dimension in the range of approximately 1–100 nm and exhibit dimension-dependent phenomena" as nanomaterials. The latter is to emphasize that bulk material of the same composition lacks many properties which are unique to nanomaterials. Dimension-wise, zero-dimension nanomaterials have all dimensions in nano-range. Nanoparticles are an example of this. One-dimension nanomaterials have two-dimension in nano-range, whereas the third one can be larger e,g., nanotubes, nanorods, and nanowires. Two-dimension nanomaterials have only one-dimension in nano-range, and are exemplified by nano-films and nano-coatings (Fan et al. 2019). This covers the term nano-structured materials as well. Nano-sized surface of a bulk is exploited in lithography or tunneling microscopes. The International Standards Organization (ISO) definition is more explicit in including nanostructured materials, in which the existence of nanostructures can be clearly identified or established. Supramolecular solids, nano-phase materials, and nano-crystalline materials, as well as structures having embedded/entrapped nano-sized materials are all also included among nanomaterials (Gleiter 2000, Kannan et al. 2014). The descriptions by British Standard Institution of nano-structures (in which interconnected constituents are in nano-range); nano-structured materials (as those which contain internal or surface nano-structures), and nano-composites (as multiphase materials with at least one phase with nano-dimensions) bring much needed clarity into this area.

Carbon Nanotubes

Carbon nanotubes (CNTs) are either single-walled nanotubes (SWNT) or multi-walled nanotubes (MWNT). Former are essentially a cylinder of graphene (see later for their description) with diameters in the range of 0.4–2.0 nm; these normally occur as bundles wherein individual tubes are held together by van der Waals interactions (Kumar 2005). The commonly used methods of preparations produce a mixture of metallic and semiconducting nanotubes; this nature depends on their crystal lattice structures. MWNTs are concentric cylinders of graphene with inner diameter in the 1–3 nm range and outer diameter in the range of 2–100 nm. CNTs are useful in many applications because of their high mechanical strength. The early common methods for preparation of CNTs were based upon either vaporization of carbon in an inert atmosphere or hydrocarbon decomposition carried out catalytically.

For most of the applications, it is necessary to dissociate the bundles of SWNTs. Oxidation with strong acids, such as sulfuric acid or nitric acid oxidizes the aromatic ring caps to introduce carboxylic acid functions. This not only leads to solubilization of SWNTs, it also provides a means of their functionalization using standard chemical protocols. The sonication of the bundles in the presence of surfactants such as octadecylamine, coats the hydrophobic sidewalls of the nanotubes with the amphiphiles and disperses them in an aqueous phase as individual nanotubes. The covalent linking of water-soluble polymers such as polyethylene glycol and polyvinyl alcohol enhances their solubility in aqueous phase considerably. Various peptides, proteins, lipids, and nucleic acids get adsorbed on the walls of nanotubes. In several cases, this makes them soluble and is also a way to prepare the bioconjugates of the nanotubes with biological molecules. An early paper by Nepal and Geckeler (2007) also describes the characterization of carbon nanotubes with few proteins by Raman spectroscopy. In another example, 1-pyrenebutanoic acid succinimidyl ester could be adsorbed to the nanotube surface due to pi-stacking interactions. The succinimdyl ester chemistry linked glucose oxidase to create a glucose biosensor (Kumar 2005). Of course, more stable bioconjugates of nanotubes (or for that matter of any nanomaterials) require covalent coupling. One obvious way is to use carbodiimide chemistry with carboxyl groups on oxidized nanotubes to link any free amino group molecule. Some early examples of applications of CNTs, such as DNA directed self-assembly of multiple CNTs and gold nanoparticles; assembly of SWNTs, polylysine and horseradish peroxidase for designing an enzyme electrode and use of CNTs as substrates for neuronal cell lines have been described by Kumar (2005). Many of the approaches mentioned above are also valid for MWNTs.

Quantum Dots

Quantum dots (QD) are nanoparticles made of (a core of) semiconducting materials. These show fluorescence, and their fluorescent properties are dictated by size and material composition. QD of 5–6 nm diameter emit at longer wavelength, smaller with 1–3 nm emit at shorter wavelength. While quantum dots are of 2–10 nm size, the self-assembled QDs are generally of 10–50 nm in size (Kluson et al. 2007).

Graphene

Graphene is another nanomaterial which in recent years has attracted considerable attention because of its interesting properties. It is essentially a single atomic plane made up of graphite. Nevertheless, it is common to talk of bilayer or multilayer graphene. Laser Raman spectroscopy is the simplest tool to determine the number of layers in a multi-layered graphene preparation. To provide some perspective, 1 mm thick graphite has about 3 million layers of graphene. Graphene is a building block of CNTs, fullerenes, and graphite (Fan et al. 2019).

Casein

Casein, the milk protein, constitutes a unique design of nanoclusters in nature. Milk caseins are both phosphorylated as well as glycosylated. A casein isoform, αs-casein is present in the milk of all mammals, and it is an intrinsically disordered protein, and like many such proteins is also a molecular chaperone (Bhattacharyya and Das 1999). The protein is prevented from forming fibrils (as many intrinsically disordered proteins do), because it sequesters nanoclusters of calcium phosphate to form casein micelle. It is worth noting that many other secreted phosphoproteins also sequester calcium phosphate and thus are able to prevent undesirable calcification in some physiological fluids and tissues (Holt et al. 2013).

The size at that level gives rise to many unique properties (Vollath 2008). One which is often mentioned is the ratio of surface area to the volume. Taking the example of a spherical nanoparticle, where volume is proportional to cube of *r* (radius, surface area is proportional to square of *r*; this makes surface area/volume to be equal to 3/*r*). So, for small particles, this works out as follows. Corresponding to a diameter of 1000 μm, the surface area/volume is 0.006; for diameter of 1 μm, it rises to 6. This high surface area per unit volume makes many technologies possible. To start with, one could load lot more drug, link much higher amounts of proteins/nucleic acids, etc. Given the simultaneous advances in technology related to working with small dimensions like microfluidics, etc., this has paved the way for diverse kinds of electronic devices. Very many other magnetic, optical, electrical, and mechanical properties unique to nano dimensions are also exploited in designing a new class of materials/devices (Vollath 2008). Just to cite an example, iron oxide nanoparticles show superparamagnetism (Wahajuddin 2012).

Corona Formation

In the context of interaction of nanomaterials with any complex system, it is important not to ignore that many different kinds of molecules/materials invariably present in such systems bind to the nanomaterials by either covalent or non-covalent bonds; in either way the binding may be specific or nonspecific (Nepal and Geckeler 2007). This coverage by the molecules present in the milieu is called corona. Hence, as pointed out by Westmeier et al. (2018), pristine (without corona) nanoparticles neither exist in our ecosystem nor inside animal bodies. In the latter case, blood, interstitial fluids, lung surfactants which contain proteins, lipids, carbohydrates, or their conjugates form the corona. Natural products varying in complexity are part of this corona formation in terrestrial and aquatic environments.

The ill-defined and often unpredictable (also because of the fact that many parameters, such as pH, ionic strength, and size/shape/composition influence binding) nature of the corona makes it difficult to understand interactions of the nanomaterials with living organisms. So, we should be cautious in over-interpreting results in this domain.

The results of Shannahan et al. (2013) demonstrate the complexity of the issue. These workers incubated 20 nm and 110 nm silver nanoparticles capped with citrate and polyvinylpyrrolidone in a culture medium. The compositions of protein coronas formed were examined by mass spectrometry. The bigger particles bound more different kinds of proteins, indicating that surface curvature influenced the binding. Another interesting result was that 20 nm particles bound more hydrophobic proteins. The corona formation inside a living organism or even a cell is a much more complex process to understand.

Measuring the Sizes

Westmeier et al. (2018) provide a fairly comprehensive listing of various techniques (analytical centrifugation, atomic force microscopy (AFM), differential light scattering (DLS), scanning electron microscopy (SEM), transmission electron microscopy (TEM), isothermal titration calorimetry (ITC), gas chromatography–mass spectrometry (GC-MS), energy-dispersive X-Ray analysis, flow cytometry, confocal laser scanning microscopy, two-photon microscopy, zeta potential measurements, 1D/2D gel electrophoresis) for detecting complex formations between nanomaterials and microorganisms. All those methods are also used for measuring the sizes of particles of sub-micron sizes. For measuring the size (and with many techniques even changes in shapes), the larger list (and adequate discussion on each technique) is available at a number of places (Horynak et al. 2008, Vollath 2008).

DLS is one of the most frequently used techniques; one must be cautious and realize that it tends to overlook the presence of small particles (as it measures the intensity of the scattered light which is proportional to the sixth power of the diameter of the particle size) if much bigger particles are present in the polydisperse population. This is not the case with AFM, SEM, TEM, Asymmetric Flow Field-Flow Fractionation (AF4), and Nanoparticle Tracking Analysis (NTA). NTA output is a video of brownian motion of the population and provides the number average size. Good advice on how to use NTA to obtain best results is provided by Kim et al. (2019). A critical discussion on various techniques is available at several places for which Kim et al. (2019) has provided references. Yohannes et al. (2011) have discussed AF4.

TEM and DLS almost always give significantly different results, with DLS giving much bigger estimates for the sizes. The reason is that unlike TEM, it measures the diameters of hydrated species. Such factors must be kept in mind while comparing sizes obtained by different techniques. Also, most of the theories for such techniques are developed by assuming spherical shapes, which is seldom the true situation.

Another misconception among non-specialists is not to realize that Ultraviolet/Visible spectra reported for metallic nanoparticles actually originate in an optical phenomenon called localized surface plasmon resonance (LSPR), which results from collective oscillations of electrons. The 'absorption' is linear to the concentration of nanoparticles (up to a limit), but the spectra also provide rich details about changes in particle size (or aggregation) and even shape in the hands of an experienced person.

WHEN NANOMATERIALS "MEET" MICROBES

Westmeier et al. (2018) believes that there has been greater attention paid to the "crosstalk" between nanomaterials and eukaryotes as compared to the interaction of former with microbial organisms. Two important points in this regard are:

- Nanomaterials and microbes have both been around even before we became concerned with their existence. So, the interactions between microbes and nanomaterials have a very long history (largely unexplored). So, when we rightly worry about the lack of data or about the long-term consequences of this interaction, it may be useful to look backwards, that is, interpolate the data. Science has often profited enormously by scientists viewing the "rear mirror" in their journey of explorations (Gupta and Roy 2018). Of course, as the example of evolutionary biology has taught us, it is seldom easy and often fraught with contentious debates. Biologists, even after so many decades, have not come to an agreement about how to define the term "species" {even though in the earlier section the word has been used in the "sense" it is routinely used by most biologists} (Taylor 2016). This point is relevant as it is good to make it clear at the outset that we don't have clear answers about whether the "good" outweighs the "bad" when we discuss the utility of nanomaterial-microbe interactions in various applied sectors.

- Our recent increased concern (about such interactions) is valid, since now much larger number and diverse kinds of engineered nanoparticles (eNP) are being released in the ecosystem. Also, their becoming a part of consumer products (toothpaste, cosmetics, etc.) has resulted in enhanced opportunities for the encounters between nanoparticles and our own microbiome. As mentioned in the earlier section, microbiome alteration now is known to affect even functions of our brains and immune systems.

Nanoparticles as Drug Carriers

Torchilin (2006) has discussed polymeric nanoparticles, niosomes (non-ionic surfactant vesicles), lipoproteins/solid lipid nanoparticles, nano-capsules, dendrimers, drug nano-crystals/nano-suspensions, ghosts of bacterial cells erythrocytes, cochelates (phospholipid-ion precipitates) nanoparticles, magnetic nanoparticles, and liposomes as drug carriers. Preparations of nano-suspensions for therapeutic purposes have been described by Verma et al. (2009). Baptista et al. (2018) provides an update of many drug delivery and targeting strategies based upon nanomaterials. The advantages listed by Baptista et al. (2018) in using the nano-carriers include increase in solubility and stability of drugs, biocompatibility, controlled release which prevents toxicity often associated with side effects because of high systemic concentration, smart release by stimuli such as light, pH, and heat, large loading possible, possibility of functionalization, and easy crossing of membranes by endocytosis. (Obviously, not all the advantages are associated with every system, so this is a list of potentials). The chemistry associated with various approaches for functionalization/conjugation of nanomaterials has been discussed at various places (Kumar 2005, Gupta and Mukherjee 2012, Ahmad and Sardar 2015). The applications of nanoparticles in both protein immobilization (Solanki and Gupta 2011, Gupta et al. 2011, Mukherjee and Gupta 2016, Sinha and Khare 2015, Goel et al. 2017) and bioseparation of proteins (Kannan et al. 2013) share the chemistry and methods for using them as drug carriers, and it may be useful to refer to the literature in these areas, which are more extensively developed and have established and proven protocols. An interesting approach is by Mukherjee and Gupta (2012), in which an enzyme chymotrypsin was coated over superparamagentic iron oxide nanoparticles of about 3.6 nm average diameter by simply precipitating the enzyme from its solution in aqueous buffer by adding to the swirling nanoparticles in n-propanol.

One critical issue related to pharmaceutical applications of nanomaterials is the strategies used for creating nano-suspensions (Verma et al. 2009). For oral administration (or by injection) of drugs (which often tend to be poorly soluble in aqueous phase), optimization of the protocols for obtaining desirable formulations is quite important. Some methods which have been used are: employing water soluble polymers/surfactant solutions/co-solvents, entrapment in liposomes, preparing their cyclodextrin conjugates, or simply attempting to adjust pH to enhance solubility. Top-down approaches aim at reducing the particle size of the drug by wet milling methods, such as media milling, microfluidization, or homogenization using high pressures. Two disadvantages of these approaches are that these are energy intensive and generate heat which may inactivate the drug (by decomposition). For protein drugs, heat may denature the molecules. The bottom-up approach broadly consists of dissolving the drug in an appropriate organic solvent and adding an anti-solvent in the presence of an additive, which acts as a stabilizer. The stabilizer is generally a surfactant which coats the drug; the efficiency of the coverage can be tracked by measurement of zeta potential if either the drug or the surfactant or both have charges. The antisolvent used for precipitation is generally an aqueous solution. While controlling particle morphology is difficult in the mechanical processes, this approach allows greater possibility of obtaining crystallinity or amorphous nature. There are various versions of this approach: solvent-antisolvent mixing, use of supercritical fluids, spray drying, and evaporation of emulsion-solvent mixtures.

Matteucci et al. (2006) described antisolvent precipitation of the drug itraconazole to obtain drug particles of below 300 nm average diameter. A non-ionic polymeric surfactant poloxamer 407 mixed with the drug in tetrahydrofuran was introduced in aqueous solution. Optimization of temperature to obtain the best nucleation and growth rate was carried out. The nozzle size for adding the drug mixture was also optimized to obtain best dissipation of mixing energy. Guo et al. (2005) have described the usefulness of ultrasonication in anti-solvent crystallization of the drug roxithromycin; ultrasonication increased the nucleation rate constant significantly. Cavitation due to ultrasonic radiation affects the process in multiple ways. Apart from the turbulence resulting in better mixing, the drug solution forms the droplets of much smaller sizes, which in turn leads to precipitation of drugs into smaller sizes.

Actually, as pointed out by Matteucci et al. (2008), drugs in amorphous forms can be about 1,600 times more soluble in amorphous forms as compared to even small crystals. These authors again precipitated itraconozole stabilized by the non-ionic polymer, but immediately added sodium sulphate to flocculate the drug particles. The salt flocculation also got rid of much of the polymer to obtain drug loading upto about 90%. The amorphous preparation could be used to prepare supersaturated drug solution. The antisolvent precipitation approach has also been described for obtaining lysozyme nanoparticles by Muhrer and Mazzotti (2003) using dimethyl sulfoxide as solvent and supercritical carbon oxide as antisolvent. Rodrigues et al. (2009) used aqueous ethanol instead as the solvent for lysozyme, and also used "assisted atomization" (increasing a liquid jet dispersion by co-pressurization with a gaseous or supercritical phase) in different ways to obtain either spherical shaped particles or fibers of the enzyme. Solanki et al. (2012) has examined the effect on biological activities of two enzymes (chymotrypsin and subtilisin) in various high activity formulations (for biocatalysis in low water media) when their aqueous solutions are mixed with n-propanol for their precipitation. Intriguingly, better activity and native structure (as measured by circular dichroism and Fourier-transform infrared spectroscopy) was retained when aqueous solutions of the enzymes were added to the organic solvent rather than vice-versa. It may be interesting to investigate whether this matters during anti-solvent precipitation.

As the drug particle size affects its solubility during formulations, let us discuss this aspect a little more. How the particle size varies during solution to solid phase transition changes has been discussed by Madras and McCoy (2003). The last stage in this transition is Ostwald ripening, a process in which smaller particles dissolve fast, and the mass is transferred to the larger particles. Ostwald ripening has been frequently referred to during the formation of nanoparticles (Nam et al. 2008, Liu et al. 2007, Simonsen et al. 2010). Beyond a critical size, the small particles just denucleate and undergo solid to solution phase transition. Ostwald ripening has been observed both for inorganic materials and biological macromolecules. Streets and Quake (2010) describe a microfluidic dynamic light scattering system to track the growth kinetics during Ostwald ripening of lysozyme to develop a model. Earlier, Ng et al. (1996) showed that Ostwald ripening is involved in crystallization of thaumatin, concanavalin A, α-amylase, and tomato bushy stunt virus. All the work on understanding growth kinetics and mechanisms is useful in designing nanoparticles of desirable size.

Nanoparticles as Bactericidal Agents

Many metal/metal oxide nanoparticles (silver, gold, and oxides of iron, copper, titanium) as such have been reported to be bactericidal (Baptista et al. 2018, Sinha and Khare 2013). The common mechanisms of bactericidal effects of nanomaterials include binding to the surfaces, followed by cellular entry, followed by generation of reactive oxygen species (ROS).

The reactive oxygen species and redox homeostasis

In view of their frequent mention in this book and their importance in the effect of nanoparticles on microorganisms, a little detailed discussion is desirable. Just to establish proper context, it should be remembered that while studying interactions of nanoparticles with microorganisms as free suspensions or part of the biofilms is important; another important setting is when these as pathogens infect animals (including humans). ROS are part of redox homeostasis, and are involved in cell growth, differentiation, signaling, and inflammation (which itself is a very complicated biological process characterized by ancients, as one accompanied by rubor (redness), calor (heat), and tubor (swelling). ROS are generated in mitochondria, endoplasmic reticulum, peroxisomes, microsomes, and enzymes such as Nicotidamide Adenine Dinucleotide Phosphate Hydrogen (NADPH) oxidases (which occur in membranes and have various isoforms), monoamine oxidase, and α-ketoglutaric dehydrogenase, lipoxygenase, and cyclooxygenase, etc. Exposure to pollutants, radiation, and nanoparticles also lead to production of ROS.

ROS can be radicals or non-radicals. ROS include singlet oxygen, superoxide anion, hydroxyl, hydroperoxyl, carbonate, peroxyl, alkoxyl, and carbon dioxide radicals. The non-radical ROS include hydrogen peroxide, hypobromous acid, hypochlorous acid, ozone, organic peroxides, peroxynitrite anion, peroxynitrate anion, peroxynitrous acid, peroxymonocarbonate, nitric oxide, and hypochlorite anion.

As a part of redox homeostasis, ROS level is controlled by superoxide dismutase (which converts superoxide anion radical into hydrogen peroxide, which in turn is decomposed by catalase into water and oxygen; animal livers being detoxifying organs are rich in catalase!) Glutathione peroxidase is another important enzyme for maintaining redox homeostasis. To clarify how redox homeostasis works, a brief discussion on the role of glutathione peroxidase provides a good illustration. Ingestion/exposure to sulpha drugs, many herbicides, antimalarials (such as primaquine and divicine) also produces superoxide radical, which by superoxide dismutase, is converted to hydrogen peroxide. Glutathione peroxidase prevents build-up of this ROS by converting it to water. Simultaneously, glutathione is converted to its oxidized form by the enzyme. The reduced form of glutathione is regenerated by glutathione reductase, which is a NADPH-dependent enzyme and generates Nicotinamide Adenine Dinucleotide Phosphate (NADP) in the process. NADPH is regenerated by the first enzyme in pentose phosphate pathway, namely glucose-6-phosphate dehydrogenase (g6pd). The reduced form of glutathione also destroys free hydroxyl radical, which may be produced from excess hydrogen peroxide. Not taking care of hydrogen peroxide/hydroxyl radical leads to damage via peroxidation of membrane lipids and oxidation of both protein and DNA constituents.

How do antimalarial drugs act? These act exactly by producing ROS, which kills malarial parasite! Why do we need glutathione peroxidase when catalase can decompose hydrogen peroxide? Given the K_m of catalase towards hydrogen peroxide, its low level is poorly decomposed by catalase. In fact, glutathione peroxidase/glutathione systems are very important at low levels of oxidative stress (Beutler 1972). The g6pd deficiency is inherited and leads to glutathione peroxidase catalyzed control of ROS levels. As an interesting related fact, g6pd deficiency is very common in regions where malaria is known to occur more commonly. The low level oxidative stress (because of the deficiency of g6pd) kills the parasite *Plasmodium falciparum*. The natural selection in malaria prone regions encourages g6pd deficiency in humans living in malaria prone regions. The redox balance varies with cellular location; the oxidative environments of endoplasmic environments allow cysteine residues to form disulphide bridges in proteins.

Another positive role of ROS is in innate immunity. One arm of innate immunity is circulating white blood cells (WBCs) called monocytes (about 5% of total WBCs in blood), which differentiate into macrophages upon reaching different tissues. For example, they become Kupffer cells in liver. The macrophages have half lives in years in tissues; those which go on circulating after differentiation become part of antigen presenting cells. Neutrophils are a more abundant class (about 70%) of total WBCs, which also end up in tissues by chemotaxis. Both kinds of cells are phagocytes, i.e., uptake of foreign cells (such as pathogenic cells) is followed by their destruction. After the uptake of foreign/invading material, these phagocytes increase their oxygen uptake and pentose phosphate pathway activity. The phagosome (the fusion vesicle formed by combination of phagocytosed material with lysosome) membrane NADPH oxidase uses increased uptake of oxygen to produce superoxide anion, which in turn gives rise to other ROS, such as hydrogen peroxide, singlet oxygen, and hydroxyl radicals. Myeloperoxidase (of lysosomes) uses hydrogen peroxide to convert halide anion into hypohalites. Murine macrophages and human neutrophils also have inducible nitric oxide synthetase, which produces nitric oxide, which in turn reacts with other ROS to form peroxynitrites. These ROS destroy the endocytosed pathogen/other foreign material. Thus, ROS form part of immune system in defense against infectious microorganisms. Again, glutathione prevents damage to the animal cells in modulating the action of these ROS. Antioxidants such as flavonoids, vitamin C, and vitamin E also reduce oxidative stress by reacting with ROS. That is why antioxidants are recommended by nutritionists.

ROS generation by nanoparticles

It starts with their internalization by cells. Their surface to volume ratio, chemical composition, and their solubility (or propensity to aggregate), all play important roles in the entire process. Quartz nanoparticles have SiO and SiO_2 radicals on their surface, which generate ROS upon entering the cells. Adsorbed nitric oxide, ozone on these nanoparticles also results in generation of oxidative stress. The nanoparticles made of Si, Fe, Cu, Cr, and V produce ROS via Fenton and/or Haber-Weiss reactions. Fenton reaction is transition metal catalyzed conversion of hydrogen peroxide into highly reactive hydroxyl radical, and produces oxidized metal ion. The latter further helps in the production of hydroxyl radicals from hydrogen peroxide. Iron oxide nanoparticles catalyze Fenton reaction, whereas nanoparticles of Cr, Co, and V catalyze both types of reactions. Cellular internalization of nanoparticles can take place with/without participation of cell surface clathrin proteins. Once inside, apart from catalyzing production of ROS by the two reactions, their interaction, notably with mitochondria (and influencing few other cellular processes referred to in this chapter and elsewhere in the book) adds to the oxidative stress.

Titanium dioxide nanomaterials and quantum dots upon irradiation with visible/ultraviolet light convert oxygen to its singlet state.

In stem cell lines, many nanoparticles (e.g., silver) are reported to have opposite effects of modulating oxidative stress. Nanoceria (cerium oxide nanoparticles) have cerium ions with +3 and +4 oxidation states on their surface and act as free radical scavengers (Abdal Dayem et al. 2017).

ROS in fact serve as important signaling molecules during normal physiological processes. The disturbed redox balance leads to metabolic dysfunction and inflammatory responses. Atheroscelorosis, diabetes mellitus, and strokes involve ROS signaling under disturbed redox balance. The involvement of ROS is different in cytoplasm, mitochondria, peroxisomes, and endoplasmic reticulum. ROS in turn produce secondary signaling products, such as 4-hydroxynonenal and oxidized phospholipids. Their production and function depend upon cell type, concentration of ROS, and cellular stress in real time (Forrester et al. 2018).

Invariably, membrane is destroyed, and apoptosis takes place (Baptista et al. 2018, Sinha and Khare 2013) There are cases wherein nanomaterials or their bioconjugates have been able to destroy the bacteria or disrupt biofilms (Baptista et al. 2018, Li et al. 2019). In the case of infections, use of targeted nanomaterials has augmented the triggering of innate as well as adaptive immunity (Baptista et al. 2018).

Zhang et al. (2018) have examined the effect of concentration and size of the silver (Ag) nanoparticles on their cytotoxicity towards *A. vinelandii*, a gram-negative bacteria which is an important part of nitrogen cycle, as it can fix atmospheric nitrogen via its nitrogenase. Figure 2.1 shows that not only was the cytotoxicity dependent on the concentration, smaller particles were significantly more efficient. The particles bind to the cell membrane, enter the cell, and generate hydroxyl radical by a Fenton-like reaction. These ROS destroy the membrane, damage DNA and proteins (including nitrogenase activity), and ultimately cause apoptosis. The amount of ROS (as estimated by electron-spin resonance spectra) generated was higher in case of smaller particles.

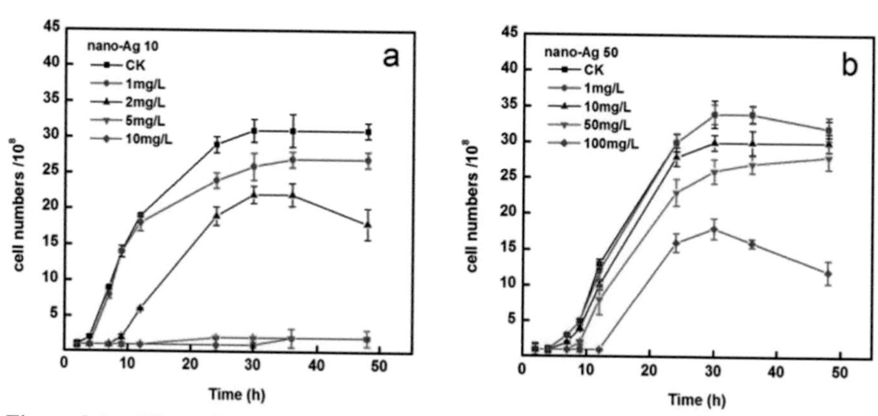

Figure 2.1 Effect of (a) 10 nm nano silver and (b) 50 nm nano silver on the growth of *A. vinelandii,* cultured for 48 hours in media with and without silver nanoparticles (Reprinted from Zhang et al. 2018).

Another nontoxic nanoparticles are those prepared from magnesium oxide (Cai et al. 2018). *R. solanacearum* is a phytopathogen and affects hundreds of plants. Figure 2.2 shows how the nanoparticles of magnesium oxide are far more effective (formed "deep craters and burst cells") against this phytopathogen as compared to the bulk form of the oxide. The nanoparticles even destroy the biofilms and affect the motility. The involvement of ROS has been shown in this case as well.

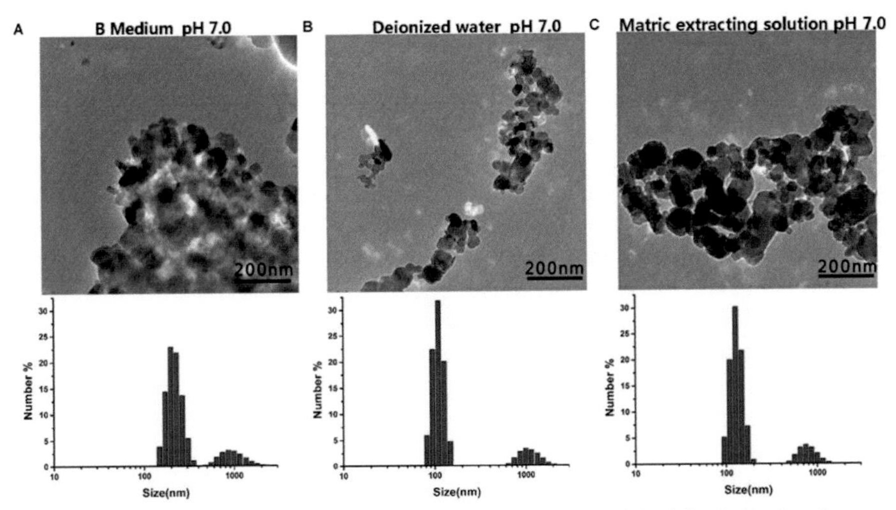

Figure 2.2 Transmission Electron Microscope and Differential Light Scattering images of the magnesium oxide nanoparticles (a) suspended in B Medium, (b) in B Medium and deionized water, and (c) in matrix extraction solution (Reprinted from Cai et al. 2018).

Wang et al. (2019) has discussed the effect of size and shape on internalization of nanomaterials by mammalian cells in the context of targeting the sub-cellular locations, wherein the drug cargo can be released "on the site" (e.g., inside the cell nucleus). It may be added that their experiments were carried out with cell lines, so the corona formation inside living organisms is not factored in. So, the discussion about the charges, zeta potential, etc. is only of academic interest, but the data on effect of sizes is a valuable first step in understanding cellular toxicity. Nanomaterials up to 50 nm pass through plasma membrane passively; for bigger ones, endocytosis is involved. In general, rods pass through easily than spherical particles. Even for crossing nuclear envelopes, a 40 nm wide rod is far better than a spherical particle of 40 nm diameter. The article also provides good information about the relative merits of various imaging tools (such as electron microscope, fluorescence microscope, confocal microscope, etc.) for tracking the fate of nanomaterials inside the cells.

Interaction of Nanomaterials with Microbial Proteins

Increased presence of nanoparticles in the environment necessitates a basic understanding about their potential impact on the biological systems. The dispersion of nanoparticles in a biological milieu results in their surfaces being immediately enveloped by a complex layer of proteins, forming a "protein corona", which may influence cellular uptake, inflammation, accumulation, degradation, and clearance of nanoparticles. Adsorption of proteins on the surface strongly depends on the nature of the protein, the surface chemistry, and physicochemical properties. Subsequent to surface adsorption, other interactions like those of electrostatic attraction, hydrogen bonding, and hydrophobic interactions provide further binding between proteins and nanoparticles. This brings about conformational changes in proteins, transformation into molten globule states, co-adsorption or release of other ions, and gain in entropy (Mahmoudi et al. 2011). Nanoparticle induced structural and conformational changes in adsorbed protein molecules may affect the overall bio-reactivity of the nanoparticle (Saptarshi et al. 2013). The attachment of nanoparticles to proteins has been used in imaging, catalysis, drug delivery, and understanding protein folding (Nie et al. 2007, Marcato and Duran 2008, Colmenares and Luque 2014, Cao-Milán and Liz-Marzán 2014).

Several studies have reported the effects of interaction between proteins and nanoparticles. The interaction of Ag nanoparticles with serum albumin was found to have a significant effect on their antibacterial activity (Gnanadhas et al. 2013). It was observed that uncapped Ag nanoparticles exhibited no antibacterial activity in the presence of serum proteins, while citrate or polyvinylpyrrolidone (PVP) capped Ag nanoparticles exhibited antibacterial properties due to minimized interactions with serum proteins. This was because PVP-Ag nanoparticles showed the least interaction with serum albumin, and hence better antibacterial activity in the presence of BSA. To confirm this finding, uncapped, citrate coated, and PVP coated Ag nanoparticles, at the same concentration, were incubated with 3% Bovine Serum Albumin (BSA), and the absorption spectrum was analyzed. The spectra indicated that uncapped Ag nanoparticles had more interaction with BSA, and PVP-Ag nanoparticles have the least. Citrate-capped Ag nanoparticles showed a moderate interaction with BSA. The authors reason out that protein adsorption onto nanoparticles occurs due to an increase in the collective entropy of the proteins on the nanoparticle surface and nonspecific interactions between the nanoparticle surface and the proteins. The increase in the entropy is due to the rearrangement of the protein structure to a more stable position on the uncoated Ag nanoparticle surface. Strong interactions occur between uncoated Ag nanoparticles and proteins due to the increase in entropy, which in turn leads to a significant decrease in the Gibbs free energy. This decrease in energy is primarily due to the relaxation of protein secondary structures.

The interactions between gold nanoparticles and BSA were investigated using fluorescence and circular dichroism (CD) spectroscopies (Wangoo et al. 2008). Fluorescence quenching of tryptophan residues of the protein molecules after conjugation, was used to determine the binding constant (K_b) of protein to gold nanoparticles. The conformational change in BSA at its native form after conjugation with gold nanoparticles confirmed that protein undergoes a more flexible conformational state on the boundary surface of nanoparticles after bioconjugation. The CD studies further showed a decrease in the helical content after conjugation. The results confirmed that the change in conformation was larger at higher concentrations of gold nanoparticles.

Sinha and Khare (2014) compared the effect of interaction of nanoparticles with halophilic and non-halophilic proteins. Halophilic proteins have highly charged surfaces, which enable them to retain structural and functional integrity under high salt concentrations (Karan et al. 2012). The effects of two commonly used Ag and zinc oxide (ZnO) nanoparticles on the biological activity and conformational transitions of halophilic *Bacillus* sp. EMB9 protease were studied and compared with a commercially available non-halophilic *Bacillus* sp. protease (subtilisin Carlsberg). The exposure of Ag and ZnO nanoparticles (1.0 mM) caused a significant decrease in the activity of non-halophilic *B. licheniformis* protease, while halophilic proteases were more stable towards both the nanoparticles. Halophilic *Bacillus* sp. protease exhibited slight enhancement in activity in the presence of 1.0 mM Ag nanoparticles as compared to its control. The frequency of acidic amino acid residues in halophilic proteases is higher than those in subtilisin Carlsberg. It may therefore be inferred that nanoparticles were adsorbed differently in halophilic and non-halophilic proteases due to the influence of charge. Greater electrostatic interaction between positive charge density of the nanoparticle with the negatively charged domain on the halophilic protein surface may have led to favorable adherence of halophilic proteases onto nanoparticle surface and influenced the conformation into functionally favorable form. Secondary structure analysis using CD spectra showed that the α-helical content of non-halophilic proteases was significantly lost in the presence of Ag and ZnO nanoparticles, while helical and sheet content of halophilic *Bacillus* sp. protease remained mostly unchanged. With regard to the tertiary structure, nanoparticles induced the quenching of the intrinsic tryptophan fluorescence in halophilic proteases, reflecting changes in microenvironment of the proteases due to nanoparticle-protein interactions. Nanoparticle protein interaction has also been investigated using other approaches, such as molecular docking, Surface Plasmon Resonance (SPR), etc. (Ranjan et al. 2018, Di Ianni et al. 2017).

Nanomaterials and Multi-drug Resistance

A penicillin resistant *Staphylococcus* strain was reported in 1940, even though penicillin was commercially introduced only around 1943. Thus, antibiotic resistance is a part of the inherent defense mechanism of microorganisms. While antibiotic resistance has been observed in cases of all kinds of microorganisms, this has become especially alarming in the case of bacteria. Overuse of antibiotics through routine prescription and extensive use as growth supplements in livestock have led to bacterial strains developing resistance to almost every available antibiotic! The selection pressure has led to bacteria either evolving new genes or selecting the existing ones. As antibiotics destroy the vulnerable population, what survives are resistance species. The resistance genes get transmitted through usual mechanism to other microbial populations, leading to spread of the antibiotic resistance.

The most notable is methicillin resistant *Staphylococcus aureus* (MRSA), which is reported to kill more Americans "each year than HIV/AIDS, Parkinson's disease, emphysema and homicide combined" (Ventola 2015). Equally alarming are the resistant strains of *Streptococcus pneumonia* and *Mycobacterium tuberculosis.* Among the gram-negative bacteria resistant *Klebsiella pneumoniae, Pseudomonas aeruginosa* and *Acinetobacter* are of serious concern in healthcare settings. *E. coli* and *N. gnorrhoeae*, on the other hand, have resistant strains which are spreading in communities. This scenario has led to practically no new antibiotics being discovered. The various ways by which the microorganism can win the war against the drug have been listed by Baptista et al. (2018).

- The hydrolase enzyme expressed by bacteria hydrolyzes the antibiotic.
- The cell permeability is decreased so that the drug can no longer enter the cell.
- The antibiotic target is protected or overproduced or altered.
- The efflux pumps are over-expressed to throw out the drug.
- Antibiotics can induce film formation.

Baptista et al. (2018) have also discussed at length how various nanoparticles as such or as carriers/vectors can overcome multi-drug resistance (MDR) (Figure 2.3).

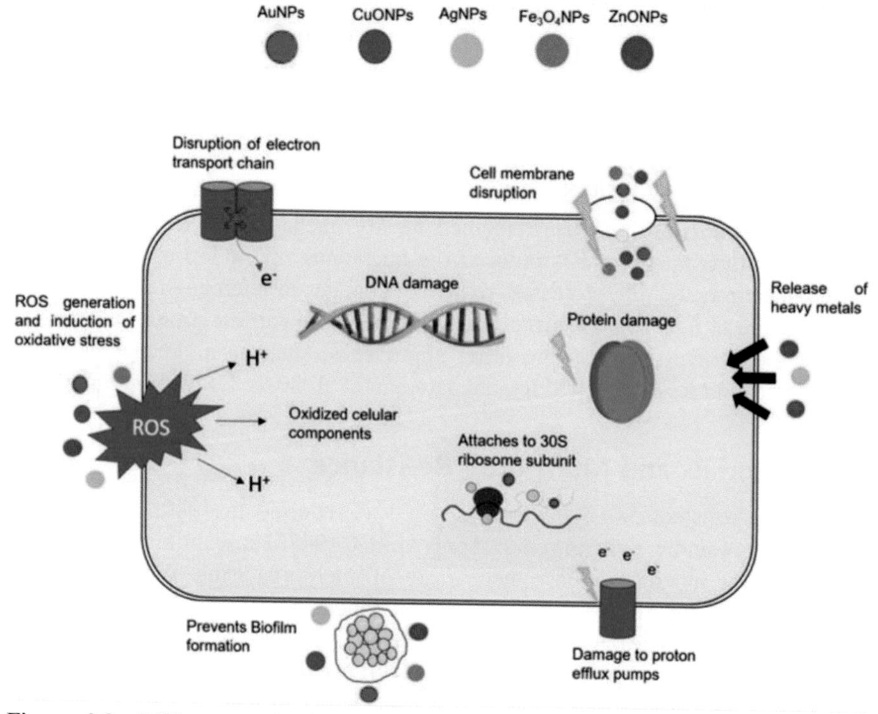

Figure 2.3 Different mechanisms of action of nanoparticles in bacterial cells. The combination in a single nanomaterial of a multitude of cellular effects may have a tremendous impact in fighting MDR bacteria. Au nanoparticles: gold nanoparticles; CuO nanoparticles: Copper oxide nanoparticles; Ag nanoparticles: silver nanoparticles; Fe_3O_4 nanoparticles: iron oxide nanoparticles; ZnO nanoparticles: zinc oxide nanoparticles. (Reprinted from Baptista et al. 2018).

CONCLUSION

In the end, it may be interesting to look at some current "leads" which may shape the future in this area. The toxicity in general and ecotoxicity in particular should continue to be of concern (Juganson et al. 2015). Techniques such as single/

multiple tracking and auto/pair correlation microscopy should help in designing nano-conjugates for sub-cellular targeting. Microbiome is emerging as an important "organ" with considerable role in human health, and early work on manipulating the microbiome as a therapeutic approach has already appeared in the literature (Hooper et al. 2012, Young 2016). More work in developing refined *in vitro* and *in vivo* models for interactions between nanomaterials and microbiota is urgently needed (Wang et al. 2019).

The nano-based approaches may become better at disrupting biofilms as we gain clearer understanding of quorum sensing (Abisado et al. 2018, Li et al. 2019). Also, bioconjugates of nanomaterials with natural products (e.g., essential oils) and antimicrobial peptides to combat MDR are likely to get continued attention (Baptista et al. 2018).

Theranostics brings diagnosis and therapy together on a single automated platform. Vergene® (a microarray system to detect gram-negative bacteria) and Vergene®-BC-GN (for identifying resistance mechanisms) are two of the several examples of nanotheranostics mentioned by Baptista et al. (2018). Nanotheranostics is definitely poised to grow.

References

Abdal Dayem, A., M.K. Hossain, S.B. Lee, K. Kim, S.K. Saha, G.M. Yang, et al. 2017. The role of reactive oxygen species (ROS) in the biological activities of metallic nanoparticles. Int. J. Mol. Sci. 18: 120–141.

Abisado, R.G., S. Benomar, J.R. Klaus, A.A. Dandekar and J.R. Chandler. 2018. Bacterial quorum sensing and microbial community interactions. MBio. 9: e02331–17.

Ahmad, R. and M. Sardar. 2015. Enzyme Immobilization: An Overview on Nanoparticles as Immobilization Matrix. Biochem. Anal. Biochem. 4: 178.

Baptista, P.V., M.P. McCusker, A. Carvalho, D.A. Ferreira, N.M. Mohan, M. Martins, et al. 2018. Nano-strategies to fight multidrug resistant bacteria—"A Battle of the Titans". Front. Microbiol. 9: 1441.

Beutler, E. 1972. Disorders due to enzyme defects in the red blood cell. Adv. Metab. Disord. 60: 131–160.

Bhattacharyya, J. and K.P. Das. 1999. Molecular chaperone-like properties of an unfolded protein, αS-casein. J. Biol. Chem. 274: 15505–15509.

Cai, L., J. Chen, Z. Liu, H. Wang, H. Yang and W. Ding. 2018. Magnesium oxide nanoparticles: Effective agricultural antibacterial agent against *Ralstonia solanacearum*. Front. Microbiol. 9: 790.

Cao-Milán, R. and L.M. Liz-Marzán. 2014. Gold nanoparticle conjugates: Recent advances toward clinical applications. Expert Opin. Drug Deliv. 11: 741–752.

Colmenares, J.C. and R. Luque. 2014. Heterogeneous photocatalytic nanomaterials: prospects and challenges in selective transformations of biomass-derived compounds. Chem. Soc. Rev. 43: 765–778.

Di Ianni, M.E., G.A. Islan, C.Y. Chain, G.R. Castro, A. Talevi and M.E. Vela. 2017. Interaction of solid lipid nanoparticles and specific proteins of the corona studied by surface plasmon resonance. J. Nanomater. 2017: Article ID 6509184, 11 p. https://doi.org/10.1155/2017/6509184

Fan, S., X. Feng, Y. Han, Z. Fan and Y. Lu. 2019. Nanomechanics of low-dimensional materials for functional applications. Nanoscale Horiz. 4: 781–788.

Forrester, S.J., D.S. Kikuchi, M.S. Hernandes, Q. Xu and K.K. Griendling. 2018. Reactive oxygen species in metabolic and inflammatory signaling. Circ. Res. 122: 877–902.

Gleiter, H. 2000. Nanostructured materials: Basic concepts and microstructure. Acta Mater. 48: 1–29.

Gnanadhas, D.P., M.B. Thomas, R. Thomas, A.M. Raichur and D. Chakravortty. 2013. Interaction of silver nanoparticles with serum proteins affects their antimicrobial activity *in vivo*. Antimicrob. Agents Chemother. 57: 4945–4955.

Goel, A., R. Sinha and S.K. Khare. 2017. Immobilization of *A. oryzae* β-galactosidase on silica nanoparticles: Development of an effective biosensor for determination of lactose in milk whey. pp. 3–18. *In*: P.Shukla [ed]. Recent Advances in Applied Microbiology. Springer, Singapore.

Guo, Z., M. Zhang, H. Li, J. Wang and E. Kougoulos. 2005. Effect of ultrasound on anti-solvent crystallization process. J. Cryst. Growth. 273: 555–563.

Gupta, M.N., M. Kaloti, M. Kapoor and K. Solanki. 2011. Nanomaterials as matrices for enzyme immobilization. Artif. Cells, Blood Substitutes, Biotechnol. 39: 98–109.

Gupta, M.N. and J. Mukherjee. 2012. Designing nano carriers for drug delivery. pp. 417–442. *In:* A.K. Mishra [ed]. Nanomedicine for Drug Delivery and Therapeutics. Wiley, New Jersey.

Gupta, M.N and I. Roy. 2018. Rear mirror view of some biological tools and processes. Proc. Indian Natl. Sci. Acad. 84: 611–624.

Holt, C., J.A. Carver, H. Ecroyd and D.C. Thorn. 2013. Invited review: Caseins and the casein micelle: Their biological functions, structures, and behavior in foods. J. Dairy Sci. 96: 6127–6146.

Hooper, L.V., D.R. Littman and A.J. Macpherson. 2012. Interactions between the microbiota and the immune system. Science. 336: 1268–1273.

Horynak, G.L., H.F. Tibbals, J. Dutta and J.J. Moore. 2008. Introduction to Nanoscience and Nanotechnology. CRC Press, New York.

Jeevanandam, J., A. Barhoum, Y.S. Chan, A. Dufresne and M.K. Danquah. 2018. Review on nanoparticles and nanostructured materials: history, sources, toxicity and regulations. Beilstein J. Nanotechnol. 9: 1050–1074.

Juganson, K., A. Ivask, I. Blinova, M. Mortimer and A. Kahru. 2015. NanoE-Tox: New and in-depth database concerning ecotoxicity of nanomaterials. Beilstein J. Nanotechnol. 6: 1788–1804.

Kannan, K., J. Mukherjee and M.N. Gupta. 2013. Use of polyethyleneimine coated Fe_3O_4 nanoparticles as an ion-exchanger for protein separation. Sci. Adv. Mater. 5: 1477–1484.

Kannan, K., J. Mukherjee and M.N. Gupta. 2014. Immobilization of a lipase on mesocellular foam of silica for biocatalysis in low-water-containing organic solvents. Chem. Lett. 43: 1064–1066.

Karan, R., S. Kumar, R. Sinha and S.K. Khare. 2012. Halophilic microorganisms as source of novel enzymes. pp. 555–579. *In:* T. Satyanarayana and B.N. Johri [eds]. Microorganisms in Sustainable Agriculture and Biotechnology. Springer, Netherlands.

Kim, A., W.B. Ng, W. Bernt and N.J. Cho. 2019. Validation of size estimation of nanoparticle tracking analysis on polydisperse macromolecule assembly. Sci. Rep. 9: 2639.

Kluson, P., M. Drobek, H. Bartkova and I. Budil. 2007. Welcome in the nanoworld. Chem. Listy. 101: 262–272.

Kumar, C.S.S.R. 2005. Biofunctionalization of Nanomaterials. Wiley-VCH, Weinheim.

Li, J., R. Nickel, J. Wu, F. Lin, J. van Lierop and S. Liu. 2019. A new tool to attack biofilms: Driving magnetic iron-oxide nanoparticles to disrupt the matrix. Nanoscale. 11: 6905–6915.

Liu, Y., K. Kathan, W. Saad and R.K. Prud'homme. 2007. Ostwald ripening of β-carotene nanoparticles. Phys. Rev. Lett. 98: 036102.

Madras, G. and B.J. McCoy. 2003. Continuous distribution theory for ostwald ripening: comparison with the LSW approach. Chem. Eng. Sci. 58: 2903–2909.

Mahmoudi, M., I. Lynch, M.R. Ejtehadi, M.P. Monopoli, F.B. Bombelli and S. Laurent. 2011. Protein-nanoparticle interactions: Opportunities and challenges. Chem. Rev. 111: 5610–5637.

Malhotra, D., J. Mukherjee and M.N. Gupta, 2013. Post-ultrasonic irradiation time is important in initiating citrate-coated α-Fe_2O_3 nanorod formation. RSC Adv. 34: 14322–14328.

Marcato, P.D. and N. Duran. 2008. New aspects of nanopharmaceutical delivery systems. J. Nanosci. Nanotechnol. 8: 2216–2229.

Matteucci, M.E., M.A. Hotze, K.P. Johnston and R.O. Williams. 2006. Drug nanoparticles by antisolvent precipitation: mixing energy versus surfactant stabilization. Langmuir. 22: 8951–8959.

Matteucci, M.E., J.C. Paguio, M.A. Miller, R.O. Williams III and K.P. Johnston. 2008. Flocculated amorphous nanoparticles for highly supersaturated solutions. Pharm. Res. 25: 2477–2487.

Muhrer, G. and M. Mazzotti. 2003. Precipitation of lysozyme nanoparticles from dimethyl sulfoxide using carbon dioxide as antisolvent. Biotechnol. Prog. 19: 549–556.

Mukherjee, J. and M.N. Gupta. 2012. Alpha chymotrypsin coated clusters of Fe_3O_4 nanoparticles for biocatalysis in low water media. Chem. Cent. J. 6: 133.

Mukherjee, J. and M.N. Gupta. 2016. Lipase coated clusters of iron oxide nanoparticles for biodiesel synthesis in a solvent free medium. Biores. Technol. 209: 166–171.

Nam, K.T., Y.J. Lee, E.M. Krauland, S.T. Kottmann and A.M. Belcher. 2008. Peptide-mediated reduction of silver ions on engineered biological scaffolds. ACS Nano. 2: 1480–1486.

Nepal, D. and K.E. Geckeler. 2007. Proteins and carbon nanotubes: Close encounters in water. Small 3: 1259–1265.

Ng, J.D., B. Lorber, J. Witz, A. Théobald-Dietrich, D. Kern and R. Giegé. 1996. The crystallization of biological macromolecules from precipitates: Evidence for ostwald ripening. J. Cryst. Growth. 168: 50–62.

Nie, S., Y. Xing, G.J. Kim and J.W. Simons. 2007. Nanotechnology applications in cancer. Annu. Rev. Biomed. Eng. 9: 257–288.

Ranjan, S., N. Dasgupta, N., C. Sudandiradoss, C. Ramalingam and A. Kumar. 2018. Titanium dioxide nanoparticle–protein interaction explained by docking approach. Int. J. Nanomedicine. 13(T-NANO 2014 Abstracts): 47–50.

Rodrigues, M.A., J. Li, L. Padrela, A. Almeida, H.A. Matos and E.G. de Azevedo. 2009. Anti-solvent effect in the production of lysozyme nanoparticles by supercritical fluid-assisted atomization processes. J. Supercrit. Fluid. 48: 253–260.

Roy, S., C.K. Ghosh and C.K. Sarkar. 2017. Nanotechnology: Synthesis to Applications. CRC Press, Boca Raton.

Saptarshi, S.R., A. Duschl and A.L. Lopata. 2013. Interaction of nanoparticles with proteins: relation to bio-reactivity of the nanoparticle. J. Nanobiotechnol. 11: 26.

Shannahan, J.H., X. Lai, P.C. Ke, R. Podila, J.M. Brown and F.A. Witzmann. 2013. Silver nanoparticle protein corona composition in cell culture media. PloS one. 8: e74001.

Shukla, A.K. and S. Iravani. 2019. Green Synthesis, Characterization and Applications of Nanoparticles. Elsevier, Amsterdam.

Simonsen, S.B., I. Chorkendorff, S. Dahl, M. Skoglundh, J. Sehested and S. Helveg. 2010. Direct observations of oxygen-induced platinum nanoparticle ripening studied by in situ TEM. J. Am. Chem. Soc. 132: 7968–7975.

Sinha, R. and S.K. Khare. 2013. Molecular basis of nanotoxicity and interaction of microbial cells with nanoparticles. Curr. Biotechnol. 2: 64–72.

Sinha, R. and S.K. Khare. 2014. Differential interactions of halophilic and non-halophilic proteases with nanoparticles. Sustain. Chem. Processes. 2: 4.

Sinha, R. and S.K. Khare. 2015. Immobilization of halophilic *Bacillus* sp. EMB9 protease on functionalized silica nanoparticles and application in whey protein hydrolysis. Bioprocess. Biosyst. Eng. 38: 739–748.

Solanki, K. and M.N. Gupta. 2011. Simultaneous purification and immobilization of *Candida rugosa* lipase on superparamagnetic Fe_3O_4 nanoparticles for catalyzing transesterification reactions. New J. Chem. 35: 2551–2556.

Solanki, K., M.N. Gupta and P.J. Halling. 2012. Examining structure–activity correlations of some high activity enzyme preparations for low water media. Biores. Technol. 115: 147–151.

Streets, A.M. and S.R. Quake. 2010. Ostwald ripening of clusters during protein crystallization. Phys. Rev. Lett. 104: 178102.

Taylor, H. 2016. What is a species? The most important concept in all of biology is a complete mystery. https://theconversation.com/what-is-a-species-the-most-important-concept-in-all-of-biology-is-a-complete-mystery-119200 [accessed on October 20, 2019].

Torchilin, V.P. 2006. Nanoparticulates as Drug Carriers. Imperial College Press, London.

Ventola, C.L. 2015. The antibiotic resistance crisis: part 1: Causes and threats. Pharm. Ther. 40: 277–283.

Verma, S., R. Gokhale and D.J. Burgess. 2009. A comparative study of top-down and bottom-up approaches for the preparation of micro/nanosuspensions. Int. J. Pharm. 380: 216–222.

Vollath, D. 2008. Nanomaterials. Wiley-VCH, Weinheim.

Wahajuddin, S.A. 2012. Superparamagnetic iron oxide nanoparticles: Magnetic nanoplatforms as drug carriers. Int. J. Nanomedicine. 7: 3445–3471.

Wang, W., K. Gaus, R.D. Tilley and J.J. Gooding. 2019. The impact of nanoparticle shape on cellular internalisation and transport: What do the different analysis methods tell us? Mater. Horiz. 6: 1538–1547.

Wangoo, N., C.R. Suri and G. Shekhawat. 2008. Interaction of gold nanoparticles with protein: A spectroscopic study to monitor protein conformational changes. Appl. Phys. Lett. 92: 133104.

Westmeier, D., A. Hahlbrock, C. Reinhardt, J. Fröhlich–Nowoisky, S. Wessler, C. Vallet, U. Pöschl, S.K. Knauer and R.H. Stauber. 2018. Nanomaterial–microbe cross–talk: physicochemical principles and (patho) biological consequences. Chem. Soc. Rev. 47: 5312–5337.

Yang, J., M. Mayer, J.K. Kriebel, P. Garstecki and G.M. Whitesides. 2004. Self-assembled aggregates of IgGs as templates for the growth of clusters of gold nanoparticles. Angew. Chem. Int. Ed. 43: 1555–1558.

Yohannes, G., M. Jussila, K. Hartonen and M.L. Riekkola. 2011. Asymmetrical flow field-flow fractionation technique for separation and characterization of biopolymers and bioparticles. J. Chromatogr. A. 1218(27): 4104–4116.

Young, W.B. 2016. Therapeutic manipulation of the microbiota: Past, present and considerations for the future. Clin. Microbiol. Infect. 22: 905–909.

Zhang, L., L. Wu, Y. Si and K. Shu. 2018. Size-dependent cytotoxicity of silver nano-particles to *Azotobacter vinelandii*: Growth inhibition, cell injury, oxidative stress and internalization. PloS one. 13: e0209020.

3

Interactions of Metal-Containing Nanomaterials with Microorganisms

Wei Liu, Isabelle A. Worms and Vera I. Slaveykova*

Environmental Biogeochemistry and Ecotoxicology
Department F.-A. Forel for Environmental and Aquatic Sciences
Faculty of Sciences, Section of the Earth and Environmental Sciences
University of Geneva, Uni Carl Vogt, Bvd Carl-Vogt 66
CH-1211 Geneva 4, Switzerland
Tel: +41 22 379 0335
Email: wei.liu@unige.ch; isabelle.worms@unige.ch; vera.slaveykova@unige.ch

INTRODUCTION

Enhanced understanding of the interactions of engineered nanomaterials (ENMs) with microorganisms is central for predicting their behavior and effects in living organisms, for grouping and categorizing of nanomaterials, and for development of "safe-by-design" ENMs (Lynch et al. 2013, Ma and Lin 2013, Pulido-Reyes et al. 2017). The present chapter focuses on the interactions of the metallic and metal oxide nanomaterials and microorganisms, such as algae, bacteria, and protozoa. The basic concepts and current progress concerning the interactions between ENMs and ambient medium, ENMs and microorganisms, and ENMs and biological fluids, once ENMs are taken up by the microorganisms, are illustrated with selected examples of our own research as well as the literature published in

*For Correspondence: vera.slaveykova@unige.ch

the past 5 years. Recent advances in studying the ENMs behavior and effects on the environment can be found in the article by Lead et al. (2018). Further information on the specific types of ENMs can be found in some recent reviews on nanoAg (Akter et al. 2018, McGillicuddy et al. 2017, Zhang et al. 2018b), TiO_2 (Joonas et al. 2019), ZnO (Kumar et al. 2017), or CuO (Joonas et al. 2019).

The interactions of ENMs with microorganisms broadly involve several consecutive steps (Figure 3.1) (Slaveykova et al. 2016, 2020, von Moos and Slaveykova 2014). Namely, ENMs suspended in the water diffuse towards the microorganism surface. This process is size-dependent and differs, especially for released ions, single ENMs, and agglomerates formed in the ambient medium.

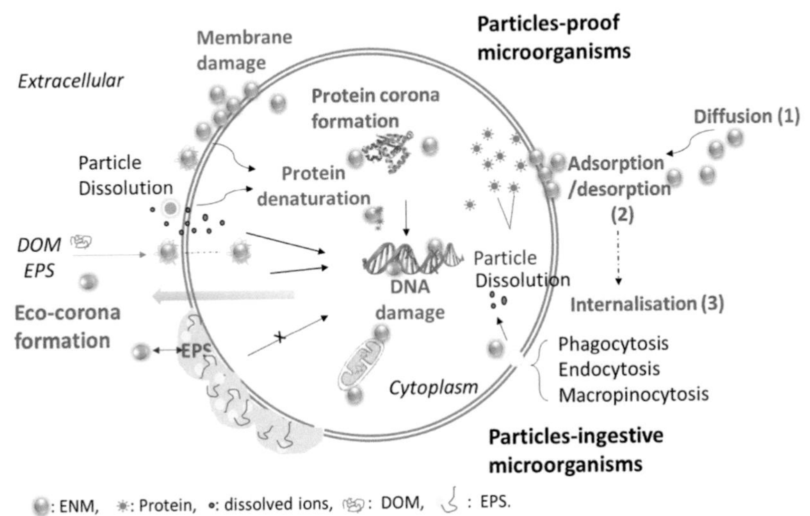

Figure 3.1 Conceptual presentation of the key processes at the interface of ambient medium-microorganisms, which govern ENM's interactions with "particle-proof" and "particle-ingestive" microorganisms; the interactions with dissolved organic matter (DOM) and extracellular polymeric substances (EPS), formation of eco-corona; and the intracellular handling of the ENMs.

Once at the vicinity of the microorganism, different ENM forms adhere on the cell walls and cell membranes, and then (but not necessarily) penetrate through the membrane. Depending on the microorganism, the ENMs can be internalized by different pathways, including endocytosis in the case of "particle-ingestive" microorganisms, alteration in cell membrane permeability for the "particle-proof" microorganism, or other possible mechanisms (Ivask et al. 2014, Ma and Lin 2013, von Moos and Slaveykova 2014). Once inside the cell, ENMs interact with different intracellular structures and biomolecules, such as proteins, lipids, and DNA, and could affect vital cellular processes; consequently ENMs can be transformed or excreted. The complex interplay between all the processes presented above will determine the overall effect of the ENMs on the microorganisms. Conversely, the microorganisms could affect the ENMs behavior in the aquatic systems directly and indirectly. An example of indirect effects is the secretion of molecules and extracellular polymeric substances (EPS), which could modify surface properties of the ENMs, their dissolution, and agglomeration. However, this aspect has received much less (or almost no) attention as compared with the significant work exploring the toxicity of the ENMs to microorganisms.

The interactions of ENMs with the microorganisms depend on: (i) the composition and physicochemical properties of the ENMs, such as size, shape, surface coating, etc.; (ii) the variables of the ambient media (e.g., pH, ionic strength, presence and type of dissolved organic matter (DOM)); and (iii) the characteristics of the microorganisms (e.g., cell wall, membrane, differentiation stage, and its pathways of particle uptake and cellular processing)(Handy et al. 2008, Ma and Lin 2013, Nel et al. 2009, von Moos et al. 2014, von Moos and Slaveykova 2014). The following sections will focus on the key interactions of the ENMs and ambient medium, ENMs and microorganisms, and ENMs and cellular biomolecules, once ENMs were taken up by the microorganisms.

INTERACTIONS OF ENMS WITH AMBIENT MEDIUM

The interactions of ENMs with different biotic and abiotic components of aquatic environment trigger their dissolution, agglomeration, aggregation, and sedimentation (Baalousha 2017, Christian et al. 2008, Lowry et al. 2012). Different physical-chemical variables, including temperature, O_2, light, ionic composition, pH, DOM concentration, etc. influence these interactions (Peng et al. 2017). In most cases, pristine ENMs form aggregates and sediment in natural waters (Smith et al. 2015, Wang et al. 2019). For dissolving ENMs, the release of ionic species correlated directly to their dispersive state, as illustrated in different natural waters (Liu et al. 2018b). Among several factors, the aggregation kinetics of nanoCuO and nanoZnO were dependent mostly on the ionic composition, strength, and were dominated by DOM stabilizing nanoparticle dispersions. The importance of the particle coating and protective role of DOM in ENMs aggregation was demonstrated for different ENMs (Clavier et al. 2019, Ellis et al. 2018, Oriekhova and Stoll 2019, Surette and Nason 2019). The size and coating of the ENMs are key factors determining their bio-interactions (Louie et al. 2016, Schaumann et al. 2015). For example, any alterations of the original coating by modification, destruction, or by molecular exchange could lead to strong changes in behavior of ENMs, and thus their toxicity (Markiewicz et al. 2018, Yu et al. 2018). Among the different ambient medium factors, we focus below on the role of the material released by the microorganisms on the behavior of the ENMs.

Microorganisms excrete various components, including nucleic acids, fatty acids, carbohydrates polymers, and proteins (which constitute a microorganism secretome) and a large part of DOM at the vicinity of the microorganisms. Despite the importance of the secretome in modulating ENMs toxicity (Chen et al. 2012a, Gao et al. 2018, Stevenson et al. 2013) and behavior (Avellan et al. 2018, Gao et al. 2018, Jiménez-Lamana and Slaveykova 2016), this topic is yet to be studied in detail. Exudate isolated from cyanobacterium *Synechocystis* sp. culture medium stabilized efficiently 20 nm citrate-nanoAg and lipoic acid-nanoAg

(Jiménez-Lamana and Slaveykova 2016). By contrast, the fibril polysaccharide enhanced the flocculation of citrate-nanoAg (Ellis et al. 2018). The exudates from green alga *Chlorella* sp. interacted with nanoZnO via electrostatic interactions and hydrogen bonding, involving amide and/or sulphate groups, which decreased Zn ion release (Chen et al. 2012b).

Interactions of the ENMs and EPS extracted either by using saline (alkaline) solutions or mechanically from the surface of planktonic microorganisms or from biofilms were also studied (Xu and Jiang 2015, Xu et al. 2016). For instance, the EPS isolated from cyanobacterial bloom, showed differences in binding rates on the nanoZnO, which decreased in the order humic-like > tryptophan-associated>fulvic-like components (Xu and Jiang 2015). Low molecular mass EPS (< 1 kDa) extracted from *S. cerevisiae* stabilized nanoCeO_2 more strongly than higher molecular mass proteins or polysaccharides (Masaki et al. 2017). EPS, such as alginate promoted nanoCuO dissolution more than EPS isolated from sludge, but with lower Cu complexation capacities (Miao et al. 2015).

The microorganisms secretome constituents are highly heterogeneous; consequently, studies able to identify molecular specificity in the adsorption are rare. Among the few examples, the extracellular DNA of bacterium *Bacillius subtilis* did not alter nanoAg, but poly-gama-glutamate (PGA) stabilized both nanoAg and nanoAg_2S (Eymard-Vernain et al. 2018b). *B. subtilis* secretome was shown to convert nanoAg to Ag_2S, because of excretion of thiol molecules and nanoAg surface oxidation, a phenomenon enhanced by the presence of bacteria (Eymard-Vernain et al. 2018a). Small ligands such as siderophores, specifically synthetized by bacteria to acquire extracellular iron, actively dissolved Al-Fe doped Ge-imogolite, rendering the initial mono-elemental nanotubes nontoxic to bacterium *Pseudomonas brassicacearum* (Avellan et al. 2016).

Although the exudate secretion is a constitutive process, the exposure of microorganism to ENMs could affect both the production and composition of the released compounds. For instance, exposure of green alga *Chlamydomonas reinhardtii* to nanoAg resulted in a decrease of the high molecular mass EPS components (Taylor et al. 2016), while the nanoCeO_2, CuO, and ZnO induced changes in the proportion of loosely bound and tightly bound EPS produced by cyanobacterium *M. aeruginosa* (Hou et al. 2017). Changes in exudates production were also shown for diatom *Thalassiosira pseudonana* exposed to CdSe/ZnS QDs (Zhang et al. 2013) and for several diatoms and green algae in the presence of nanoTiO_2, SiO_2, and CeO_2 (Chiu et al. 2017). In addition, nanoAg_2S impaired extracellular enzymatic activities involved in important biological functions, such as those of β-glucosidase (involved in carbon-cycling), L-leucine aminopeptidase (involved in nitrogen-cycling), and alkaline phosphatase (Liu et al. 2018a).

ADSORPTION AND INTERNALISATION OF ENMS BY MICROORGANISMS

ENMs adsorption and internalisation by microorganisms are key processes determining the toxicological outcome. An informative list on the studies concerning the adsorption and uptake of metal-containing ENMs in various organisms, including microorganisms, can be found in various articles (Ma and Lin 2013, von Moos et al. 2014). Hence, below we will provide some selected recent examples. For the sake of clarity, we will discuss separately adsorption and internalisation processes, although the adsorption is considered as a prerequisite for ENMs internalisation in the cells (von Moos et al. 2014).

Numerous examples show the adsorption of ENMs to microorganisms, such as microalgae and bacteria. ENMs adsorption on the microalgae was demonstrated for nanoPt and algae *Pseudokirchneriella subcapitata* and *C. reinhardtii* (Sørensen et al. 2016), nanoAg, Au, Pd, and Pt obtained (Nguyen et al. 2018a), and four iron-containing nanomaterials (Nguyen et al. 2018b), and *C. reinhardtii*. NanoAg adsorbed, but did not penetrate into alga *Euglena gracilis* with no rigid cell wall, but a flexible glycoprotein-containing pellicle (Yue et al. 2017). The positively charged nanoAu interacted with negatively charged surfaces of diatom *Eolimna minima* in a way similar to that of metal cations (Gonzalez et al. 2018), however the interactions were strongly dependent on pH and presence of exo-metabolites. Capturing of the algal cells within nanoTiO_2 agglomerates was observed in case of green algae *Raphidocelis subcapitata* and *C. reinhardtii*, a diatom *Fistulifera pelliculosa*, and a cyanobacterium *Synechocystis* sp. (Joonas et al. 2019). Similarly, various nano-sized oxides were shown previously to adsorb on different microalgae: nanoSiO_2 and CeO_2 on *P. subcapitata* (Rodea-Palomares et al. 2011, Van Hoecke et al. 2008, 2009); nanoTiO_2 (Chen et al. 2012a), nanoCuO (Cheloni et al. 2016) on *C. reinhardtii.* Negatively charged carboxyl CdSe/ZnS QDs only adsorbed onto cells of the wall-less strain, but not onto wild type *C. reinhardtii* (Worms et al. 2012).

NanoCeO_2 adhered to *E. coli*, *Synechocystis* sp. (Limbach et al. 2008, Thill et al. 2006), and *Anabaena* CPB4337 (Rodea-Palomares et al. 2011). The adsorption of nanoTiO_2 and Al_2O_3 onto *Cupriavidus metallidurans* and *E. coli* (Simon-Deckers et al. 2009), as well as for nanoAl_2O_3, SiO_2, TiO_2, ZnO onto *B. subtilis*, *E. coli* and *Pseudomonas fluorescens* (Jiang et al. 2009) was also demonstrated. NanoAg induced damage of the thylakoids and disruption of the endocytosis process of *M. aeruginosa,* but no nanoAg internalization into the cytoplasm or absorption on the membrane surface was found (Zhang et al. 2018a). To the authors' knowledge, no similar studies concerning the adhesion mechanisms of ENMs to other micro-organisms, such as protozoa and fungi in the environmental context exist.

The main forces governing the processes at the nano-bio interfaces and the adsorption of ENMs on microorganisms involve van der Waals forces, hydro-phobic forces, hydrogen bonding, electrostatic attraction, or specific chemical interactions including receptor-ligand interactions (Ma and Lin 2013, von Moos et al. 2014). Nonetheless, a rather limited number of studies clearly demonstrated the importance of a given type of interaction. Some evidences were provided about the role of the electrostatic interactions. Notably, the positive charge of nanoAg facilitated their attachment to the negatively charged cell membrane of the microorganisms (Abbaszadegan et al. 2015). The electrostatic association of the nanoAu with negatively charged lipopolysaccharides in the cell envelope of *Shewanella oneidensis* was demonstrated by manipulating the lipopolysaccharide content in the bacterial cell walls (Jacobson et al. 2015). However, the electrostatic interactions were suggested to be negligible, given similar biological response of *C. metallidurans* towards carboxyl- and amine-group functionalized CdSe/ZnS QDs (Slaveykova et al. 2009).

The adsorption of ENMs onto microorganisms depended on the thickness and composition of the cell wall (Dakal et al. 2016, Ma and Lin 2013, von Moos et al. 2014). For example, the gram-negative bacteria (~3–4 nm thickness of the cell wall), such as *E. coli*, were more sensitive to nanoAg than gram-positive bacteria, characterized by thicker (30 nm) cell walls made up of peptidoglycan, such as *S. aureus* (Feng et al. 2000). Carboxyl-CdSe/ZnS QDs adsorbed to the wall-less alga *C. reinhardtii*, while reduced adsorption was observed in the strain with wall containing glycoproteins and no adsorption was found in case of *Chlorella kesslerii* possessing a cellulosic cell wall (Worms et al. 2012). Agglomerates of nanoPt of approximately 100 nm were identified on the surface of *P. subcapitata* and *C. reinhardtii* (Sørensen et al. 2016), with a higher accumulation in *P. subcapitata*, most likely due to favored binding of Pt to the polysaccharide containing cell wall of this algal species. Overall, the diversity of the cell envelope and difference in proteins and lipopolysaccharides content could strongly affect the ENMs-microorganisms interactions. The association of nanoTiO_2 with EPS resulted in an enhancement of their accumulation by *Chlorella pyrenoidosa*, but the particles were mainly located on the cell surfaces (Gao et al. 2018). NanoCuO were attached onto the surface of green alga *C. pyrenoidosa*, however in spite of the EPS layer, they were internalized into the algae (Zhao et al. 2016). By contrast, the cell surface exopolysaccharides protected cyanobacterium *Synechocystis* against the deleterious effects of nanoTiO_2 in natural and artificial waters (Planchon et al. 2013), possibly by reducing the contact between ENMs and bacterial cell membrane.

The interaction of ENMs with proteins and polysaccharides of the cell wall and membrane is recognized as a very important aspect of the interactions between ENMs and microorganisms; however, it is almost unexplored for bacteria and algae and to be subjects of future research. Among the few examples, nanoAg caused irreversible changes in cell wall structure, resulting in its disruption due to the interaction with the sulfur-containing proteins present in the cell wall (Ghosh et al. 2012). The surface tertiary amine groups or hybrid NH_4^+/COO^- terminated surfaces were suggested as possible charged groups of diatoms adsorbing nanoAu at the cell surfaces (Gonzalez et al. 2018). Van der Waals forces were found to play a role in the interactions of nanoSiO_2 with a dioleoyl phosphatidylcholine monolayer (Vakurov et al. 2012). Hydrogen bonding and hydrophobic forces were also considered to affect bio-nano interactions (Nel et al. 2009, von Moos et al. 2014), however no specific studies have been carried out with the aquatic microorganisms so far. The only recent example showed that the capture of nanoAg with different surface modifications (oxalate, arginine, or cysteine) by *Pseudomonas putida* was related more to hydrophobic and hydrophilic forces rather than electrostatic interactions (Figueredo et al. 2018).

Endocytosis is accepted as the major entry pathway of ENMs in cells (Ma and Lin 2013, von Moos et al. 2014, von Moos and Slaveykova 2014).

Several mechanisms, including clathrin-mediated, caveolae-mediated, clathrin/caveolae-independent endocytosis, were demonstrated (Iversen et al. 2011, Mayor and Pagano 2007). Carboxylic CdSe/ZnS entered cells of ciliate *Tetrahymena thermophyla* by phagocytosis and clathrin-mediated endocytosis, as well as by using an alternative uptake pathway (Mortimer et al. 2014b). Similarly, nanoAg, Au, CuO, and TiO_2, accumulated inside *T. thermophyla* (Mortimer et al. 2014a). Endocytic internalisation was demonstrated for nanoAg (Miao et al. 2010) and thioglycolic acid stabilized CdTe QDs (Wang et al. 2013) into golden-brown alga *Ochromonasdanica* as well as nanoTiO_2 into *Anabaena variabilis* (Cherchi et al. 2011). Macropinocytosis was pointed out as the main route of internalisation of thioglycolic acid stabilized CdTe QDs in case of *O. danica* (Wang et al. 2013).

Whether the majority of algae and bacteria have specific mechanisms of particle uptake, such as endocytosis, is still a matter of debate. Nevertheless, various examples illustrate the possible internalisation of ENMs in microorganisms. For example, nanoTiO_2 were found in the cell wall and cytoplasm of the microalga *C. reinhardtii* (Chen et al. 2012a) and inside cells of the blue-green alga *Anabaena variabilis* (Cherchi et al. 2011). Bare and polymer coated nanoCuO also entered cells of the microalga *C. reinhardtii* with a higher penetrating capability in case of the polymer-coated nanoCuO (Perreault et al. 2012b). NanoAu internalized in both the wild-type and the cell wall-deficient mutant *C. reinhardtii* (Perreault et al. 2012a) and adsorbed onto the cellulosic layer of the cell walls of the microalga *S. subspicatus*, but did not enter its cytoplasm (Renault et al. 2008). Adsorption and internalization of nanoAg were suggested to cause the toxicity to *Phaeodactylum tricornutum* (Sendra et al. 2017), whereas dissolved Ag ions were the main toxicity factor in *C. reinhardtii* (Piccapietra et al. 2012, Pillai et al. 2014, Sendra et al. 2017). NanoAu dose and time dependent uptake was recently demonstrated in alga *Cryptomonas ovate*, however only 40–50% of cells contained ENMs (Merrifield et al. 2018). NanoCuO were internalized by *M. aeruginosa* (Wang et al. 2011). NanoZnO was found to be taken up by *E. coli* and *S. aureus* in a size-dependent manner, with the small sized particles accumulating more (Applerot et al. 2009). NanoZnO and TiO_2 (Kumar et al. 2011) and nanoAg were also observed inside *E. coli* cells and *B. subtilis* (Kim et al. 2011). Peptide conjugated nanoAu with a size of 5 nm penetrated cells of *E. coli* and *S. aureus* opposite to the larger 10 nm particles did not internalize. This difference was supposed to be due to the large size of the nanoAu as compared to the pore size of the bacterial cell wall (Kumar et al. 2018). CdTe/CdS QDs were also taken up by *C. reinhardtii* cells but the observed bioaccumulation was greatest for the largest QDs (Domingos et al. 2011). The above examples demonstrated that the ENMs of different chemical composition and various sizes are capable of penetrating the cell walls and plasma membranes of bacteria and microalgae. However, determining the exact mechanisms needs further attention.

In addition to the endocytosis, other possible mechanisms of internalisation of ENMs through the cell membrane/walls of bacterium and microalgae may include (Kloepfer et al. 2005, Ma and Lin 2013, von Moos et al. 2014): uptake by non-specific diffusion, uptake by non-specific membrane damage, specific uptake, or adhesive interactions. *Non-specific diffusion* is considered for particles smaller than e.g., 2 nm, size corresponding to the largest globular protein that can pass through the intact bacterium cell walls (Kloepfer et al. 2005). Although theoretically possible, this pathway has not yet been experimentally demonstrated. Exposure of *liposomes* (used as models for biological membranes to study the potential uptake by passive diffusion through the phospholipid bilayer) to 5 nm polyvinylpyrrolidone-coated nanoAg showed no significant uptake of nanoAg, nor their invagination (Guilleux et al. 2018). Study of the interaction of nanoAg functionalized with citrate and phytomolecules and liposomes revealed that the phytomolecules capped nanoAg had significantly higher affinity presumably via the insertion of aromatic/hydrophobic moieties of capping agents on the surface of nanoAg into the lipid bilayer (Maillard et al. 2018). *Specific uptake* is another possible transport way for entry of particles in the cells. In most cases, the pore sizes are unknown, but the largest pore size for specific transport in bacterium is *ca*. 6 nm (Wang et al. 2003), therefore this pathway will be possible for particles smaller than about 5 nm. This uptake route was demonstrated for the adenine and AMP labelled CdSe and CdSe/ZnS quantum dots smaller than 5 nm that entered the *B. subtilis* cells walls *via purine dependent mechanisms*, as mutant lacking single enzymes demonstrated different signal than wild-type (Kloepfer et al. 2005). However, the uptake of ENMs via transport proteins for ions is supposed as unlikely, because they are much larger than ions (ranging 30 pm to $\geq$ 200 pm), and therefore most likely do not fit the respective binding sites (Handy et al. 2008). In a case of dissolving ENMs, the internalisation occurs by the same pathways as for trace metals, including active and passive transport mechanisms via (non-) specific transporter (carrier) proteins and channels, co-transport, or simple diffusion (Newman 2010, Tessier and Turner 1995). The cell entry of the ENMs via *non-specific cell wall and membrane damage*, which causes significant changes in the cell membrane structure and permeability, is considered as a plausible mechanism favoring the internalisation of ENMs in algal and bacterial cells (Ma and Lin 2013, von Moos et al. 2014). Indeed, adsorption of nanoAg to the cell wall resulted in membrane permeability alteration and cell wall disruption for different bacteria (Dakal et al. 2016, Kim et al. 2007). For example, nanoAg attached to the cell membrane of *E. coli* cells induced the numerous electron dense pits (Sondi and Salopek-Sondi 2004). Alteration of the membrane permeability of bacteria by nanoAg was shown to result in (i) an alteration of the cellular ability to properly regulate transport activity through the plasma membrane, and (ii) induce leakage of cellular contents, including ions, proteins, reducing sugars, and cellular energy reservoir (Dakal et al. 2016). Similarly, adsorption of nanoTiO_2 on alga *P. subcapitata* could interfere with membrane, nutrient uptake, and other cellular processes involving the cell surface (Hartmann et al. 2010). Indeed, nanoTiO_2 accumulated in the cell wall and membrane of *C. reinhardtii*, causing membrane and organelles damage, as well as plasmolysis (Chen et al. 2012a). NanoTiO_2 induced the membrane deformation in green alga *R. subcapitata*, as well as altered cellular membrane permeability of *C. reinhardtii* (Sottocasa 2012). Similarly, more than 20% of the cells of *C. reinhardtii* had damaged membranes after 1 hour of exposure to 200 mg/L nanoTiO_2 (von Moos et al. 2016). ENMs adsorption to the surface of photosynthetic microorganism could also induce shading, reducing light received by the algae, thus indirectly affecting the photosynthesis efficiency and growth. NanoPt adhesion to the algal surface of *P. subcapitata* and *C. reinhardtii* affected their membrane permeability (Sørensen et al. 2016), as did the nanoCuO for *C. reinhardtii* (Cheloni et al. 2016, von Moos et al. 2015). It is also worth noting that the cell walls contain pores with diameters ranging from 5–20 nm (Bhatt and Tripathi 2011, Zemke-White et al. 2000), which could be potential entry ports for small ENMs. The possibility that interactions of ENMs with cells may even induce the formation of larger, new pores permeable to bigger ENMs was also pointed out (Navarro et al. 2008). In addition, the permeability of cell walls changes throughout cell cycling, but especially during the cell division, in the course of which cell walls are newly synthesized (Navarro et al. 2008, Wang et al. 2011).

It is also worth noting that the cell walls of microorganisms are considered efficient barriers which prevent internalisation of ENMs, and thus majority of the bacteria and algae are often considered as *particle-proof* microorganisms, as previously reviewed (Ma et al. 2013, von Moos et al. 2014). Indeed, Au-mannose ENMs were shown to penetrate the cell wall-deficient strain of *C. reinhardtii* more easily (Perreault et al. 2012a). Similarly, the gram-positive bacterium *B. subtilis* with thick outer peptidoglycan layer may have better protection from the toxic effects of nanoAg as compared with *E. coli* (Kim et al. 2011). No uptake of functionalized CdSe/ZnS QDs was observed in the cell wall-deficient *C. reinhardtii* strain, despite obvious interactions between the algae and QDs (Worms et al. 2012). By contrast, the same QDs were found in the periplasmic space and on the surface of the gram-negative bacterium *C. metallidurance* (Slaveykova et al. 2013).

INTRACELLULAR INTERACTIONS OF ENMS

Once ENMs are internalised into the cells, they interact with diverse biomolecules, including protein, DNA, lipids, hormones, and other small molecules (Lundqvist et al. 2008, Medlin and Snyder 2009, Lynch et al. 2013, von Moos et al. 2014). Understanding the interaction between ENMs and these biomolecules has attracted interest due to the key role of the biomolecules in the cellular function and pathways (Lead et al. 2018). ENMs could affect the biomolecules via direct interactions or indirectly via formation of reactive oxygen species (ROS).

The interaction of ENMs with cellular proteins is considered as an important mechanism of toxicity (Monopoli et al. 2012, Nel et al. 2009). Proteins adsorb on the surface of ENMs, forming the "protein corona", which gives "new signature" to ENMs, but also could alter protein conformation and activity, and thus affect cellular function (Saptarshi et al. 2013, Zarei and Aalaie 2019). Despite the importance of protein corona in intracellular fate and toxicity of ENMs to microorganisms, this topic is still largely unexplored. Among the few examples, tryptophanase protein from *E. coli* was shown to exclusively associate with carbonate-coated nanoAg and not with bare nanoAg, suggesting preferential binding of some protein fragments depending on ENM surface coating (Wigginton et al. 2010). Metallothionein rapidly surrounded the 20 nm citrate coated nanoAg with the transient reticulate corona, thus promoting their dissolution associated with the metal substitution, while ceruloplasmin formed stable corona with a limited substitution of Cu and a decrease in its ferroxidase activity (Liu et al. 2017). These two *in vitro* studies of the interactions between ENMs and single protein could facilitate the understanding of the processes underlying protein corona formation.

Proteomic studies on bacteria provided an overview of the changes in the proteomics profile in cells and the function of those proteins upon exposure to ENMs (Zhang and Wu 2015). For example, nanoAg-induced up-regulation of formate acetyltransferase, which promoted anaerobic condition by the termination of electron transport chain functions in bacterium *S. aureus* (Li et al. 2011). Both bare and graphene oxide coated nanoAg induced distinct changes in the proteomic profile of *Pseudomonas aeruginosa*, affecting eight responsive proteins, involved in translation (He et al. 2014). NanoTiO_2 induced changes in the expression of proteins involved in responses to osmotic stress, metabolism of cell envelope components, and uptake/metabolism of endogenous and exogenous compounds (Sohm et al. 2015).

Interactions of ENMs and DNA result in a formation of DNA/ENMs complexes, DNA conformation changes, and DNA degradation (Kumar et al. 2011, Liu 2012, Prado-Gotor and Grueso 2011). Particularly, study of DNA/ENMs complex structures suggested that there are different kinds of complexation, which occur as ENMs size decreases in relation to DNA length (Zinchenko et al. 2007). *In vitro* studies demonstrated that DNA binds to nanoAu through the DNA bases and depends on the particle size, as well as salt concentration, pH, temperature, and DNA secondary structures (Liu 2012). NanoAu capped with glycine formed a stable DNA/Au nanoparticles complex via hydrophilic groups of the tiopronin (2-(2-sulfanylpropanoylamino) acetic acid) and the DNA grooves, thus inducing DNA conformational change (Prado-Gotor and Grueso 2011). Similarly, binding to grooves was observed with plasmid DNA isolated from *E. coli* in the presence of Cu-containing particles, which resulted in a dose-dependent degradation of plasmid DNA pET11a (Giannousi et al. 2014) and pUC19 (Chatterjee et al. 2014).

NanoAg bound with DNA induced deformation of *S. aureus* cells (Grigor'eva et al. 2013). Similarly, nanoAg caused DNA damage, resulting in a loss of replication and degradation of DNA, thereby inhibiting the growth of bacteria *Proteus* sp. and *Klebsiella.* sp. isolates from microbial polluted water (Ouda 2014). The comparison between nanoAg (8 nm) on *E. coli* wild type strains and strains with mutation in genes responsible for the repair of DNA containing oxidative lesion (*mutY, mutS, mutM, mutT, nth*) indicated that the oxidative DNA damage plays an important role in the toxicity of nanoAg (Radzig et al. 2013). Exposure to nanoAl_2O_3 induced stronger DNA oxidative damage to *P. aeruginosa* than as compared to *Bacillus altitudinis*, demonstrating that the excessive generation of ROS could act as a major factor of DNA damage in the bacteria cells (Bhuvaneshwari et al. 2016). NanoCu caused DNA degradation, triggering the ROS (superoxide anions, hydrogen peroxide) and single-stranded DNA responses in *E. coli* (Bondarenko et al. 2012). NanoCu and nanoCuO internalized as intact particles caused complete degradation of plasmid DNA in *E. coli* and *Lactobacillus brevis* (Kaweeteerawat et al. 2015). NanoCuO induced DNA strand breaks in a concentration-dependent manner due to the oxidative stress in aquatic fungi collected from streams, too (Pradhan et al. 2015).

The potential of nanoTiO_2, SiO_2, and ZnO to damage DNA and the underlying mechanisms were investigated in microalga *Dunaliella tertiolecta* (Schiavo et al. 2016). It was found that nanoZnO induced significant DNA strand breaks close to environmental concentration through the release of Zn^{2+}, while nanoSiO_2 induced DNA strand breaks via oxidative stress; nanoTiO_2 indirectly affected the DNA structure by activation of cellular signals during cell division or cell wall destruction.

CONCLUSION

ENMs-microorganisms interactions are highly complex and result from dynamic interplay of the physicochemical properties of ENMs, variables of the ambient medium, and the type of microorganisms. Most of the research has been focused on understanding how ENMs affect microorganisms, while the question of how microorganisms influence ENMs has received much less attention. Therefore, future work should tackle the interaction of ENMs with secretome of the microorganism.

A significant progress has been made in the understanding of the interactions of the metal-containing ENMs and microorganisms, providing various evidences about the capability of the ENMs to be adsorbed and internalized in a way depending on the characteristics of the particles, the nature of the microorganisms, and the variables of the ambient medium. However, a quantitative characterization of the kinetics of the uptake and excretion processes, assessment of the importance of internalization with respect to adsorption to surfaces, vis-a-vis the intracellular fate of ENMs in aquatic microorganisms, are still lacking and should be one of the important areas of future research. In addition, the majority of the current knowledge is based on bioassays with the model microalgae and bacteria, or ciliate in well-controlled media with relatively high concentrations of ENMs. It is yet to be demonstrated that it is applicable for other groups of microorganisms, such as fungi and archaea. Various possible internalisation mechanisms of ENMs through the cell membrane/walls of microorganisms including the endocytosis, uptake by non-specific diffusion, non-specific membrane damage, and specific uptake/adhesive interactions have been suggested. Some of these are shown to be relevant for the microorganisms. However, it is still unclear, if endocytic pathways are common in microorganisms, such as bacteria and microalgae, or rather are an exception.

The interactions of ENMs with intracellular molecules, such as protein, DNA, and cellular membrane component all contribute to their toxicity to microorganisms. However, it is still unclear what the key proteins and signalling pathways involved in ENMs toxicity are. Therefore, further work is necessary to address this knowledge gap. Finally most of the available information concerning the ENMs and microorganisms interactions originated from the bioassays after short-term exposure with ENMs of high concentrations. Further systematic studies are thus necessary to verify if the current concepts are applicable at long-term exposure settings and lower concentrations of ENMs.

ACKNOWLEDGMENTS

VS is grateful for the financial support of the Swiss National Science Foundation grant IZSEZ0-180186.

References

Abbaszadegan, A., Y. Ghahramani, A. Gholami, B. Hemmateenejad, S. Dorostkar, M. Nabavizadeh, et al. 2015. The effect of charge at the surface of silver nanoparticles on antimicrobial activity against Gram-positive and Gram-negative bacteria: A preliminary study. J. Nanomater. 8: 720654.

Akter, M., M.T. Sikder, M.M. Rahman, A. Ullah, K.F.B. Hossain, S. Banik, et al. 2018. A systematic review on silver nanoparticles-induced cytotoxicity: Physicochemical properties and perspectives. J. Adv. Res. 9: 1–16.

Applerot, G., A. Lipovsky, R. Dror, N. Perkas, Y. Nitzan, R. Lubart, et al. 2009. Enhanced antibacterial activity of nanocrystalline ZnO due to increased ROS-mediated cell injury. Adv. Funct. Mater. 19: 842–852.

Avellan, A., M. Auffan, A. Masion, C. Levard, M. Bertrand, J. Rose, et al. 2016. Remote biodegradation of Ge-Imogolite nanotubes controlled by the iron homeostasis of *Pseudomonas brassicacearum*. Environ. Sci. Technol. 50: 7791–7798.

Avellan, A., M. Simonin, E. McGivney, N. Bossa, E. Spielman-Sun, J.D. Rocca, et al. 2018. Gold nanoparticle biodissolution by a freshwater macrophyte and its associated microbiome. Nat. Nanotechnol. 13: 1072–1077.

Baalousha, M., 2017. Effect of nanomaterial and media physicochemical properties on nanomaterial aggregation kinetics. Nanoimpact. 6: 55–68.

Bhatt, I. and B.N. Tripathi. 2011. Interaction of engineered nanoparticles with various components of the environment and possible strategies for their risk assessment. Chemosphere 82: 308–317.

Bhuvaneshwari, M., S. Bairoliya, A. Parashar, N. Chandrasekaran and A. Mukherjee. 2016. Differential toxicity of Al_2O_3 particles on Gram-positive and Gram-negative sediment bacterial isolates from freshwater. Environ. Sci. Pollut. Res. 23: 12095–12106.

Bondarenko, O., A. Ivask, A. Käkinen and A. Kahru. 2012. Sub-toxic effects of CuO nanoparticles on bacteria: Kinetics, role of Cu ions and possible mechanisms of action. Environ. Pollut. 169: 81–89.

Chatterjee, A.K., R. Chakraborty and T. Basu. 2014. Mechanism of antibacterial activity of copper nanoparticles. Nanotechnology 25: e135101.

Cheloni, G., E. Marti and V.I. Slaveykova. 2016. Interactive effects of copper oxide nanoparticles and light to green alga *Chlamydomonas reinhardtii*. Aquat. Toxicol. 170: 120–128.

Chen, L.Z., L.N. Zhou, Y.D. Liu, S.Q. Deng, H. Wu and G.H. Wang. 2012a. Toxicological effects of nanometer titanium dioxide (nano-TiO_2) on *Chlamydomonas reinhardtii*. Ecotoxicol. Environ. Saf. 84: 155–162.

Chen, P., B.A. Powell, M. Mortimer and P.C. Ke. 2012b. Adaptive interactions between zinc oxide nanoparticles and *Chlorella* sp. Environ. Sci. Technol. 46: 12178–12185.

Cherchi, C., T. Chernenko, M. Diem and A.Z. Gu. 2011. Impact of nano titanium dioxide exposure on cellular structure of *Anabaena variabilis* and evidence of internalization. Environ. Toxicol. Chem. 30: 861–869.

Chiu, M.H., Z.A. Khan, S.G. Garcia, A.D. Le, A. Kagiri, J. Ramos, et al. 2017. Effect of engineered nanoparticles on exopolymeric substances release from marine phytoplankton. Nanoscale. Res. Lett. 12: 620.

Christian, P., F. Von der Kammer, M. Baalousha and T.J.E. Hofmann. 2008. Nanoparticles: Structure, properties, preparation and behaviour in environmental media. Ecotoxicology 17: 326–343.

Clavier, A., A. Praetorius and S. Stoll. 2019. Determination of nanoparticle heteroaggregation attachment efficiencies and rates in presence of natural organic matter monomers. Monte Carlo modelling. Sci. Total. Environ. 650: 530–540.

Dakal, T.C., A. Kumar, R.S. Majumdar and V. Yadav. 2016. Mechanistic basis of antimicrobial actions of silver nanoparticles. Front. Microbiol. 7: 1831.

Domingos, R.F., D.F. Simon, C. Hauser and K.J. Wilkinson. 2011. Bioaccumulation and effects of CdTe/CdS quantum dots on *Chlamydomonas reinhardtii* – nanoparticles or the free ions? Environ. Sci. Technol. 45: 7664–7669.

Ellis, L.A., M. Baalousha, E. Valsami-Jones and J.R. Lead. 2018. Seasonal variability of natural water chemistry affects the fate and behaviour of silver nanoparticles. Chemosphere 191: 616–625.

Eymard-Vernain, E., Y. Coute, A. Adrait, T. Rabilloud, G. Sarret and C. Lelong. 2018a. The poly-gamma-glutamate of *Bacillus subtilis* interacts specifically with silver nanoparticles. PLoS One 13: e0197501.

Eymard-Vernain, E., C. Lelong, A.E. Pradas Del Real, R. Soulas, S. Bureau, V. Tardillo Suarez, et al. 2018b. Impact of a model soil microorganism and of its secretome on the fate of silver nanoparticles. Environ. Sci. Technol. 52: 71–78.

Feng, Q.L., J. Wu, G.Q. Chen, F.Z. Cui, T.N. Kim and J.O. Kim. 2000. A mechanistic study of the antibacterial effect of silver ions on *Escherichia coli* and *Staphylococcus aureus*. J. Biomed. Mater. Res. 52: 662–668.

Figueredo, F., A. Saavedra, E. Corton and V.E. Diz. 2018. Hydrophobic forces are relevant to bacteria-nanoparticle interactions: *Pseudomonas putida* capture efficiency by using arginine, cysteine or oxalate wrapped magnetic nanoparticles. Colloid. Interface. 2: 29.

Gao, X., K.J. Zhou, L.Q. Zhang, K. Yang and D.H. Lin. 2018. Distinct effects of soluble and bound exopolymeric substances on algal bioaccumulation and toxicity of anatase and rutile TiO_2 nanoparticles. Environ. Sci. Nano. 5: 720–729.

Ghosh, S., S. Patil, M. Ahire, R. Kitture, S. Kale, K. Pardesi, et al. 2012. Synthesis of silver nanoparticles using *Dioscorea bulbifera* tuber extract and evaluation of its synergistic potential in combination with antimicrobial agents. Int. J. Nanomed. 7: 483–496.

Giannousi, K., K. Lafazanis, J. Arvanitidis, A. Pantazaki and C. Dendrinou-Samara. 2014. Hydrothermal synthesis of copper based nanoparticles: antimicrobial screening and interaction with DNA. J. Inorg. Biochem. 33: 24–32.

Gonzalez, A.G., O.S. Pokrovsky, I.S. Ivanova, O. Oleinikova, A. Feurtet-Mazel, S. Mornet, et al. 2018. Interaction of freshwater diatom with gold nanoparticles: Adsorption, assimilation, and stabilization by cell exometabolites. Minerals 8: 99.

Grigor'eva, A., I. Saranina, N. Tikunova, A. Safonov, N. Timoshenko, A. Rebrov, et al. 2013. Fine mechanisms of the interaction of silver nanoparticles with the cells of *Salmonella typhimurium* and *Staphylococcus aureus*. Biometals 26: 479–488.

Guilleux, C., P.G.C. Campbell and C. Fortin. 2018. Interactions between silver nanoparticles/silver Ions and liposomes: Evaluation of the potential passive diffusion of silver and effects of speciation. Arch. Environ. Con. Tox. 75: 634–646.

Handy, R.D., R. Owen and E. Valsami-Jones. 2008. The ecotoxicology of nanoparticles and nanomaterials: Current status, knowledge gaps, challenges, and future needs. Ecotoxicol. 17: 315–325.

Hartmann, N.B., F. Von der Kammer, T. Hofmann, M. Baalousha, S. Ottofuelling and A. Baun. 2010. Algal testing of titanium dioxide nanoparticles—Testing considerations, inhibitory effects and modification of cadmium bioavailability. Toxicology 269: 190–197.

He, T., H. Liu, Y. Zhou, J. Yang, X. Cheng and H. Shi. 2014. Antibacterial effect and proteomic analysis of graphene-based silver nanoparticles on a pathogenic bacterium *Pseudomonas aeruginosa*. Biometals 27: 673–682.

Hou, J., Y. Yang, P. Wang, C. Wang, L. Miao, X. Wang, et al. 2017. Effects of CeO_2, CuO, and ZnO nanoparticles on physiological features of *Microcystis aeruginosa* and the production and composition of extracellular polymeric substances. Environ. Sci. Pollut. Res. 24: 226–235.

Ivask, A., K. Juganson, O. Bondarenko, M. Mortimer, V. Aruoja, K. Kasemets, et al. 2014. Mechanisms of toxic action of Ag, ZnO and CuO nanoparticles to selected ecotoxicological test organisms and mammalian cells in vitro: A comparative review. Nanotoxicology 8: 57–71.

Iversen, T.G., T. Skotland and K. Sandvig. 2011. Endocytosis and intracellular transport of nanoparticles: Present knowledge and need for future studies. Nano. Today 6: 176–185.

Jacobson, K.H., I.L. Gunsolus, T.R. Kuech, J.M. Troiano, E.S. Melby, S.E. Lohse, et al. 2015. Lipopolysaccharide density and structure govern the extent and distance of nanoparticle interaction with Actual and Model Bacterial Outer Membranes. Environ. Sci. Technol. 49: 10642–10650.

Jiang, W., H. Mashayekhi and B.S. Xing. 2009. Bacterial toxicity comparison between nano- and micro-scaled oxide particles. Environ. Pollut. 157: 1619–1625.

Jiménez-Lamana, J. and V.I. Slaveykova. 2016. Silver nanoparticle behaviour in lake water depends on their surface coating. Sci. Total. Environ. 573: 946–953.

Joonas, E., V. Aruoja, K. Olli and A. Kahru. 2019. Environmental safety data on CuO and TiO_2 nanoparticles for multiple algal species in natural water: Filling the data gaps for risk assessment. Sci. Total. Environ. 647: 973–980.

Kaweeteerawat, C., C.H. Chang, K.R. Roy, R. Liu, R. Li, D. Toso, et al. 2015. Cu nanoparticles have different impacts in *Escherichia coli* and *Lactobacillus brevis* than their microsized and ionic analogues. ACS Nano 9: 7215–7225.

Kim, J.S., E. Kuk, K.N. Yu, J.H. Kim, S.J. Park, H.J. Lee, et al. 2007. Antimicrobial effects of silver nanoparticles. Nanomed. Nanotechnol. 3: 95–101.

Kim, S.W., Y.W. Baek and Y.J. An. 2011. Assay-dependent effect of silver nanoparticles to *Escherichia coli* and *Bacillus subtilis*. Appl. Microbiol. Biotechnol. 92: 1045–1052.

Kloepfer, J.A., R.E. Mielke and J.L. Nadeau. 2005. Uptake of CdSe and CdSe/ZnS quantum dots into bacteria via purine-dependent mechanisms. Appl. Environ. Microbiol. 71: 2548–2557.

Kumar, A., A.K. Pandey, S.S. Singh, R. Shanker and A. Dhawan. 2011. Engineered ZnO and TiO_2 nanoparticles induce oxidative stress and DNA damage leading to reduced viability of *Escherichia coli*. Free. Radic. Biol. Med. 51: 1872–1881.

Kumar, M., W. Tegge, N. Wangoo, R. Jain and R.K. Sharma. 2018. Insights into cell penetrating peptide conjugated gold nanoparticles for internalization into bacterial cells. Biophys. Chem. 237: 38–46.

Kumar, R., A. Umar, G. Kumar and H.S. Nalwa. 2017. Antimicrobial properties of ZnO nanomaterials: A review. Cermat. Int. 43: 3940–3961.

Lead, J.R., G.E. Batley, P.J.J. Alvarez, M.N. Croteau, R.D. Handy, M.J. McLaughlin, et al. 2018. Nanomaterials in the environment: Behavior, fate, bioavailability, and effects. An updated review. Environ. Toxicol. Chem. 37: 2029–2063.

Li, W.R., X.B. Xie, Q.S. Shi, S.S. Duan, Y.S. Ouyang and Y.B. Chen. 2011. Antibacterial effect of silver nanoparticles on *Staphylococcus aureus*. Biometals. 24: 135–141.

Limbach, L.K., R. Bereiter, E. Müller, R. Krebs, R. Gälli and W.J. Stark. 2008. Removal of oxide nanoparticles in a model wastewater treatment plant: Influence of agglomeration and surfactants on clearing efficiency. Environ. Sci. Technol. 42: 5828–5833.

Liu, J. 2012. Adsorption of DNA onto gold nanoparticles and graphene oxide: Surface science and applications. Phys. Chem. Chem. Phys. 14: 10485–10496.

Liu, W., I.A. Worms, N. Herlin-Boime, D. Truffier-Boutry, I. Michaud-Soret, E. Mintz, et al. 2017. Interaction of silver nanoparticles with metallothionein and ceruloplasmin: Impact on metal substitution by Ag(I), corona formation and enzymatic activity. Nanoscale. 9: 6581–6594.

Liu, S., C. Wang, J. Hou, P. Wang, L. Miao and T. Li. 2018a. Effects of silver sulfide nanoparticles on the microbial community structure and biological activity of freshwater biofilms. Environ. Sci. Nano. 5: 2899–2908.

Liu, Z., C. Wang, J. Hou, P. Wang, L. Miao, B. Lv, et al. 2018b. Aggregation, sedimentation, and dissolution of CuO and ZnO nanoparticles in five waters. Environ. Sci. Pollut. Res. 25: 31240–31249.

Louie, S.M., R.D. Tilton and G.V. Lowry. 2016. Critical review: Impacts of macromolecular coatings on critical physicochemical processes controlling environmental fate of nanomaterials. Environ. Sci. Nano. 3: 283–310.

Lowry, G.V., K.B. Gregory, S.C. Apte and J.R. Lead. 2012. Transformations of nanomaterials in the environment. Environ. Sci. Technol. 46: 6893–6899.

Lundqvist, M., J. Stigler, G. Elia, I. Lynch, T. Cedervall and K.A. Dawson. 2008. Nanoparticle size and surface properties determine the protein corona with possible implications for biological impacts. Proceedings of the National Academy of Sciences 105(38): 14265–14270.

Lynch, I., A. Ahluwalia, D. Boraschi, J. Byrne Hugh, B. Fadeel, P. Gehr, et al. 2013. The bio-nano-interface in predicting nanoparticle fate and behaviour in living organisms: towards grouping and categorising nanomaterials and ensuring nanosafety by design. BioNanoMaterial. 14: 195–216.

Ma, H., P.L. Williams and S.A. Diamond. 2013. Ecotoxicity of manufactured ZnO nanoparticles—A review. Environ. Pollut. 172: 76–85.

Ma, S. and D. Lin. 2013. The biophysicochemical interactions at the interfaces between nanoparticles and aquatic organisms: adsorption and internalization. Environ. Sci. Processes Impacts. 14: 145–160.

Maillard, A., P.R. Dalmasso, B.A.L. de Mishima and A. Hollmann. 2018. Interaction of green silver nanoparticles with model membranes: possible role in the antibacterial activity. Colloid. Surface. 171: 320–326.

Markiewicz, M., J. Kumirska, I. Lynch, M. Matzke, J. Köser, S. Bemowsky, et al. 2018. Changing environments and biomolecule coronas: consequences and challenges for the design of environmentally acceptable engineered nanoparticles. Green Chem. 20: 4133–4168.

Masaki, S., Y. Nakano, K. Ichiyoshi, K. Kawamoto, A. Takeda, T. Ohnuki, et al. 2017. Adsorption of extracellular polymeric substances derived from *S. cerevisiae* to ceria nanoparticles and the effects on their colloidal stability. Environments. 4: 48.

Mayor, S and R.E. Pagano. 2007. Pathways of clathrin-independent endocytosis. Nat. Rev. Mol. Cell Biol. 8: 603–612.

McGillicuddy, E., I. Murray, S. Kavanagh, L. Morrison, A. Fogarty, M. Cormican, et al. 2017. Silver nanoparticles in the environment: Sources, detection and ecotoxicology. Sci. Total. Environ. 575: 231–246.

Medlin, D. and G.J. Snyder. 2009. Interfaces in bulk thermoelectric materials: A review for current opinion in colloid and interface science. Curr. Opin. Colloid Interface Sci. 14: 226–235.

Merrifield, R.C., C. Stephan and J.R. Lead. 2018. Quantification of Au nanoparticle biouptake and freshwater algae using single cell - ICP-MS. Environ. Sci. Technol. 52: 2271–2277.

Miao, A.-J., Z. Luo, C.-S. Chen, W.-C. Chin, P.H. Santschi and A. Quigg. 2010. Intracellular uptake: A possible mechanism for silver engineered nanoparticle toxicity to a freshwater alga *Ochromonas danica*. PLoS ONE. 5: e15196.

Miao, L.Z., C. Wang, J. Hou, P.F. Wang, Y.H. Ao, Y. Li, et al. 2015. Enhanced stability and dissolution of CuO nanoparticles by extracellular polymeric substances in aqueous environment. J. Nanoparticle. Res. 17: 404.

Monopoli, M.P., C. Aberg, A. Salvati and K.A. Dawson. 2012. Biomolecular coronas provide the biological identity of nanosized materials. Nat. Nanotech. 7: 779–786.

Mortimer, M., A. Gogos, N. Bartolomé, A. Kahru, T.D. Bucheli and V.I. Slaveykova. 2014a. Potential of hyperspectral imaging microscopy for semi-quantitative analysis of nanoparticle uptake by protozoa. Environ. Sci. Technol. 48: 8760–8767.

Mortimer, M., A. Kahru and V.I. Slaveykova. 2014b. Uptake, localization and clearance of quantum dots in ciliated protozoa *Tetrahymena thermophila*. Environ. Pollut. 190: 58–64.

Navarro, E., A. Baun, R. Behra, N.B. Hartmann, J. Filser, A.J. Miao, et al. 2008. Environmental behavior and ecotoxicity of engineered nanoparticles to algae, plants, and fungi. Ecotoxicology. 17: 372–386.

Nel, A.E., L. Mädler, D. Velegol, T. Xia, E.M.V. Hoek, P. Somasundaran, et al. 2009. Understanding biophysicochemical interactions at the nano-bio interface. Nat. Mater. 8: 543–557.

Newman, M., 2010. Fundamentals of Ecotoxicology, 3rd Ed. CRC Press, Boca Raton, FL USA.

Nguyen, N.H.A., V.V.T Padil, V.I. Slaveykova, M. Černík and A. Ševců. 2018a. Green synthesis of metal and metal oxide nanoparticles and their effect on the unicellular alga *Chlamydomonas reinhardtii*. Nanoscale. Res. Lett. 13: 159.

Nguyen, N.H.A., N.R. Von Moos, V.I. Slaveykova, K. Mackenzie, R.U. Meckenstock, S. Thűmmler, et al. 2018b. Biological effects of four iron-containing nanoremediation materials on the green alga *Chlamydomonas* sp. Ecotoxicol. Environ. Saf. Safety. 154: 36–44.

Oriekhova, O. and S. Stoll. 2019. Heteroaggregation of CeO_2 nanoparticles in presence of alginate and iron (III) oxide. Sci. Total. Environ. 648: 1171–1178.

Ouda, S.M. 2014. Some nanoparticles effects on *Proteus* sp. and *Klebsiella* sp. isolated from water. Am. J. Infect. Dis. Microbiol. 2: 4–10.

Peng, C., W. Zhang, H. Gao, Y. Li, X. Tong, K. Li, et al. 2017. Behavior and potential impacts of metal-Based engineered nanoparticles in aquatic environments. Nanomaterials. 7: 1.

Perreault, F., N. Bogdan, M. Morin, J. Claverie and R. Popovic. 2012a. Interaction of gold nanoglycodendrimers with algal cells (*Chlamydomonas reinhardtii*) and their effect on physiological processes. Nanotoxicology. 6: 109–120.

Perreault, F., A. Oukarroum, S.P. Melegari, W.G. Matias and R. Popovic. 2012b. Polymer coating of copper oxide nanoparticles increases nanoparticles uptake and toxicity in the green alga *Chlamydomonas reinhardtii*. Chemosphere. 87: 1388–1394.

Piccapietra, F., C.G. Allué, L. Sigg and R. Behra. 2012. Intracellular silver accumulation in *Chlamydomonas reinhardtii* upon exposure to carbonate coated silver nanoparticles and silver nitrate. Environ. Sci. Technol. 46: 7390–7397.

Pillai, S., R. Behra, H. Nestler, M.J-F. Suter, L. Sigg and K. Schirmer. 2014. Linking toxicity and adaptive responses across the transcriptome, proteome, and phenotype of *Chlamydomonas reinhardtii* exposed to silver. Proc. Natl. Acad. Sci. 111: 3490–3495.

Planchon, M., T. Jittawuttipoka, C. Cassier-Chauvat, F. Guyot, A. Gelabert, M.F. Benedetti, et al. 2013. Exopolysaccharides protect *Synechocystis* against the deleterious effects of titanium dioxide nanoparticles in natural and artificial waters. J. Colloid Interf. Sci. 405: 35–43.

Pradhan, A., S. Seena, D. Schlosser, K. Gerth, S. Helm, M. Dobritzsch, et al. 2015. Fungi from metal-polluted streams may have high ability to cope with the oxidative stress induced by copper oxide nanoparticles. Environ. Toxicol. Chem. 34: 923–930.

Prado-Gotor, R. and E. Grueso. 2011. A kinetic study of the interaction of DNA with gold nanoparticles: Mechanistic aspects of the interaction. Phys. Chem. Chem. Phys. 13: 1479–1489.

Pulido-Reyes, G., F. Leganes, F. Fernandez-Pinas and R. Rosal. 2017. Bio-nano interface and environment: A critical review. Environ. Toxicol. Chem. 36: 3181–3193.

Radzig, M., V. Nadtochenko, O. Koksharova, J. Kiwi, V. Lipasova and I. Khmel. 2013. Antibacterial effects of silver nanoparticles on gram-negative bacteria: Influence on the growth and biofilms formation, mechanisms of action. Colloids. Surf. B. Biointerfaces. 102: 300–306.

Renault, S., M. Baudrimont, N. Mesmer-Dudons, P. Gonzalez, S. Mornet and A. Brisson. 2008. Impacts of gold nanoparticle exposure on two freshwater species: A phytoplanktonic alga (*Scenedesmus subspicatus*) and a benthic bivalve (*Corbicula fluminea*). Gold Bull. 41: 116–126.

Rodea-Palomares, I., K. Boltes, F. Fernandez-Pinas, F. Leganes, E. Garcia-Calvo, J. Santiago, et al. 2011. Physicochemical characterization and ecotoxicological assessment of CeO_2 nanoparticles using two aquatic microorganisms. Toxicol. Sci. 119: 135–145.

Saptarshi, S.R., A. Duschl and A.L. Lopata. 2013. Interaction of nanoparticles with proteins: Relation to bio-reactivity of the nanoparticle. J. Nanobiotechnol. 11: 26.

Schaumann, G.E., A. Philippe, M. Bundschuh, G. Metreveli, S. litzke, D. Rakcheev, et al. 2015. Understanding the fate and biological effects of Ag- and TiO_2-nanoparticles in the environment: The quest for advanced analytics and interdisciplinary concepts. Sci. Total. Environ. 535: 3–19.

Schiavo, S., M. Oliviero, M. Miglietta, G. Rametta and S. Manzo. 2016. Genotoxic and cytotoxic effects of ZnO nanoparticles for *Dunaliella tertiolecta* and comparison with SiO_2 and TiO_2 effects at population growth inhibition levels. Sci. Total. Environ. 550: 619–627.

Sendra, M., M.P. Yeste, J.M Gatica, I. Moreno-Garrido and J. Blasco. 2017. Direct and indirect effects of silver nanoparticles on freshwater and marine microalgae (*Chlamydomonas reinhardtii* and *Phaeodactylum tricornutum*). Chemosphere. 179: 279–289.

Simon-Deckers, A., S. Loo, M. Mayne-L'Hermite, N. Herlin-Boime, N. Menguy, N. Reynaud, et al. 2009. Size-, composition- and shape-dependent toxicological impact of metal oxide nanoparticles and carbon nanotubes toward bacteria. Environ. Sci. Technol. 43: 8423–8429.

Slaveykova, V.I., K. Startchev and J. Roberts. 2009. Amine- and carboxyl-quantum dots affect membrane integrity of bacterium *Cupriavidus metallidurans* CH34. Environ. Sci. Technol. 43: 5117–5122.

Slaveykova, V.I., J.P. Pinheiro, M. Floriani and M. Garcia 2013. Interactions of core–shell quantum dots with metal resistant bacterium *Cupriavidus metallidurans*: Consequences for Cu and Pb removal. J. Hazard. Mater. 261: 123–129.

Slaveykova, V.I., B. Sonntag and J.C. Gutiérrez. 2016. Stress and Protists: No life without stress. Eur. J. Protistol. 55: 39–49.

Slaveykova, V.I., M. Li, I.A. Worms and W. Liu. 2020. When environmental chemistry meets ecotoxicology: Bioavailability of inorganic nanoparticles to phytoplankton. Chimia. 74: 115–121.

Smith, B.M., D.J. Pike, M.O. Kelly and J.A. Nason. 2015. Quantification of heteroaggregation between citrate-stabilized gold nanoparticles and hematite colloids. Environ. Sci. Technol. 49: 12789–12797.

Sohm, B., F. Immel, P. Bauda and C.J.P. Pagnout. 2015. Insight into the primary mode of action of TiO_2 nanoparticles on *Escherichia coli* in the dark. Proteomics. 15: 98–113.

Sondi, I. and B. Salopek-Sondi. 2004. Silver nanoparticles as antimicrobial agent: A case study on *E. coli* as a model for Gram-negative bacteria. J. Colloid. Interface. Sci. 275: 177–182.

Sørensen, S.N., C. Engelbrekt, H.-C.H. Lützhøft, J. Jiménez-Lamana, J.S. Noori, F.A. Alatraktchi, et al. 2016. A Multimethod approach for investigating algal toxicity of platinum nanoparticles. Environ. Sci. Technol. 50: 10635–10643.

Sottocasa, B. 2012. Étude de l'impact de nanoparticules sur des organismes unicellulaires: Le cas du dioxyde de titane sur l'algue *Chlamydomonas reinhardtii*. University of Geneva, Geneva.

Stevenson, L.M., H. Dickson, T. Klanjscek, A.A. Keller, E. McCauley and R.M. Nisbet. 2013. Environmental feedbacks and engineered nanoparticles: Mitigation of silver nanoparticle toxicity to *Chlamydomonas reinhardtii* by algal-produced organic compounds. PLoS One. 8: e74456.

Surette, M.C. and J.A. Nason. 2019. Nanoparticle aggregation in a freshwater river: The role of engineered surface coatings. Environ. Sci. Nano. 6: 540–553.

Taylor, C., M. Matzke, A. Kroll, D.S. Read, C. Svendsen and A. Crossley. 2016. Toxic interactions of different silver forms with freshwater green algae and cyanobacteria and their effects on mechanistic endpoints and the production of extracellular polymeric substances. Environ. Sci. Nano. 3: 396–408.

Tessier, A. and D.R. Turner (eds). 1995. Metal Speciation and Bioavailability in Aquatic Systems. Wiley, Chichester, UK.

Thill, A., O. Zeyons, O. Spalla, F. Chauvat, J. Rose, M. Auffan, et al. 2006. Cytotoxicity of CeO_2 nanoparticles for *Escherichia coli*. Physico-chemical insight of the cytotoxicity mechanism. Environ. Sci. Technol. 40: 6151–6156.

Vakurov, A., R. Brydson and A. Nelson. 2012. Electrochemical modeling of the silica nanoparticle–biomembrane interaction. Langmuir. 28: 1246–1255.

Van Hoecke, K., K.A.C. De Schamphelaere, P. Van der Meeren, S. Lucas and C.R. Janssen. 2008. Ecotoxicity of silica nanoparticles to the green alga *Pseudokirchneriella subcapitata*: Importance of surface area. Environ. Toxicol. Chem. 27: 1948–1957.

Van Hoecke, K., J.T.K. Quik, J. Mankiewicz-Boczek, K.A.C. De Schamphelaere, A. Elsaesser, P. Van der Meeren, et al. 2009. Fate and effects of CeO_2 nanoparticles in aquatic ecotoxicity tests. Environ. Sci. Technol. 43: 4537–4546.

Von Moos, N., P. Bowen and V.I. Slaveykova. 2014. Bioavailability of inorganic nanoparticles to planktonic bacteria and aquatic microalgae in freshwater. Environ. Sci. Nano. 1: 214–232.

Von Moos, N. and V.I. Slaveykova. 2014. Oxidative stress induced by inorganic nanoparticles in bacteria and aquatic microalgae-state of the art and knowledge gaps. Nanotoxicology. 8: 605–630.

Von Moos, N., L. Maillard and V.I. Slaveykova. 2015. Dynamics of sub-lethal effects of nano-CuO on the microalga *Chlamydomonas reinhardtii* during short-term exposure. Aquat. Toxicol. 161: 267–275.

Von Moos, N., V.B. Koman, C. Santschi, O.J.F. Martin, L. Maurizi, A. Jayaprakash, et al. 2016. Pro-oxidant effects of nano-TiO_2 on *Chlamydomonas reinhardtii* during short-term exposure. RSC. Advances.6: 115271–115283.

Wang, C.Y., D.Y. Lyon, J.B. Hughes and G.N. Bennett. 2003. Role of hydroxylamine intermediates in the phytotransformation of 2,4,6-trinitrotoluene by *Myriophyllum aquaticum*. Environ. Sci. Technol. 37: 3595–3600.

Wang, Z.Y., J. Li, J. Zhao and B.S. Xing. 2011. Toxicity and internalization of CuO nanoparticles to prokaryotic alga *Microcystis aeruginosa* as affected by dissolved organic matter. Environ. Sci. Technol. 45: 6032–6040.

Wang, Y., A.J. Miao, J. Luo, Z.B. Wei, J.J. Zhu, and L.Y. Yang. 2013. Bioaccumulation of CdTe quantum dots in a freshwater alga *Ochromonas danica*: A kinetics study. Environ. Sci. Technol. 47: 10601–10610.

Wang, R., F. Dang, C. Liu, D.J. Wang, P.X. Cui, H.J. Yan, et al. 2019. Heteroaggregation and dissolution of silver nanoparticles by iron oxide colloids under environmentally relevant conditions. Environ. Sci. Nano. 6: 195–206.

Wigginton, N.S., A.D. Titta, F. Piccapietra, J. Dobias, V.J. Nesatyy, M.J. Suter, et al. 2010. Binding of silver nanoparticles to bacterial proteins depends on surface modifications and inhibits enzymatic activity. Environ. Sci. Technol. 44: 2163–2168.

Worms, I.A.M., J. Boltzman, M. Garcia and V.I. Slaveykova. 2012. Cell-wall-dependent effect of carboxyl-CdSe/ZnS quantum dots on lead and copper availability to green microalgae. Environ. Pollut. 167: 27–33.

Xu, H. and H. Jiang. 2015. Effects of cyanobacterial extracellular polymeric substances on the stability of ZnO nanoparticles in eutrophic shallow lakes. Environ. Pollut. 197: 231–239.

Xu, H., C. Yang and H. Jiang. 2016. Aggregation kinetics of inorganic colloids in eutrophic shallow lakes: Influence of cyanobacterial extracellular polymeric substances and electrolyte cations. Water. Res. 106: 344–351.

Yu, S., J. Liu, Y. Yin and M. Shen. 2018. Interactions between engineered nanoparticles and dissolved organic matter: A review on mechanisms and environmental effects. J. Environ. Sci (China). 63: 198–217.

Yue, Y., X.M. Li, L. Sigg, M.J.F. Suter, S. Pillai, R. Behra, et al. 2017. Interaction of silver nanoparticles with algae and fish cells: A side by side comparison. J. Nanobiotechnol. 15: 1–16.

Zarei, M. and J. Aalaie. 2019. Profiling of nanoparticle-protein interactions by electrophoresis techniques. Anal. Bioanal. Chem. 411: 79–96.

Zemke-White, W.L., K.D. Clements and P.J. Harris. 2000. Acid lysis of macroalgae by marine herbivorous fishes: Effects of acid pH on cell wall porosity. J. Exp. Marine Biol. 45: 57–68.

Zhang, S., Y. Jiang, C.S. Chen, D. Creeley, K.A. Schwehr, A. Quigg, et al. 2013. Ameliorating effects of extracellular polymeric substances excreted by *Thalassiosira pseudonana* on algal toxicity of CdSe quantum dots. Aquat. Toxicol. 126: 214–223.

Zhang, H. and R.A. Wu. 2015. Proteomic profiling of protein corona formed on the surface of nanomaterial. Sci. China. Chem. 58: 780–792.

Zhang, J.L., Z.P. Zhou, Y. Pei, Q.Q. Xiang, X.X. Chang, J. Ling, et al. 2018a. Metabolic profiling of silver nanoparticle toxicity in *Microcystis aeruginosa*. Environ. Sci. Nano. 5: 2519–2530.

Zhang, W.C., B.D. Xiao and T. Fang. 2018b. Chemical transformation of silver nanoparticles in aquatic environments: Mechanism, morphology and toxicity. Chemosphere. 191: 324–334.

Zhao, J., X. Cao, X. Liu, Z. Wang, C. Zhang, J.C. White, et al. 2016. Interactions of CuO nanoparticles with the algae *Chlorella pyrenoidosa*: Adhesion, uptake, and toxicity. Nanotoxicology. 10: 1297–1305.

Zinchenko, A.A., T. Sakaue, S. Araki, K. Yoshikawa and D. Baigl. 2007. Single-chain compaction of long duplex DNA by cationic nanoparticles: Modes of interaction and comparison with chromatin. J. Phys. Chem. B. 111: 3019–3031.

4

Challenges in the Risk Assessment of Nanomaterial Toxicity Towards Microbes

Jaison Jeevanandam[1], Sharadwata Pan[2], Akula Harini[3] and Michael K. Danquah[4*]

[1]CQM – Centro de Química da Madeira, MMRG, Universidade da Madeira, Campus da Penteada, 9020-105 Funchal, Portugal.
Email: jaison.jeevanandam@staff.uma.pt

[2]School of Life Sciences Weihenstephan
Technical University of Munich, 85354 Freising, Germany
Email: sharadwata.pan@tum.de

[3]Department of Physics
Osmania University, Amberpet, Hyderabad, Telangana 500007, India
Email: akulaharini22@gmail.com

[4]Chemical Engineering Department
University of Tennessee, Chattanooga, TN 37403, United States
Email: michael-danquah@utc.edu

INTRODUCTION

Nanosized materials are extensively used in various fields, due to their exceptional properties and ability to alter their exclusive characteristics, based on the desired application. Past studies have shown that nanosized materials possess enhanced biological properties, which tend to be beneficial in biomedical and pharmaceutical

*For Correspondence: michael-danquah@utc.edu

applications. This has been instrumental in the inception and emergence of domains, such as nano-biotechnology and nanomedicine (Jeevanandam et al. 2018a). In biomedical applications, numerous nanosized materials are proposed to be valuable as antimicrobial agents, mainly due to their targeted and effective microbial growth inhibition property (Inam et al. 2019). In general, nanosized particles are fabricated via toxic chemicals as precursors, reducing and stabilizing agents, and their toxicity towards pathogenic microbes are attributed to the synthesis approach (Khan et al. 2019). Further, the size, shape, morphology, surface charge, and texture of nanosized materials were also emphasized to be the factors that cause an inhibitory effect towards microbes, which eventually could be attributed to the fabrication procedure (Le et al. 2019). Besides, these chemically synthesized nanoparticles are proven to be toxic, towards both pathogenic as well as favorable microbes, and normal and healthy cells of other macro-organisms (Jeevanandam et al. 2019a, Jeevanandam et al. 2019d). In order to reduce the adverse hazardous effects of the chemically fabricated nanoparticles, green synthesis methods using biological organisms were introduced (Jeevanandam et al. 2016). However, these biosynthesized nanoparticles also exhibited toxic reactions towards pathogenic as well as beneficial microbes, based on dose and concentration (Ibrahim 2015, Nayak et al. 2016, Jeevanandam et al. 2020). Thus, a stringent assessment of the risk associated with utilizing nanoparticles in commercial materials, is the need of the hour.

It is noteworthy that nanosized materials can inhibit the growth of microbes, including bacteria, fungi, algae, and viruses (Durán et al. 2016). Subsequently, they are widely included in the latest consumer products, such as lotions, soaps, paints and dresses, to prevent microbial growth (Mikiciuk et al. 2017). However, an extensive usage of nanosized materials with toxicity towards microbes, has raised the need for evaluating their toxic reactions towards other beneficial microbes and other organisms (Huo et al. 2015). This has led to the emergence of risk assessment strategies, to evaluate the toxicity of nanoparticles, and formulate them to reduce their cellular interaction and toxic reactions to other organisms (Bove et al. 2017). The antimicrobial efficacy of nanosized materials is assessed using conventional disc diffusion, minimum inhibitory concentration, and cell counting methods (Logeswari et al. 2015). Besides, *in vitro*, *in vivo*, and *in silico* methods are introduced to unveil the toxicity of nanosized materials towards other organisms and the environment (Raja et al. 2017). Furthermore, novel methods, such as hybrid approach, machine learning method, and molecular dynamic studies, are employed currently to assess the risks of using nanomaterials and their toxic mechanisms towards microbes and other organisms (Melagraki and Afantitis 2015). The present chapter attempts to lay an overview of various common risk assessment procedures that are available to evaluate the toxicity of nanomaterials, limitations and challenges of conventional risk evaluation assays of nanomaterials towards microbes, and some endorsements to develop exclusive nanomaterial toxicity evaluation assays are also provided. The chapter discusses nanomaterial toxicity in general and especially towards microbes, the risk assessment of nanomaterial toxicity towards microbes, and the associated shortcomings, future perspectives, and salient inferences.

NANOMATERIAL TOXICITY

To date, numerous reports have highlighted the benefits of nanoparticles and nanomaterials. However, increasing reports in recent times predominantly reveal the toxicity of nanomaterials towards various organisms (Tan et al. 2018). It is noteworthy that several characteristics of nanoparticles, such as size, shape, morphology, texture, surface charge, concentration, dose and functionalization, play a crucial role in exhibiting toxicity towards other organisms (Andra et al. 2019, Jeevanandam et al. 2019a). Fishes and other aquatic organisms are widely reported in the literature to be the most affected group, due to the toxic reactions of nanomaterials. It has been reported that ~5–10 nm sized silver and gold particles are cytotoxic towards *Oncorhynchus mykiss* (rainbow trout), even at low concentrations (Farkas et al. 2010). Similarly, Ji et al. (2017) fabricated oxides of zinc particles using leaf extract of *Argemone maxicana* and evaluated their toxicity against cardiac cells of Sahul India *Catla catla* fish. The results showed that the nanosized oxide particle exhibited concentration dependent cytotoxicity against fish cardiac cells (Ji et al. 2017). Likewise, Abramenko et al. (2018) evaluated the toxicity of flat and spherical shaped nanosized silver particles towards the embryos of zebrafish. The study revealed that the flat nanosized plate-shaped silver particles are highly toxic to zebrafish embryos than nanospheres (Abramenko et al. 2018). Further, iron and oxides of iron particles, synthesized via aqueous *Ficus natalensis* extract and chemical approaches, are proved to be toxic towards larva and pupa of *C. quinquefasciatus* mosquito species. In addition, the study demonstrated that the chemically synthesized nanoparticles magnified *Poecilia reticulata* guppy fish predation efficiency, which reduces mosquito population by eating eggs and larvae of mosquito, rather than biosynthesized nanoparticles (Murugan et al. 2018). Furthermore, Iswarya et al. (2016a) evaluated the individual and the binary toxicity of anatase and rutile crystal phase, nanosized titania particles towards *Ceriodaphnia dubia*. The results revealed that the individual exposure to both the phases of titania are toxic to the fish cells, upon irradiation of ultraviolet-A, compared to the visible light. Moreover, the binary mixture of both phases showed a distinct uptake and agglomeration pattern to exhibit toxic effects, depending upon the type of light irradiation towards *C. dubia* (Iswarya et al. 2016a). Contrarily, Sohn et al. (2015) showed that single walled carbon nanotubes (CNTs) exhibit acute toxicity towards freshwater microalgae, such as *Chlorella vulgaris* and *Raphidocelis subcapitata*, whereas remain majorly nontoxic towards zebrafish and Medaka fish (*Oryzias latipes*) (Sohn et al. 2015). Two recent studies also revealed that the carboxylated cadmium selenide-zinc sulfide quantum dots, and tungsten carbide nanoparticles, are toxic towards the embryos of *Oncorhynchus mykiss* (rainbow trout), *Dania reria* (zebrafish), *Trichogaster leerii* (Pearl gourami) (Rotomskis et al. 2018), and rainbow trout gill cell line, respectively (Kühnel et al. 2009).

Besides the aquatic ecosystem, nanoparticles and nanosized materials are reported to be toxic towards several organisms in each trophic level of the terrestrial ecosystem. Trophic levels in the terrestrial ecosystem consist of plants as the producers, herbivorous animals as the primary consumers, followed by carnivorous animals as the secondary consumers, and omnivorous animals as the tertiary consumers (Lefcheck et al. 2015), whereas microbes serve as a decomposer in all the ecosystems (Tlili et al. 2017). Iswarya et al. (2016b) evaluated the toxicity of 30 and 40 nm sized, as well as polyvinyl pyrrolidone (PVP) capped gold particles towards bacteria (decomposer), algae (producer), mouse and human cell lines (primary and secondary consumer) as model organisms, representing different trophic levels. The study showed that 30 and 40 nm sized particles are toxic towards bacteria and algae, compared to the capped nanoparticles. Further, the study emphasized that the nanosized gold particles induce DNA damage, depending upon their size and dose to cause toxic reactions in mouse hepatocytes, and reduce cell viability of human cervix cell lines (SiHa) via exposure and dose dependent toxic reactions (Iswarya et al. 2016b). Hossain et al. (2015) stated that plants can negatively respond to the stress induced by nanosized particles, especially nanosized silver and zinc oxide particles, by modulating the composition of the proteome (Hossain et al. 2015). Further, Reddy et al. (2016) emphasized that the chemically engineered nanosized particles are toxic towards plants, which can be attributed to their properties, soil dynamics, plant species, and the soil microbial community (Reddy et al. 2016). In our previous review article, we have discussed that even phytosynthesized nanoparticles are toxic towards animal models, such as mouse, *Daphnia magna* and mussels (Andra et al. 2019). Similarly, it has been reported in recent studies that nanoparticles are toxic towards animals and birds, such as hen (Freire et al. 2015), rabbit (Hanini et al. 2016), non-human primates (Xu et al. 2018), bull (Vinita et al. 2017), tadpole, *Lithobates sylvaticus* (wood frog), and *L. catesbeianus* (bullfrog) (Thompson et al. 2017). Mostly, nanoparticles are toxic to pathogenic insects (Benelli 2018), including mosquitoes, flies, and bugs (Suresh et al. 2019). Besides, several studies also showed that nanoparticles exhibit toxicity towards beneficial microbes in the soil (Schlich and Hund-Rinke 2015, Xu et al. 2015). Thus, it is evident that nanosized particles are toxic to all the trophic levels in the terrestrial environment. Moreover, the toxicity of nanosized polystyrene towards marine *Halomonas alkaliphila* bacteria (Sun et al. 2018), nanosized plastic towards green algae (Nolte et al. 2017), and zebrafishes (Chen et al. 2017), were also reported in recent studies, apart from chemically or green fabricated nanoparticles. Figure 4.1 shows the toxicity of nanoparticles towards various organisms across trophic levels in terrestrial ecosystems.

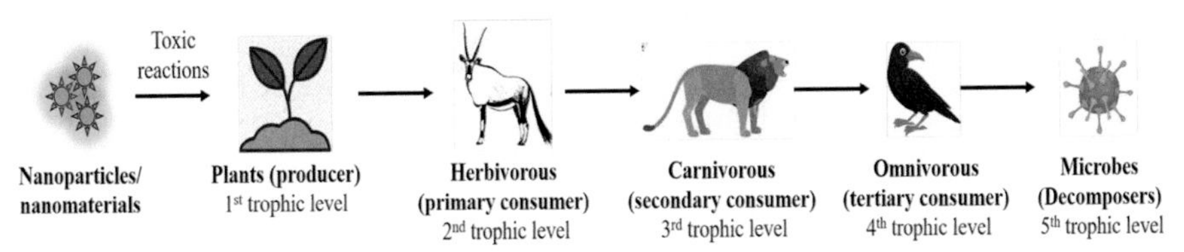

Figure 4.1 Toxicity of nanomaterials towards various trophic levels in ecosystems.

NANOMATERIAL TOXICITY TOWARDS MICROBES

Among other organisms, numerous reports are available to show the toxicity of nanomaterials towards microbes. Nanomaterials exhibit effective toxicity towards microbes, since they are mostly unicellular organisms, and are prone to utilize nutrients from the materials that are available in the environment, compared to the larger, macroscopic organisms. Almost all the unicellular and multicellular microbes, such as bacteria, algae, fungi, cyanobacteria, and viruses, are reported to be affected by the toxic reactions exhibited by various types of nanomaterials, as shown in Figure 4.2.

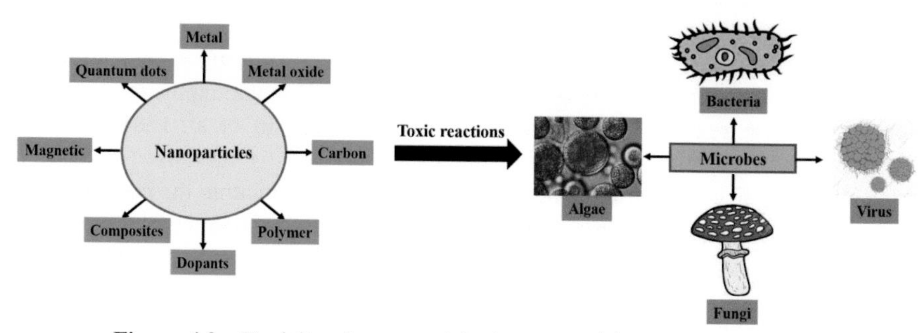

Figure 4.2 Toxicity of nanoparticles/nanomaterials towards microbes.

Toxicity Towards Bacteria

Several studies have reported that nanoparticles exhibit toxic reactions towards bacteria. Generally, nanoparticles are designed as antibacterial agents and are mostly considered as a beneficial entity that are helpful in industrial and commercial applications (Zheng et al. 2018). The size of nanoparticles is typically below 100 nm, which can easily penetrate 0.2 to 2 μm sized bacteria to exhibit toxic reactions (Vimbela et al. 2017). Further, the surface charge and high surface to volume ratio are the positive aspects of nanoparticles to be an effective antibacterial agent (Abbaszadegan et al. 2015, Dizaj et al. 2015). Silver nanoparticles are widely used as nano-anti-bacterial agents, and have been recently incorporated in several commercial products, to protect them from bacterial infestation (Franci et al. 2015). Nanosized silver exhibited an enhanced antibacterial activity against common pathogenic bacterial species, such as *Escherichia coli* and *Staphylococcus aureus* (Kubo et al. 2018), as well as environmentally significant bacterial species, such as *Cupriavidus metallidurans* CH 34 (Billen 2018). These silver nanoparticles are reported to possess antibacterial activities against various types of bacteria, such as *Micrococcus* (Ashkarran et al. 2016), *Staphylococcus* (Yuan et al. 2017), *Bacillus* (Rafińska et al. 2019), and *Pseudomonas* (Kora and Arunachalam 2011). In addition, silver nanoparticles are also effective in inhibiting the growth of psychrophiles (bacteria that grows in cold temperature) (Marchiore et al. 2017), mesophiles (bacteria that grows in moderate temperature) (Gloria et al. 2017), and thermophiles (bacteria that grows in high temperature) (Nthunya et al. 2019). Likewise, morphologies, such as hexagonal, triangular (El-Zahry et al. 2015), nanorods (Shaheen and Fouda 2018), nanoflowers (Molina et al. 2019), nanocubes (Alshareef et al. 2017), and nanowires (Hong et al. 2016) of nanosized silver are also beneficial in inhibiting pathogenic microbes. Hence, the nanosized silver particles are perceived as a potential broad-spectrum antibacterial agent. Moreover, metallic gold nanoparticles are also used as an antibacterial agent against certain bacterial species. Recently, it was reported that gold nanoparticles possess enhanced antibacterial activity against *E. coli*, *Klebsiella pneumonia*, methicillin-resistant *Staphylococcus aureus* (MRSA), *S. aureus*, *Pseudomonas aeruginosa* (Abdel-Raouf et al. 2017), *Corynebacterium pseudotuberculosis* (Mohamed et al. 2017), *Bacillus subtilis*, *Proteus vulgaris* (Muthukumar et al. 2016), *Staphylococcus epidermis*, *Streptococcus bovis*, *Enterobacter aerogenes*, *P. aeruginosa* PA01, *P. aeruginosa* UNC-D, and *Yersinia pestis* (Payne et al. 2016). Similar to silver, nanosized gold also exhibited morphology dependent antibacterial activity towards various bacterial species (Mahmoud et al. 2017, Penders et al. 2017). Apart from these common metal nanoparticles, platinum (Tahir et al. 2017), copper (Khatami et al. 2017), and selenium (Muthu et al. 2019) are the other nanosized metals that are currently under extensive investigations as enhanced antibacterial agents.

Metal oxide nanoparticles, such as zinc, copper, aluminum, iron, magnesium, manganese, titanium, and other rare earth oxides, also exhibit toxic reactions towards harmful bacterial species. Several studies revealed that zinc oxide (ZnO) nanoparticles possess enhanced antibacterial properties (Sirelkhatim et al. 2015), especially against multidrug resistant bacterial strains (Jesline et al. 2015, Kadiyala et al. 2018). Likewise, copper oxide (CuO) nanoparticles are proved to possess an improved ability to inhibit a wide variety of pathogenic bacteria (Kumar et al. 2015, Meghana et al. 2015). Similarly, aluminum oxide (Al_2O_3) nanoparticles are reported to possess antibacterial activities against multidrug resistant *Pseudomonas aeruginosa* (Ansari et al. 2015) and other common pathogenic bacterial strains (Mukherjee et al. 2011). Further, the nanosized magnetic oxides of iron particles also exhibited enhanced antibacterial activities against *B. subtilis*, *E. coli* (Arakha et al. 2015), *P. aeruginosa* (Irshad et al. 2017), *S. aureus*, and *Serratia marcescens* (Ismail et al. 2015). These nanoparticles are widely used in inhibiting bacterial pathogens during wastewater treatment, attributing to their magnetic property, which helps to purify water (Herlekar et al. 2015). It was recently reported that magnesium oxide (MgO) nanoparticles are highly beneficial in inhibiting foodborne pathogenic bacteria (He et al. 2016), bacteria causing wilt disease in tomato (Imada et al. 2016), and other bacterial species (Jeevanandam et al. 2019a). Furthermore, manganese dioxide (MnO_2) nanoparticles were also demonstrated to be a beneficial antibacterial agent against food- and waterborne pathogens (Krishnaraj et al. 2016), as well as certain gram-positive and gram-negative bacterial strains (Azhir et al. 2015, Cherian et al. 2016). Moreover, titanium dioxide (TiO_2) nanocrystals have proved to exhibit antibacterial activity against cariogenic bacteria (Sodagar et al. 2016) as well as *E. coli* and *S. aureus* when used along with sunlight irradiation (Liu et al. 2017b). In addition, metal oxide nanoparticles, such as cerium oxide (Arumugam et al. 2015), tin oxide (Vidhu and Philip 2015), nickel oxide (Ezhilarasi et al. 2018), silicon dioxide (Chai et al. 2017), and yttrium oxide (Mariano-Torres et al. 2018), have been under extensive exploration of late, and are revealed to possess specific antibacterial activities.

Dopant included metallic nanoparticles are another unique set of nanosized materials that are proven to possess enhanced potential in inhibiting pathogenic bacterial species, compared to free metal and metal oxide nanoparticles (Wen et al. 2018). Dopants are added to the crystal structure of the metal-based nanoparticles, to alter them for exhibiting improved biological properties (Figueroba et al. 2017). It is noteworthy that metals are doped with metal oxides, in order to alter their electronic configurations and improve their properties at the crystal level, to exhibit enhanced antibacterial activity (Pathak et al. 2018). Yadav et al. (2014) reported that copper doped anatase titanium dioxide nanoparticles exhibited visible light mediated photocatalytic activity. They also proved that the doping process with metallic copper assists the antibacterial titanium dioxide nanoparticles, to exhibit an enhanced inhibition ability, in the presence of visible light (Yadav et al. 2014). Further, Dutta et al. (2010) stated that the susceptibility of *E. coli* varies with respect to variations in the transition metals, which are doped with ZnO nanoparticles (Dutta et al. 2010). The doping process of metals with metal oxide nanoparticles are highly beneficial in reducing the band gap, by being at the intermediate position, which converts a normal antibacterial nanoparticle into a photocatalytic antibacterial agent (Mashitah et al. 2016). In recent times, reduced graphene oxide supported metal doped TiO_2 nanoparticles (Dhanasekar et al. 2018), neodymium doped ZnO nanoparticles (Hameed et al. 2016), copper doped silver sulphide nanoparticles (Fakhri et al. 2015), nickel doped (Vijayaprasath et al. 2016), and iron-doped ZnO nanoparticles (Khatir et al. 2016), are rigorously proven to possess antibacterial activities, which were improved due to the doping process.

Carbon-based nanoparticles are gaining more significance than metal nanoparticles, as they exhibit unique mechanisms in inhibiting bacteria. The antibacterial activity of carbon nanoparticles is highly dependent on its dimensionality and its reactivity (Dizaj et al. 2015). Carbon nanoparticles are classified into graphite (3D), graphene (2D), carbon nanotube (1D), and carbon dots (0D), based on the dimensionality (Jeevanandam et al. 2019c). Several studies have demonstrated that 3D graphite in its bulk form can be loaded with metal or metal oxide nanoparticles, to exhibit enhanced antibacterial activity. Silver (Hou et al. 2017), silver oxide (Chen and Liu 2016), metal oxides (Hung et al. 2017), titanium dioxide (Dědková et al. 2015b), and zinc oxide nanoparticles (Dědková et al. 2015a) are some of the nanosized particles that are loaded into graphite, to elevate their antibacterial efficacy. Graphene is a 2D nanofilm that makes up carbon atoms, and thus, can accommodate several nanoparticles to exhibit antibacterial activities (Ji et al. 2016). Generally, nanoparticles are embedded in the graphene structures and are used as an antibacterial film, which will extend their efficacy compared to the free nanoparticles. Silver nanoparticles (Shao et al. 2015), cuprous oxide (Yang et al. 2019), zinc oxide, strontium (Ravichandran et al. 2016), and cobalt oxide nanoparticle decorated graphene oxides (Alsharaeh et al. 2016) are under exhaustive research, to utilize them effectively to inhibit pathogenic bacteria. Likewise, carbon nanotubes (CNTs) are 1D carbon nanoparticles, where the ends can be functionalized with nanomaterials, such as metals or polymers to enhance their antimicrobial properties. The hollow tube-like structure can be used to encapsulate antibacterial nanomaterials, and the ends can be functionalized to exhibit antibacterial activity against specific pathogenic bacterial species (Mocan et al. 2017, Ma et al. 2018). It may be noted that multi-walled CNTs are under extensive investigations for antibacterial activities, compared to the single-walled counterparts, as the preparation process of single-walled CNTs is highly tedious (Entezari et al. 2015). Certain studies also proved that single-walled CNTs also possess antibacterial properties, and are highly beneficial in increasing the bacterial inhibition efficiency of conventional antibiotics, when antibiotics are incorporated with CNTs via covalent functionalization (Assali et al. 2017). Similarly, carbon nanorods and nanofibers are rod-shaped and fibrous 1D carbon-based nanomaterials, respectively, without the hollow portion that are available in CNTs (Pachfule et al. 2016, Gopinath and Krishna 2019). These nanorods and nanofibers do not possess antibacterial activities, whereas, antibacterial nanomaterials are embedded on these nanorods in order to elevate their bacterial inhibition efficacy (Chou et al. 2016, Sathe et al. 2016). Furthermore, carbon nanodots are zero-dimensional nanostructure that are highly useful as antibacterial agents (Zhu et al. 2019). It is noteworthy that these nanodots are photoluminescent in nature, which is highly beneficial to demonstrate photoinduced antibacterial activity (Liu et al. 2017a, Venkateswarlu et al. 2018). Similar to metals, doping in carbon nanostructure also elevated their electronic as well as antibacterial properties. Numerous studies have confirmed that doping metal or metal oxide nanoparticles with carbon nanostructures drastically enhances their antibacterial properties (Mihailescu et al. 2016, Mohammad et al. 2019, Mohammed et al. 2019).

Polymer nanoparticles constitute another unique set of nanoparticles that are employed as antibacterial agents (Fu et al. 2016). Natural polymeric nanoparticles act as direct antibacterial agents (Wu et al. 2017), whereas, synthetic nanosized polymers are used to encapsulate antibacterial nanomaterials and deliver them at the target sites (Darvishi et al. 2015). Natural polymers, such as chitosan (Piras et al. 2015), chitin (Jiang et al. 2017), lignin (Yang et al. 2016), and cellulose (Zheng et al. 2016), whereas synthetic polymers such as poly lactic acid (PLA) (Khan et al. 2016), poly (lactic-co-glycolic acid) (PLGA) (Arasoglu et al. 2017), and poly ethylene glycol (PEG) (Jayaramudu et al. 2016) are commonly fabricated as nanoparticles to exhibit antibacterial activities. These synthetic nanoparticles are further used to formulate conventional antibacterial agents in the form of liposomes (Hu et al. 2019), micelles (Hisey et al. 2017), and dendrimers (Sardana et al. 2018). Likewise, the quantum dots are proven to have potential both for being a nanocarrier of antibiotics, and as an individual antibacterial nanomaterial. Sulphur and nitrogen doped carbon quantum dots (Sardana et al. 2018), PEGylated silver-graphene quantum dots (Habiba et al. 2015), core-shell (Shariati et al. 2018b), ZnO (Garcia et al. 2018), and molybdenum disulfide quantum dots (Tian et al. 2019), are some of the recently identified quantum dots with antibacterial activities. Nanocomposites are the combination of two nanosized materials into one to elevate their properties (Zare and Shabani 2016). Zinc-lanthanide-iron oxide-nickel-titanium trioxide (Sobhani-Nasab et al. 2017), polyacrylamide-zirconium vanadophosphate (Sharma et al. 2016), copper-bioactive glass-egg shell membrane (Li et al. 2016), cross-linked chitosan-palladium (Dhanavel et al. 2018), and pectin-zirconium silicophosphate (Pathania et al. 2015), are some of the nanocomposites that are reported to exhibit enhanced antibacterial activities.

Even though nanosized particles are toxic towards pathogenic bacteria, certain nanoparticles are also known to be toxic towards commercially beneficial bacteria, which may affect the microbiome ecology (Su et al. 2015). For instance, *Bacillus subtilis* is a well-known Generally Recognized As Safe (GRAS) bacterial species that is mostly found in soil and in the gastrointestinal (GI) tract in humans and ruminants (Borriss et al. 2018). This bacterial species is highly beneficial in stimulating the immune system for the treatment of GI and urinary tract infections (Hamdy et al. 2018). Similarly, *Pseudomonas fluorescens* is an obligate aerobic bacteria that is extensively found in soil and water (Garrido-Sanz et al. 2016). This bacterial species is reported to protect the roots of the plant from parasitic fungal infections, such as *Fusarium* (Habiba et al. 2016). Auger et al. (2018) reported that zinc-magnesium-oxide nanoparticles are toxic towards *Bacillus subtilis* and macrophages (Auger et al. 2018). Likewise, Xie et al. (2019) stated that *B. subtilis* causes dissolution of ceria nanoparticles, which may lead to toxic effects (Xie et al. 2019). Even though there are no reports on the toxicity of nanoparticles towards *P. fluorescens*, several studies demonstrated their antibacterial activities towards *P. aeruginosa* (Bhargava et al. 2018, Mohan et al. 2019). Toxicity of nanomaterials towards these useful bacteria has led to the emergence of novel nanoparticles that exhibit specific toxic reactions towards harmful and disease-causing bacteria. Such novel nanoparticles are highly beneficial in maintaining the integrity of microbiota without any noteworthy side-effects (Pietroiusti et al. 2016).

Toxicity Towards Fungi and Algae

Numerous fungal and algal species are also affected by the toxic reactions exhibited by nanomaterials, similar to bacterial strains, out of which, most of them are either harmful to human health or affect the economy. Gold and silver are the most common metal nanoparticles that show effective antifungal and antialgal activities. Recently, Swain et al. (2016) reported a novel green synthesis method to fabricate gold nanoparticles via leaf and root extracts of *Vertiveria zizanioides* and *Cannabis sativa*. The synthesized nanosized gold particles are in the size range of 10–35 nm, and showed antifungal efficacy against fungal pathogens, including *Penicillium* species, *Aspergillus fumigates*, *A. flavus*, *Fusarium*, and *Mucor* species (Swain et al. 2016). Likewise, Eskandari-Nojedehi et al. (2018) fabricated nanosized gold particles (25 nm) via *Agaricus bisporus* mushroom extract, which exhibited toxicity against fungal strains, such as *A. flavus* and *A. terreus* (Eskandari-Nojedehi et al. 2018). Similarly, *Ziziphus zizyphus* leaf extract mediated nanosized gold particles (~50 nm) were synthesized recently by Aljabali et al. (2018), and revealed that nanosized green gold particles are highly toxic towards the *Candida albicans* fungal strain (Aljabali et al. 2018). Further, Renault et al. (2008) showed that exposure of 10 nm sized gold particles for 24 hours towards phytoplanktonic freshwater algal species, namely *Scenedesmus subspicatus* and a benthic bivalve *Corbicula fluminea*, led to ~20% of mortality in both the algal species, which has increased up to 50%, while increasing concentrations. The study further revealed that the toxicity of the nanoparticle is due to the high absorption ability of algae in their cell walls, which eventually led to progressive disturbances in their intracellular regions and the walls (Renault et al. 2008). Furthermore, nanosized gold rods embedded with chitosan-based hydrogel exhibited enhanced antifungal activities against *C. albicans*, *Cryptococcus neoformans*, and *Trichophyton mentagrophytes* (Bermúdez-Jiménez et al. 2019).

Silver nanoparticles are known for their enhanced antimicrobial properties. Several past studies have reported on the antifungal and antialgal activities of nanosized silver particles. Recently, Medda et al. (2015) demonstrated that nanosized silver particles fabricated via *Aloe vera* leaf extract possess effective antifungal activity against *Rhizopus* and *Aspergillus* species (Medda et al. 2015). Likewise, Xia et al. (2016) revealed the antifungal activity of commercially available, 5–20 nm sized silver particle solution against the pathogenic *Trichosporon asahii* fungi (Xia et al. 2016). Similarly, Bonilla et al. (2015) investigated the *in vitro* antifungal efficacy of nanosized silver particles against *C. krusei and C.* glabrata, which are resistant to fluconazole, and revealed that the nanosized silver are toxic to fungal strains at low concentrations, and are nontoxic to Murine fibroblast cells (CC_{50}), even at high concentrations (Bonilla et al. 2015). Additionally, silver nanoparticles also exhibited antialgal activities against microalgae *Protheca* genus (Jagielski et al. 2018, Nowakowska et al. 2018) *Chlorella vulgaris*, *C. pyrenoidosa* (Kumari et al. 2017), and bloom forming cyanobacteria (Chaturvedi and Verma 2015). Recently, novel, nanosized silver particles were fabricated via *Nannochloropsis oculate* and *Tetraselmis tetrathele* cultures, and their toxicity towards *Oscillaroria simplicissima* algae was evaluated by El-Kassas and Ghobrial (2017). The study demonstrated that the biosynthesized, nanosized silver particles were highly toxic towards the algal strain, which can be utilized as a potential antialgal agent to reduce neurotoxic effects secreted by these algae that can affect aquatic organisms (El-Kassas and Ghobrial 2017). Moreover, bimetallic gold-silver nanoparticles that are synthesized via nanoreactors are proven to possess high antifungal activity against *Candida* species, such as *C. glabrata*, *C. parapsilosis*, *C. albicans*, *C. krusei*, and *C. guillermondii* (Gutiérrez et al. 2018). Apart from gold and silver nanoparticles, there are other metal nanoparticles that are reported to possess efficient antifungal activities. These include nanosized platinum fabricated from *Prunus*^{x} *yedoensis* tree gum (Velmurugan et al. 2016), chemically reduced nanosized copper (Viet et al. 2016), palladium nanoparticles synthesized with *Melia azedarach* leaf extract (Bhakyaraj et al. 2017), biosynthesized nanosized silver-gold alloy from cell free *Bacillus safensis* LAU13 (Ojo et al. 2016), nanosized copper extracellularly synthesized from *Streptomyces griseus* (Ponmurugan et al. 2016), and nanosized copper fabricated from *Citrus medica* juice (Shende et al. 2015). In addition, nanosized lead (Gandhi et al. 2018), as well as several metal-doped metal oxide nanoparticles, such as titanium dioxide doped with silver and copper (Graziani et al. 2016), have exhibited potential antialgal activity.

Nanosized metal oxide particles, such as zinc, copper, iron, and titanium are the most common nanoparticles that are demonstrated to possess potential antifungal and antialgal activities. Nanosized oxides of zinc particles (ZnO) prepared from *Nyctanthes arbortristis* (Jamdagni et al. 2018), petal extracts of *Rosa indica*

(Tiwari et al. 2016), leaf extract of *Limonia acidissima* (Patil and Taranath 2016), homogenous precipitation (Sharma et al. 2016) and stem bark extract of *Boswellia ovalifoliolata* (Supraja et al. 2016), are proven to possess antifungal activities in recent times. Likewise, nanosized ZnO particles fabricated via secondary extracellular metabolite of *Pseudomonas aeruginosa* (Barsaiya and Singh 2018), ZnO nanorod-based antimicrobial coating (Al-Fori et al. 2014), and ZnO nanorods with photocatalytic property in visible light region (Sathe et al. 2016), are experimentally proven to exhibit antialgal activities. Further, nanosized oxides of copper particles fabricated via brown seaweed *Sargassum polycystum* (Ramaswamy et al. 2016) by electrochemical method (Katwal et al. 2015), commercial 40 nm sized copper nanoparticles (Amiri et al. 2017), and *Cissus quandrangularis* (Devipriya and Roopan 2017), showed antifungal efficacy against several common fungal strains. These nanosized particles synthesized by grafting with 3-glycidyloxypropyl) trimethoxysilane and coupled with 4-hydroxyphenylboronic acid (Halbus et al. 2019) and zeolite as composite (Du et al. 2017), also possess effective antialgal activities. Nanosized iron oxide particles with magnetic properties, that are prepared with tannic acid in alkaline medium (Parveen et al. 2018), surface functionalized with gallic acid (Shah et al. 2017), commercially synthesized with chemicals (Seddighi et al. 2017), and fabricated using aqueous *Sageretia thea* extracts (Khalil et al. 2017), are proven to possess the ability to inhibit pathogenic fungal strains. Likewise, Baniamerian et al. (2020) showed that the photocatalytic iron oxide-titanium dioxide nanocomposite possesses antialgal activity against *Chlorella vulgaris* under irradiation of visible light (Baniamerian et al. 2020). Nanosized titanium dioxide particles are unique nanoparticles that have exhibited effective antifungal (Shibata et al. 2007, Arakha et al. 2015, Durairaj et al. 2015) and antialgal activities (Goffredo et al. 2017, Nored et al. 2018, Goffredo et al. 2019). Karimiyan et al. (2015) revealed that nanosized oxides of metals, such as magnesium, copper, silicon and zinc, possess effective antifungal activity against *C. albicans* (Karimiyan et al. 2015). In addition, other nanosized oxides of metal particles, such as cerium (Zhang et al. 2017, Javadi et al. 2019), silica (Derbalah et al. 2018, Verma et al. 2018a), magnesium oxide (Sierra-Fernandez et al. 2017), and palladium oxide (Wang et al. 2015), are proven to demonstrate efficient antialgal and antifungal properties.

Nanosized carbon materials are another set of nanoparticles that are revealed to possess toxicity against several strains of algae and fungi. Sawangphruk et al. (2012) demonstrated that the reduced nanosized oxides of graphene sheets, fabricated via the modified 'Hummers method', possess enhanced antifungal activities against *Aspergillus niger*, *A. oryzae*, and *Fusarium oxysporum* (Sawangphruk et al. 2012). Further, graphene oxide blended with nanosized silver particles in the form of nanocomposites, was also emphasized to possess effective antifungal and antialgal activities in the literature (Chen et al. 2016, Whitehead et al. 2017). Furthermore, efficient antifungal activity was exhibited by surface modified (Wang et al. 2017), single-walled (Foo et al. 2018), and multi-walled CNTs (Mohamed and El-Ghany 2018), towards a broad spectrum of fungal strains. Moreover, multi-walled CNTs that are decorated with metals, such as copper, silver, and oxide of zinc particles, also exhibited improved antifungal properties (Fosso-Kankeu et al.). However, to our knowledge, there is no literature available to prove the toxicity of CNTs against algal species. Zero-dimensional carbon dots derived from deep eutectic solvents (Gao et al. 2019), carbon dots functionalized with red-emissive guanylated polyene (Li et al. 2019), and nano-conjugated carbon dots with gold (Priyadarshini et al. 2018), are proven to possess potential toxicity towards pathogenic fungal species. Recently, Li et al. (2018) stated that multifunctional carbon dots are highly beneficial as lifetime thermal sensors, for imaging nucleolus and possessing antialgal activity by inhibiting RuBisCO in *Anabaena* species (Li et al. 2018). Further, Zhang et al. (2019) showed that degradable carbon dots, fabricated using one-step electrolytic method, are toxic towards *Chlorella vulgaris* algae via the generation of reactive oxygen species (ROS) mediated oxidative stress, and disrupting their photosynthetic system (Zhang et al. 2019).

In general, it is striking that the individual polymeric nanoparticles do not possess antifungal or antialgal activities, except those that are prepared via biological origins (biopolymers) and with biocompatible property. Yien et al. (2012) showed that nanosized chitosan biopolymer particles exhibited effective antifungal activities against *C. albicans*, *Fusarium solani*, and *A. niger* (Yien et al. 2012). Recently, biodegradable nanosized polymeric particles, coated with phytochemicals of *Syzygium cumini*, were reported to have antifungal efficacy against *C. guilliermondii* and *C. haemulonii*, along with an effective antioxidant property (Bitencourt et al. 2016). Likewise, biocompatible cationic polyelectrolyte poly-(diallyldimethylammonium chloride) surfaces functionalized cross-linked with acrylate copolymer nanogel particles, were fabricated to be beneficial as a potential nanocarrier for chlorhexidine, to exhibit antialgal activity against *C. reinhardtii* and antifungal efficacy against *S. cerevisiae* (yeast) (Al-Awady et al. 2018). Apart from biopolymers, metal or metal oxide-based polymer nanocomposites, are widely used as potential antifungal and antialgal agents. Ifuku et al. (2015) showed that nanosized silver particles embedded on the surface of chitin nanofibers, possess antifungal activities against *Alternaria brassicae*, *A. alternata*, *A. brassicicola*, *Penicillium digitatum*, *F. oxysporum*, *Bipolaris oryzae*, *Botrytis cinerea*, *Colletotrichum higginsianum*, and *C. orbiculare*, compared to standalone chitin nanofibers (Ifuku et al. 2015). Similarly, nanosized, solid lipid particles based on polyethylene glycol (PEG) 40 stearate, have been reported for the efficient delivery of antifungal drugs to inhibit vaginal pathogens (Cassano et al. 2016). Additionally, nanocomposites, such as sulfur-chitosan (Yela et al. 2016), copper-silver-zinc oxide (Ghosh et al. 2020), magnetic iron oxide-zinc oxide-silver bromide (Hoseinzadeh et al. 2016), corn oil-gelatin (Sahraee et al. 2017), and copper oxide-carbon (Surendra et al. 2019), were recently reported to possess the ability to inhibit pathogenic fungal species, while being nontoxic to beneficial fungal strains. Moreover, silica-titania in core-shell structure (Verma et al. 2018b), chitosan-silver-titania (Natarajan et al. 2018), silver-activated charcoal-titania (Caro et al. 2015), visible light irradiated iron oxide-titania (Baniamerian et al. 2020), are the other nanocomposites that are under extensive research to emphasize their antialgal efficacy. Besides, copper-doped ZnO (Khan et al. 2017), Zn-doped silica (Arshad et al. 2018), iron-doped ceria (Rahdar et al. 2019), lithium, magnesium, and strontium-doped ZnO (Shanthi et al. 2018), Zn-doped tungsten oxide (Arshad et al. 2019), and cobalt-doped tin oxide (Chandran et al. 2015) nanoparticles were demonstrated to possess toxicity, due to the doping process against pathogenic fungal species. Further, studies involving palladium oxide modified with nitrogen-doped titanium (Wang et al. 2015), indium-doped ZnO (Shariati et al. 2018a), and cerium-doped copper (Zhang et al. 2017) nanoparticles, are the few that have proven that the doping process increases the antialgal efficacy of stand-alone nanosized particles.

Toxicity Towards Viruses

Majority of viruses are exclusively detrimental towards both micro and macro-organisms, unlike other microbes that are both beneficial and pathogenic in nature (Foster et al. 2018, Tennant et al. 2018). Since viruses belong to the same size range as nanoparticles, novel nanosized materials are being widely utilized to exhibit effective toxic reactions towards viruses, and consequently inhibiting their growth (Galdiero et al. 2011). It may be noted that virus capsids, such as virus-like and viral nanoparticles, are used to encapsulate and deliver nanomedicines or genes at the target site, for several unique and rare disease treatments (Jeevanandam et al. 2018b). Similar to other microbes, nanoparticles, including nanosized metals, metal oxides, polymers, carbon-based doped particles and composites, are reported in the literature as enhanced antiviral agents. Recently, nanosized gold particles coated with anionic carbosilane dendrons, and decorated with sulfonate functions, as well as single thiol moiety at the focal point, were fabricated to exhibit effective antiviral activity against the Human Immunodeficiency Virus-1 (HIV-1) (Peña-González et al. 2016). Furthermore, recent reports have revealed the antiviral activity of nanosized silver particles fabricated via electrochemical method against polio virus (Huy et al. 2017), chemically reduced nano-silver against norovirus (Castro-Mayorga et al. 2017), radiochemical process against Influenza A, Feline Calicivirus (Seino et al. 2016), and plant derived tannic acid polyphenol modified nano-silver against the vaginal herpes simplex virus (Orłowski et al. 2018). Among nanosized oxides of metals, nanosized oxides of copper particles that are prepared using *Syzygium alternifolium* fruit extract, are proven to possess efficient antiviral activity against the Newcastle Disease Virus (NDV) (Yugandhar et al. 2018). Likewise, titanium dioxide nanostructures fabricated using a self-assembly process, were shown to be a potential antiviral agent against broad bean strain virus that attacks and inhibits the growth of the faba bean plant (Elsharkawy and Derbalah 2019). Surprisingly, there is no reported antiviral property pertaining to the nanosized ZnO particles. Further, glycine coated iron oxide nanoparticles against H1N1 influenza virus (Kumar et al. 2016), PEG coated ZnO nanoparticles against type 1 herpes simplex virus (Tavakoli et al. 2018), and magnesium oxide nanoparticles against anti-foot-and-mouth-disease virus (Rafiei et al. 2015), are the other nanosized particles that are under exhaustive investigations to constrain the growth of the disease-causing viruses. Carbon-based nanoparticles, such as carbon dots (Dong et al. 2017), C60 fullerene (Dostalova et al. 2016), graphene oxide (Ye et al. 2015), and tyrosine supported CNTs (Botta et al. 2015) showed effective antiviral activity against a wide range of DNA and RNA viruses, including norovirus, bacteriophage λ plaques, porcine epidemic diarrhea virus, pseudorabies virus. Furthermore, polymeric nanoparticles, such as polyrhodanine (Nazaktabar et al. 2017), nanolipogels (Ramanathan et al. 2016), and nanoparticulate vacuolar ATPase blocker (Hu et al. 2017) encapsulated with polymer and biopolymers (Randazzo et al. 2018), were also demonstrated to possess improved antiviral properties. Moreover, graphene oxide-silver (Du et al. 2018), peptide-nucleic acid-titania (Amirkhanov et al. 2015), and titania-polylysine containing oligonucleotide (Levina et al. 2016), are the most recent nanocomposites that are fabricated to exhibit a heightened antiviral activity. However, most of these studies have focused only on evaluating either the antimicrobial effect/toxicity towards microbes, or the toxicity towards macro-organisms. This makes it tedious to compare, whether their toxicity is either favourable or destructive. Thus, it is recommended that the toxicity evaluation of nanoparticles towards both microbes, and towards the environment (plants, animals, and humans), will be necessary in the future to predict their actual or beneficial toxicity in inhibiting the disease-causing pathogenic microbes.

RISK ASSESSMENT OF NANOMATERIAL TOXICITY TOWARDS MICROBES

Novel risk assessment methods, targeting the toxicity prediction of nanomaterials, are highly essential towards various micro and macro-organisms, as discussed in the previous sections. *In vitro*, *in vivo*, and *in silico* are the three significant approaches, which are used to assess the toxicity risk of nanomaterials towards microbes, as it is not possible to use the conventional toxicity analysis methods directly, unlike for evaluating the toxicity of common drugs and soluble chemicals.

In vitro Risk Assessment of Nanomaterial Toxicity

Initially, the *in vitro* risk assessment of nanomaterials must be performed using a normal cell line, to confirm their nontoxic nature towards humans and other animals. Guadagnini et al. (2015) evaluated the toxicity of various nanomaterials, including titania, silica, uncoated iron oxide, oleic acid coated iron oxide, and PLGA-PEO (poly (lactic-co-glycolic acid)-polyethylene oxide), and examined the effects of composition, coatings, size, and agglomeration in modifying their toxic reactions. The study utilized toxicity analysis approaches, such as *in vitro* dye based cytotoxicity assay, namely water-soluble tetrazolium salt-1 (WST-1), 3-(4,5-dimethylthiazol-2-yl)-2,5-diphenyl tetrazolium bromide (MTT), lactate dehydrogenase, propidium iodide, neutral red, ^{3}H incorporation of thymidine, and cell counting methods. Further, the study also involved methods, such as enzyme-linked immunosorbent assay (ELISA), to evaluate the quantity of interleukin 6 and 8 (IL-6 and IL-8), and granulocyte-macrophage colony-stimulating factor (GM-CSF), as well as mono Bromobimane, nitric oxide (NO) and dichlorofluorescein assays, to detect oxidative stress. The study provided an insight that the interferences in the assays between nanomaterials, are nanomaterial-specific or assay-specific. The study thus emphasized that proper physical and chemical characterizations of nanoparticles may be used as a preliminary method, to screen the number of risk assessment assays. Once the samples are screened for toxicity analysis, quantitative reverse transcription polymerase chain reaction (RT-qPCR) and flow cytometry, even though time consuming, may be helpful in predicting the toxicity mechanism of nanoparticles, to reduce their toxicity (Guadagnini et al. 2015).

Similarly, Hofmann-Amtenbrink et al. (2015) exposed the current challenges in assessing the risk and using standardized inorganic nanoparticles. They reported that although *in vitro* assays can be used as a preliminary method to screen toxic nanoparticles, the *in vivo* methods are required to analyze their systematic level toxic reactions and interactions (Hofmann-Amtenbrink et al. 2015). Generally, conventional antimicrobial assays, such as disc diffusion, minimum inhibitory concentration, and cell counting methods, are used to evaluate the inhibitory effects of nanoparticles towards pathogenic microbes (Ncube et al. 2008, Balouiri et al. 2016). However, it is also necessary to evaluate the toxicity of nanoparticles towards beneficial soil or environmental microbes. Hegde et al. (2016) listed several nanoparticles that are toxic to beneficial microbes, and revealed that conventional identification and hazard characterization, assessment of exposure, and risk characterization are highly complicated in the case of evaluating nanoparticle toxicity. Further, the authors emphasized that engineered nanoparticle exposure towards the environment in the entire production life cycle must be monitored, to reduce their toxicity towards microbes. The article also demonstrated that *in vitro* risk assessment helps to predict the inhibitory mechanisms of nanoparticles towards microbes. The proposed toxic mechanisms of engineered nanoparticles in the environment, via aggregation, dissolution, sorption, agglomeration, sedimentation, and surface reaction that leads to impact on microbes, such as cell lysis, via membrane damage, nucleic acid or protein interaction, mitochondria damage, and ROS production, are predicted using *in vitro* risk assessment methods (Hegde et al. 2016).

In vivo and *In silico* Risk Assessment of Nanomaterial Toxicity

In vivo and *in silico* risk assessment methods are required to evaluate the actual toxic mechanisms of nanoparticles and their interactions between cells in live animals, apart from *in vitro* studies. *In vivo* risk assessment involves animal models to evaluate the toxicity of nanomaterials towards live, normal animal cells (Chakraborty et al. 2016, Jeevanandam et al. 2019b). There are several cases in which the nanoparticles exhibited effective antimicrobial and anticancer efficacy, while being toxic to normal cells (Jeevanandam et al. 2019a). Even certain green synthesized nanoparticles showed toxicity towards normal cells depending upon concentration, dose, and their cellular interactions (Jeevanandam et al. 2019d). However, utilization of animal models for risk assessment of toxicity, involves various ethical clearance and regulatory issues (Varga et al. 2010), which has led to the emergence of *in silico*-based computational methods to evaluate nanomaterial toxicity. Lin et al. (2015) developed a potential computational framework via pharmaceutical models, with independent multiple data sets, to simulate the distribution of nanosized gold particles in the tissues of several species. The study revealed that the computational model may assist the translation of nanomaterials from animal model to clinical trials (Lin et al. 2015). Likewise, Burden et al. (2015) utilized replacement, refinement, and reduction of animals (3Rs), as a framework for assessing the safety of nanomaterials (Burden et al. 2017). Similarly, Oomen et al. (2015) showed that grouping and read-across approaches within the broader perspective of MARINA risk assessment strategy, will be beneficial in evaluating the toxicity of nanomaterials (Oomen et al. 2015). Recently, Marvin et al. (2017) demonstrated that Bayesian network can be applied to rank nanomaterials, based on their hazard profiles, to support a human health risk assessment (Marvin et al. 2017). In addition, Costa and Fadeel (2016) showed that systems biology approaches can be utilized for the mechanism based understanding of nanomaterial hazards and risks in the nanotoxicological field (Costa and Fadeel 2016). Moreover, Pang et al. (2017) showed that probabilistic approaches can be highly beneficial in assessing the health risk of ingesting nanosized silver particles that are released from consumer products towards infants (Pang et al. 2017). Thus, it is evident that *in silico* and non-conventional methods are gaining significance, for the risk assessment to evaluate nanomaterial toxicity and replace the conventional *in vitro* and *in vivo* approaches.

LIMITATIONS, FUTURE PERSPECTIVE AND CONCLUSION

Even though risk assessment methods help to evaluate and identify the toxicity of nanomaterials, there are certain drawbacks, which hurdle the clinical translation of nanomaterials in the pharmaceutical field. *In vitro* studies are beneficial only to predict the cellular interactions, toxicity of nanomaterials, and their toxic mechanisms, for an initial screening of toxic nanomaterials (Kroll et al. 2009). However, *in vivo* studies using live animal models, are essential to confirm their effective toxicity towards normal healthy cells (Jahnel 2015). Besides, ethical issues involved in using animal models constitute a major challenge, which acts as a limitation for *in vivo* risk assessment approaches (Kwon et al. 2014). *In silico*-based computational models are used to evaluate the toxicity of nanomaterials via simulations, which is a relatively new approach, compared to the *in vitro* and *in vivo* methods. In spite of being beneficial in the risk assessment of nanomaterials, a reliability issue exists in utilizing *in silico* models for toxicity analysis (Fröhlich and Salar-Behzadi 2014). Several *in silico* studies have reported that the computational models can be useful for the preliminary screening of nanomaterials, based on their toxicity, and *in vivo* methods are necessary to confirm their actual adverse toxic effects (Hristozov et al. 2016). These risk assessment approaches are beneficial to produce commercial nanomaterials, which are toxic to microbes, namely antimicrobial creams and soaps, without toxic reactions towards humans, the environment, and beneficial microbes.

The current limitations of risk assessment methods can be avoided via hybrid risk assessment strategies, including *in vitro*, *in vivo*, *ex vivo*, and *in silico* methods (Linkov and Satterstrom 2008, Laux et al. 2018). Recently, molecular dynamics (MD) simulations were utilized by researchers to analyze the physical and chemical characteristics of nanoparticles, prior to obtaining experimental data. Baweja et al. (2013) showed that MD simulations can be beneficial in identifying the effect of helical protein stabilization via hydration pattern, while fabricating graphene-based nanoparticles embedded with proteins (Baweja et al. 2013). The same research group has recently demonstrated that MD simulation studies are useful in identifying the effects of amyloid beta peptide to modify the structure and conformational transitions of graphene oxide (Baweja et al. 2015). Further, Naicker et al. (2005) has identified the potential of MD simulations in characterizing the properties of nanosized dioxides of titanium particles (Naicker et al. 2005). Thus, it is evident that MD simulations can be beneficial in predicting the physicochemical properties of nanoparticles, which eventually help in predicting their toxicity. Besides, machine learning is gaining much applicational importance in the field of biomedical sciences, and it is expanding towards the nano-biotechnology field in recent times. Machine learning approaches are recently introduced in the risk assessment of nanoparticle toxicity. Furxhi et al. (2019) showed that classifiers and ensemble classifiers via Copeland index will be beneficial in predicting the *in vitro* toxicity of nanoparticles using machine learning approach (Furxhi et al. 2019). Likewise, Concu et al. (2017) demonstrated that a unified *in silico* machine learning model, based on the perturbation theory, can be used to probe the toxicity of nanoparticles (Concu et al. 2017). Further, Goldberg et al. (2015) stated that the machine learning approach can provide insights for predicting nanoparticle transport behavior from the physicochemical properties, to guide the next generation of transport models (Goldberg et al. 2015). In addition, Pikula et al. (2020) recently showed the efficiency of utilizing bioinformatics, machine and deep learning approaches for the risk assessment of nanotoxicology (Pikula et al. 2020). Thus, it is noteworthy that computational approaches will alter the conventional risk assessment strategies in the future, to unveil nanomaterial toxicities towards microbes, animals, and humans.

ACKNOWLEDGMENTS

The authors would like to thank their respective departments for their extensive support and motivation.

References

Abbaszadegan, A., Y. Ghahramani, A. Gholami, B. Hemmateenejad, S. Dorostkar, M. Nabavizadeh, et al. 2015. The effect of charge at the surface of silver nanoparticles on antimicrobial activity against gram-positive and gram-negative bacteria: A preliminary study. J. Nanomater. 16: 53.

Abdel-Raouf, N., N.M. Al-Enazi and I.B.M. Ibraheem. 2017. Green biosynthesis of gold nanoparticles using Galaxaura elongata and characterization of their antibacterial activity. Arabian J. Chem. 10: S3029–S3039.

Abramenko, N.B., T.B. Demidova, E.V. Abkhalimov, B.G. Ershov, E.Y. Krysanov and L.M. Kustov. 2018. Ecotoxicity of different-shaped silver nanoparticles: Case of zebrafish embryos. J. Hazard. Mater. 347: 89–94.

Al-Awady, M.J., P.J. Weldrick, M.J. Hardman, G.M. Greenway and V.N. Paunov. 2018. Amplified antimicrobial action of chlorhexidine encapsulated in PDAC-functionalized acrylate copolymer nanogel carriers. Mater. Chem. Front. 2: 2032–2044.

Al-Fori, M., S. Dobretsov, M.T.Z. Myint and J. Dutta. 2014. Antifouling properties of zinc oxide nanorod coatings. Biofouling. 30: 871–882.

Aljabali, A.A.A., Y. Akkam, S.M. Al Zoubi, M.K. Al-Batayneh, B. Al-Trad, O. Abo Alrob, et al. 2018. Synthesis of gold nanoparticles using leaf extract of ziziphus zizyphus and their antimicrobial activity. Nanomaterials. 8(3): 174.

Alsharaeh, E., Y. Mussa, F. Ahmed, Y. Aldawsari, M. Al-Hindawi and G.K. Sing. 2016. Novel route for the preparation of cobalt oxide nanoparticles/reduced graphene oxide nanocomposites and their antibacterial activities. Ceram. Int. 42: 3407–3410.

Alshareef, A., K. Laird and R.B.M. Cross. 2017. Shape-dependent antibacterial activity of silver nanoparticles on *Escherichia coli* and *Enterococcus faecium* bacterium. Appl. Surf. Sci. 424: 310–315.

Amiri, M., Z. Etemadifar, A. Daneshkazemi and M. Nateghi. 2017. Antimicrobial Effect of Copper Oxide Nanoparticles on Some Oral Bacteria and Candida Species. J. Dent. Biomater. 4: 347–352.

Amirkhanov, R.N., N.A. Mazurkova, N.V. Amirkhanov and V.F. Zarytova. 2015. Composites of peptide nucleic acids with titanium dioxide nanoparticles. IV. Antiviral activity of nanocomposites containing DNA/PNA duplexes. Russ. J. Bioorg. Chem. 41: 140–146.

Andra, S., S.K. Balu, J. Jeevanandham, M. Muthalagu, M. Vidyavathy, Y. San Chan, et al. 2019. Phytosynthesized metal oxide nanoparticles for pharmaceutical applications. Naunyn-Schmiedeberg's Arch. Pharmacol. 392: 755–771.

Ansari, M.A., H.M. Khan, M.A. Alzohairy, M. Jalal, S.G. Ali, R. Pal, et al. 2015. Green synthesis of Al_2O_3 nanoparticles and their bactericidal potential against clinical isolates of multi-drug resistant *Pseudomonas aeruginosa*. World J. Microbiol. Biotechnol. 31: 153–164.

Arakha, M., S. Pal, D. Samantarrai, T.K. Panigrahi, B.C. Mallick, K. Pramanik, et al. 2015. Antimicrobial activity of iron oxide nanoparticle upon modulation of nanoparticle-bacteria interface. Sci. Rep. 5: 14813.

Arasoglu, T., S. Derman, B. Mansuroglu, G. Yelkenci, B. Kocyigit, B. Gumus, et al. 2017. Synthesis, characterization and antibacterial activity of juglone encapsulated PLGA nanoparticles. J. Appl. Microbiol. 123: 1407–1419.

Arshad, M., A. Qayyum, G.A. Shar, G.A. Soomro, A. Nazir, B. Munir, et al. 2018. Zn-doped SiO_2 nanoparticles preparation and characterization under the effect of various solvents: Antibacterial, antifungal and photocatlytic performance evaluation. J. Photochem. Photobiol. B. 185: 176–183.

Arshad, M., S. Ehtisham-ul-Haque, M. Bilal, N. Ahmad, A. Ahmad, M. Abbas, et al. 2019. Synthesis and characterization of Zn doped WO_3 nanoparticles: photocatalytic, antifungal and antibacterial activities evaluation. Mater. Res. Express.7: 015407.

Arumugam, A., C. Karthikeyan, A.S.H. Hameed, K. Gopinath, S. Gowri and V. Karthika. 2015. Synthesis of cerium oxide nanoparticles using Gloriosa superba L. leaf extract and their structural, optical and antibacterial properties. Mater. Sci. Eng. C. 49: 408–415.

Ashkarran, A.A., S. Davoudi and S. Ahmady-Asbchin. 2016. A comparative study of silver nanoparticles and corona discharge for environmental and antibacterial applications. J. Environ. Biotechnol. Res. 4: 17–23.

Assali, M., A.N. Zaid, F. Abdallah, M. Almasri and R. Khayyat. 2017. Single-walled carbon nanotubes-ciprofloxacin nanoantibiotic: Strategy to improve ciprofloxacin antibacterial activity. Int. J. Nanomed. 12: 6647.

Auger, S., C. Henry, C. Péchoux, S. Suman, N. Lejal, N. Bertho, et al. 2018. Exploring multiple effects of $Zn_{0.15}Mg_{0.85}O$ nanoparticles on *Bacillus subtilis* and macrophages. Sci. Rep. 8: 12276.

Azhir, E., R. Etefagh, M. Mashreghi and P. Pordeli. 2015. Preparation, characterization and antibacterial activity of manganese oxide nanoparticles. Phys. Chem. Res. 3: 197–204.

Balouiri, M., M. Sadiki and S.K. Ibnsouda. 2016. Methods for in vitro evaluating antimicrobial activity: A review. J. Pharm. Anal. 6: 71–79.

Baniamerian, H., P. Tsapekos, M. Alvarado-Morales, S. Shokrollahzadeh, M. Safavi and I. Angelidaki. 2020. Anti-algal activity of Fe_2O_3–TiO_2 photocatalyst on *Chlorella vulgaris* species under visible light irradiation. Chemosphere. 242: 125119.

Barsaiya, M. and D.P. Singh. 2018. Green Synthesis of Zinc Oxide Nanoparticles by Pseudomonas aeruginosa and their Broad-Spectrum Antimicrobial Effects. J. Pure. Appl. Microbiol. 12: 2123–2134.

Baweja, L., K. Balamurugan, V. Subramanian and A. Dhawan. 2013. Hydration patterns of graphene-based nanomaterials (GBNMs) play a major role in the stability of a helical protein: a molecular dynamics simulation study. Langmuir. 29: 14230–14238.

Baweja, L., K. Balamurugan, V. Subramanian and A. Dhawan. 2015. Effect of graphene oxide on the conformational transitions of amyloid beta peptide: A molecular dynamics simulation study. J. Mol. Graphics Modell. 61: 175–185.

Benelli, G. 2018. Mode of action of nanoparticles against insects. Environ. Sci. Pollut. Res. 25: 12329–12341.

Bermúdez-Jiménez, C., M.G. Romney, S.A. Roa-Flores, G. Martínez-Castañón and H. Bach. 2019. Hydrogel-embedded gold nanorods activated by plasmonic photothermy with potent antimicrobial activity. Nanomedicine. 22: 102093.

Bhakyaraj, K., S. Kumaraguru, K. Gopinath, V. Sabitha, P.R. Kaleeswarran, V. Karthika, et al. 2017. Eco-friendly synthesis of palladium nanoparticles using Melia azedarach leaf extract and their evaluation for antimicrobial and larvicidal activities. J. Cluster Sci. 28: 463–476.

Bhargava, A., V. Pareek, S. Roy Choudhury, J. Panwar and S. Karmakar. 2018. Superior Bactericidal Efficacy of Fucose-Functionalized Silver Nanoparticles against *Pseudomonas aeruginosa* PAO1 and Prevention of Its Colonization on Urinary Catheters. ACS Appl. Mater. Interfaces. 10: 29325–29337.

Billen, M. and R. Van Houdt. 2018. The influence of copper and silver ions on Cupriavidus metallidurans biofilm formation and development. Master of Science, Uhasselt-Hasselt University.

Bitencourt, P.E.R., L.M. Ferreira, L.O. Cargnelutti, L. Denardi, A. Boligon, M. Fleck, et al. 2016. A new biodegradable polymeric nanoparticle formulation containing Syzygium cumini: Phytochemical profile, antioxidant and antifungal activity and in vivo toxicity. Ind. Crops Prod. 83: 400–407.

Bonilla, J.J.A., D.J.P. Guerrero, C.I.S. Suárez, C.C.O. López and R.G.T. Sáez. 2015. *In vitro* antifungal activity of silver nanoparticles against fluconazole-resistant Candida species. World J. Microbiol. Biotechnol. 31: 1801–1809.

Borriss, R., A. Danchin, C.R. Harwood, C. Medigue, E.P.C. Rocha, A. Sekowska, et al. 2018. *Bacillus subtilis*, the model Gram-positive bacterium: 20 years of annotation refinement. Microb. Biotechnol. 11: 3–17.

Botta, G., B.M. Bizzarri, A. Garozzo, R. Timpanaro, B. Bisignano, D. Amatore, et al. 2015. Carbon nanotubes supported tyrosinase in the synthesis of lipophilic hydroxytyrosol and dihydrocaffeoyl catechols with antiviral activity against DNA and RNA viruses. Bioorg. Med. Chem. 23: 5345–5351.

Bove, P., M.A. Malvindi, S.S. Kote, R. Bertorelli, M. Summa and S. Sabella. 2017. Dissolution test for risk assessment of nanoparticles: A pilot study. Nanoscale. 9: 6315–6326.

Burden, N., K. Aschberger, Q. Chaudhry, M.J.D. Clift, S.H. Doak, P. Fowler, et al. 2017. The 3Rs as a framework to support a 21st century approach for nanosafety assessment. Nano Today. 12: 10–13.

Caro, C., F. Gámez, M.J. Sayagues, R. Polvillo and J.L. Royo. 2015. $AgACTiO_2$ nanoparticles with microbiocide properties under visible light. Mater. Res. Express. 2: 055002.

Cassano, R., T. Ferrarelli, M.V. Mauro, P. Cavalcanti, N. Picci and S. Trombino. 2016. Preparation, characterization and *in vitro* activities evaluation of solid lipid nanoparticles based on PEG-40 stearate for antifungal drugs vaginal delivery. Drug delivery 23: 1037–1046.

Castro-Mayorga, J.L., W. Randazzo, M.J. Fabra, J.M. Lagaron, R. Aznar and G. Sánchez. 2017. Antiviral properties of silver nanoparticles against norovirus surrogates and their efficacy in coated polyhydroxyalkanoates systems. LWT-Food Sci. Technol. 79: 503–510.

Chai, Q., Q. Wu, T. Liu, L. Tan, C. Fu, X. Ren, et al. 2017. Enhanced antibacterial activity of silica nanorattles with ZnO combination nanoparticles against methicillin-resistant Staphylococcus aureus. Sci. Bull. 62: 1207–1215.

Chakraborty, C., A.R. Sharma, G. Sharma and S.S. Lee. 2016. Zebrafish: A complete animal model to enumerate the nanoparticle toxicity. J. Nanobiotechnol. 14: 65.

Chandran, D., L.S. Nair, S. Balachandran, K.R. Babu and M. Deepa. 2015. Structural, optical, photocatalytic, and antimicrobial activities of cobalt-doped tin oxide nanoparticles. J. Sol-Gel Sci. Technol. 76: 582–591.

Chaturvedi, V. and P. Verma. 2015. Fabrication of silver nanoparticles from leaf extract of *Butea monosperma* (Flame of Forest) and their inhibitory effect on bloom-forming cyanobacteria. Bioresour. Bioprocess. 2: 18.

Chen, H. and W. Liu. 2016. Cellulose-based photocatalytic paper with Ag_2O nanoparticles loaded on graphite fibers. J. Bioresour. Bioprod. 1: 192–198.

Chen, J., L. Sun, Y. Cheng, Z. Lu, K. Shao, T. Li, et al. 2016. Graphene oxide-silver nanocomposite: Novel agricultural antifungal agent against *Fusarium graminearum* for crop disease prevention. ACS Appl. Mater. Interfaces 8: 24057–24070.

Chen, Q., M. Gundlach, S. Yang, J. Jiang, M. Velki, D. Yin, et al. 2017. Quantitative investigation of the mechanisms of microplastics and nanoplastics toward zebrafish larvae locomotor activity. Sci. Total Environ. 584: 1022–1031.

Cherian, E., A. Rajan and G. Baskar. 2016. Synthesis of manganese dioxide nanoparticles using co-precipitation method and its antimicrobial activity. Int. J. Modern Sci. Technol. 1: 17–22.

Chou, T.M., Y.Y. Ke, Y.H. Tsao, Y.C. Li and Z.H. Lin. 2016. Fabrication of Te and Te-Au Nanowires-Based Carbon Fiber Fabrics for Antibacterial Applications. Int. J. Environ. Res. Public Health.13: 202.

Concu, R., V.V. Kleandrova, A. Speck-Planche and M.N.D.S. Cordeiro. 2017. Probing the toxicity of nanoparticles: A unified in silico machine learning model based on perturbation theory. Nanotoxicology. 11: 891–906.

Costa, P.M. and B. Fadeel. 2016. Emerging systems biology approaches in nanotoxicology: Towards a mechanism-based understanding of nanomaterial hazard and risk. Toxicol. Appl. Pharmacol. 299: 101–111.

Darvishi, B., S. Manoochehri, G. Kamalinia, N. Samadi, M. Amini, S.H. Mostafavi, et al. 2015. Preparation and antibacterial activity evaluation of 18-β-glycyrrhetinic acid loaded PLGA nanoparticles. Iran. J. Pharm. Res. 14: 373–383.

Dědková, K., B. Janíková, K. Matějová, K. Čabanová, R. Váňa, A. Kalup, et al. 2015a. ZnO/graphite composites and its antibacterial activity at different conditions. J. Photochem. Photobiol. B. 151: 256–263.

Dědková, K., J. Lang, K. Matějová, P. Peikertová, J. Holešinský, V. Vodárek, et al. 2015b. Nanostructured composite material graphite/TiO_2 and its antibacterial activity under visible light irradiation. J. Photochem. Photobiol. B. 149: 265–271.

Derbalah, A., M. Shenashen, A. Hamza, A. Mohamed and S. El Safty. 2018. Antifungal activity of fabricated mesoporous silica nanoparticles against early blight of tomato. Egypt. J. Basic Appl. Sci. 5: 145–150.

Devipriya, D. and S.M. Roopan. 2017. Cissus quadrangularis mediated ecofriendly synthesis of copper oxide nanoparticles and its antifungal studies against *Aspergillus niger*, *Aspergillus flavus*. Mater. Sci. Eng. C. 80: 38–44.

Dhanasekar, M., V. Jenefer, R.B. Nambiar, S.G. Babu, S.P. Selvam, B. Neppolian, et al. 2018. Ambient light antimicrobial activity of reduced graphene oxide supported metal doped TiO_2 nanoparticles and their PVA based polymer nanocomposite films. Mater. Res. Bull. 97: 238–243.

Dhanavel, S., N. Manivannan, N. Mathivanan, V.K. Gupta, V. Narayanan and A. Stephen. 2018. Preparation and characterization of cross-linked chitosan/palladium nanocomposites for catalytic and antibacterial activity. J. Mol. Liq. 257: 32–41.

Dizaj, S.M., A. Mennati, S. Jafari, K. Khezri and K. Adibkia. 2015. Antimicrobial activity of carbon-based nanoparticles. Adv. Pharm. Bull. 5: 19–23.

Dong, X., M.M. Moyer, F. Yang, Y.-P. Sun and L. Yang. 2017. Carbon dots' antiviral functions against noroviruses. Sci. Rep. 7: 519.

Dostalova, S., A. Moulick, V. Milosavljevic, R. Guran, M. Kominkova, K. Cihalova, et al. 2016. Antiviral activity of fullerene C60 nanocrystals modified with derivatives of anionic antimicrobial peptide maximin H5. Monatsh. Chem. 147: 905–918.

Du, B.D., D.V. Phu, L.A. Quoc and N.Q. Hien. 2017. Synthesis and investigation of antimicrobial activity of Cu_2O nanoparticles/zeolite. J. Nanopart. 1–6.

Du, T., J. Lu, L. Liu, N. Dong, L. Fang, S. Xiao, et al. 2018. Antiviral activity of graphene oxide–silver nanocomposites by preventing viral entry and activation of the antiviral innate immune response. ACS Appl. Bio Mater. 1: 1286–1293.

Durairaj, B., S. Muthu and T. Xavier. 2015. Antimicrobial activity of *Aspergillus niger* synthesized titanium dioxide nanoparticles. Adv. Appl. Sci. Res. 6: 45–48.

Durán, N., M. Durán, M.B. De Jesus, A.B. Seabra, W.J. Fávaro and G. Nakazato. 2016. Silver nanoparticles: A new view on mechanistic aspects on antimicrobial activity. Nanomedicine. 12: 789–799.

Dutta, R.K., P.K. Sharma, R. Bhargava, N. Kumar and A.C. Pandey. 2010. Differential susceptibility of *Escherichia coli* cells toward transition metal-doped and matrix-embedded ZnO nanoparticles. J. Phys. Chem. B. 114: 5594–5599.

El-Kassas, H.Y. and M.G. Ghobrial. 2017. Biosynthesis of metal nanoparticles using three marine plant species: Anti-algal efficiencies against "*Oscillatoria simplicissima*". Environ. Sci. Pollut. Res. 24: 7837–7849.

El-Zahry, M.R., A. Mahmoud, I.H. Refaat, H.A. Mohamed, H. Bohlmann and B. Lendl. 2015. Antibacterial effect of various shapes of silver nanoparticles monitored by SERS. Talanta. 138: 183–189.

Elsharkawy, M.M. and A. Derbalah. 2019. Antiviral activity of titanium dioxide nanostructures as a control strategy for broad bean strain virus in faba bean. Pest Manage. Sci. 75: 828–834.

Entezari, M., Z.G. Tabatabaei, A. Azarioun, S. Sarabian and G.T. Farahani. 2015. *In vitro* evaluation of the antibacterial activity of modified multi-walled carbon nanotubes with phenolic extracts. J Biomater. Tissue Eng. 2: 17–23.

Eskandari-Nojedehi, M., H. Jafarizadeh-Malmiri and J. Rahbar-Shahrouzi. 2018. Hydrothermal green synthesis of gold nanoparticles using mushroom (*Agaricus bisporus*) extract: Physico-chemical characteristics and antifungal activity studies. Green Process. Synth. 7: 38–47.

Ezhilarasi, A.A., J.J. Vijaya, K. Kaviyarasu, L.J. Kennedy, R.J. Ramalingam and H.A. Al-Lohedan. 2018. Green synthesis of NiO nanoparticles using Aegle marmelos leaf extract for the evaluation of *in vitro* cytotoxicity, antibacterial and photocatalytic properties. J. Photochem. Photobiol. B. 180: 39–50.

Fakhri, A., M. Pourmand, R. Khakpour and S. Behrouz. 2015. Structural, optical, photoluminescence and antibacterial properties of copper-doped silver sulfide nanoparticles. J. Photochem. Photobiol. B. 149: 78–83.

Farkas, J., P. Christian, J.A.G. Urrea, N. Roos, M. Hassellöv, K.E. Tollefsen, et al. 2010. Effects of silver and gold nanoparticles on rainbow trout (*Oncorhynchus mykiss*) hepatocytes. Aquat. Toxicol. 96: 44–52.

Figueroba, A., A. Bruix, G. Kovács and K.M. Neyman. 2017. Metal-doped ceria nanoparticles: Stability and redox processes. Phys. Chem. Chem. Phys. 19: 21729–21738.

Foo, M.E., P. Anbu, S.C.B. Gopinath, T. Lakshmipriya, C.G. Lee, H.S. Yun, et al. 2018. Antimicrobial activity of functionalized single-walled carbon nanotube with herbal extract of Hempedu bumi. Surf. Interface Anal. 50: 354–361.

Fosso-Kankeu, E., C.M. De Klerk, T.A. Botha, F. Waanders, J. Phoku and S. Pandey. 2016. The antifungal activities of multi-walled carbon nanotubes decorated with silver, copper and zinc oxide particles. International Conference in Science, Engineering, Technology and Natural Resources. 55–59.

Foster, J.E., J.A. Mendoza and J. Seetahal. 2018. Viruses as pathogens: animal viruses, with emphasis on human viruses. pp. 157–187. *In:* P. Tennant, G. Fermin and J.E. Foster [eds]. Viruses: Molecular Biology, Host Interactions, and Applications to Biotechnology. Academic Press.

Franci, G., A. Falanga, S. Galdiero, L. Palomba, M. Rai, G. Morelli, et al. 2015. Silver nanoparticles as potential antibacterial agents. Molecules. 20: 8856–8874.

Freire, P.L.L., T.C.M. Stamford, A.J.R. Albuquerque, F.C. Sampaio, H.M.M. Cavalcante, R.O. Macedo, et al. 2015. Action of silver nanoparticles towards biological systems: Cytotoxicity evaluation using hen's egg test and inhibition of Streptococcus mutans biofilm formation. Int. J. Antimicrob. Agents. 45: 183–187.

Fröhlich, E. and S. Salar-Behzadi. 2014. Toxicological assessment of inhaled nanoparticles: Role of *in vivo*, *ex vivo*, *in vitro*, and *in silico* studies. Int. J. Mol. Sci. 15: 4795–4822.

Fu, Y., J. Jiang, Q. Zhang, X. Zhan and F. Chen. 2016. Robust liquid-repellent coatings based on polymer nanoparticles with excellent self-cleaning and antibacterial performances. J. Mater. Chem. A 5: 275–284.

Furxhi, I., F. Murphy, M. Mullins and C.A. Poland. 2019. Machine learning prediction of nanoparticle *in vitro* toxicity: A comparative study of classifiers and ensemble-classifiers using the Copeland Index. Toxicol. Lett. 312: 157–166.

Galdiero, S., A. Falanga, M. Vitiello, M. Cantisani, V. Marra and M. Galdiero. 2011. Silver nanoparticles as potential antiviral agents. Molecules. 16: 8894–8918.

Gandhi, N., D. Sirisha and S. Asthana. 2018. Microwave mediated green synthesis of lead (Pb) nanoparticles and its potential applications. Int. J. Eng. Sci. & Res. Tech. 7: 623–644.

Gao, Z., X. Li, L. Shi and Y. Yang. 2019. Deep eutectic solvents-derived carbon dots for detection of mercury (II), photocatalytic antifungal activity and fluorescent labeling for *C. albicans*. Spectrochim Acta A Mol. Biomol. Spectrosc. 220: 117080.

Garcia, I.M., V.C.B. Leitune, F. Visioli, S.M.W. Samuel and F.M. Collares. 2018. Influence of zinc oxide quantum dots in the antibacterial activity and cytotoxicity of an experimental adhesive resin. J. Dent. 73: 57–60.

Garrido-Sanz, D., J.P. Meier-Kolthoff, M. Göker, M. Martin, R. Rivilla and M. Redondo-Nieto. 2016. Genomic and genetic diversity within the *Pseudomonas fluorescens* complex. PLoS One. 11: e0150183.

Ghosh, M., S. Mandal, A. Roy, S. Chakrabarty, G. Chakrabarti and S.K. Pradhan. 2020. Enhanced antifungal activity of fluconazole conjugated with Cu-Ag-ZnO nanocomposite. Mater. Sci. Eng. C. 106: 110160.

Gloria, E.C., V. Ederley, M. Gladis, H. César, O. Jaime, A. Oscar, et al. 2017. Synthesis of silver nanoparticles (AgNPs) with antibacterial activity. J. Phys. Conf. Ser. 850: 012023.

Goffredo, G.B., S. Accoroni and C. Totti. 2019. Nanotreatments to inhibit microalgal fouling on building stone surfaces. pp. 619–647. *In:* F. Pacheco-Torgal, M.V. Diamanti, A. Nazari, C. Goran-Granqvist, A. Pruna and S. Amirkhanian [eds]. Nanotechnology in Eco-efficient Construction. Elsevier.

Goffredo, G.B., S. Accoroni, C. Totti, T. Romagnoli, L. Valentini and P. Munafo. 2017. Titanium dioxide based nanotreatments to inhibit microalgal fouling on building stone surfaces. Build. Sci. 112: 209–222.

Goldberg, E., M. Scheringer, T.D. Bucheli and K. Hungerbühler. 2015. Prediction of nanoparticle transport behavior from physicochemical properties: Machine learning provides insights to guide the next generation of transport models. Environ. Sci. Nano. 2: 352–360.

Gopinath, A. and K. Krishna. 2019. Dual role of chemically functionalized activated carbon fibres: Investigation of parameters influencing the degradation of organophosphorus compounds and antibacterial behaviour. J. Chem. Technol. Biotechnol. 94: 611–617.

Graziani, L., E. Quagliarini and M. D'Orazio. 2016. The role of roughness and porosity on the self-cleaning and anti-biofouling efficiency of TiO_2-Cu and TiO_2-Ag nanocoatings applied on fired bricks. Constr. Build. Mater. 129: 116–124.

Guadagnini, R., B. Halamoda Kenzaoui, L. Walker, G. Pojana, Z. Magdolenova, D. Bilanicova, et al. 2015. Toxicity screenings of nanomaterials: Challenges due to interference with assay processes and components of classic *in vitro* tests. Nanotoxicology. 9: 13–24.

Gutiérrez, J.A., S. Caballero, L.A. Díaz, M.A. Guerrero, J. Ruiz and C.C. Ortiz. 2018. High antifungal activity against candida species of monometallic and bimetallic nanoparticles synthesized in nanoreactors. ACS Biomater. Sci. Eng. 4: 647–653.

Habiba, K., D.P. Bracho-Rincon, J.A. Gonzalez-Feliciano, J.C. Villalobos-Santos, V.I. Makarov, D. Ortiz, et al. 2015. Synergistic antibacterial activity of PEGylated silver–graphene quantum dots nanocomposites. Appl. Mater. Today. 1: 80–87.

Habiba, R.N., S.A. Ali, V. Sultana, J. Ara and S. Ehteshamul-Haque. 2016. Evaluation of biocontrol potential of epiphytic fluorescent Pseudomonas associated with healthy fruits and vegetables against root rot and root knot pathogens of mungbean. Pak. J. Bot. 48: 1299–1303.

Halbus, A.F., T.S. Horozov and V.N. Paunov. 2019. Self-grafting copper oxide nanoparticles show a strong enhancement of their anti-algal and anti-yeast action. Nanoscale Adv. 1: 2323–2336.

Hamdy, A.A., N.A. Elattal, M.A. Amin, A.E. Ali, N.M. Mansour, G.E.A. Awad, et al. 2018. *In vivo* assessment of possible probiotic properties of *Bacillus subtilis* and prebiotic properties of levan. Biocatal. Agric. Biotechnol. 13: 190–197.

Hameed, A.S.H., C. Karthikeyan, A.P. Ahamed, N. Thajuddin, N.S. Alharbi, S.A. Alharbi, et al. 2016. *In vitro* antibacterial activity of ZnO and Nd doped ZnO nanoparticles against ESBL producing *Escherichia coli* and *Klebsiella pneumoniae*. Sci. Rep. 6: 24312.

Hanini, A., M. El Massoudi, J. Gavard, K. Kacem, S. Ammar and O. Souilem. 2016. Nanotoxicological study of polyol-made cobalt-zinc ferrite nanoparticles in rabbit. Environ. Toxicol. Pharmacol. 45: 321–327.

He, Y., S. Ingudam, S. Reed, A. Gehring, T.P. Strobaugh and P. Irwin. 2016. Study on the mechanism of antibacterial action of magnesium oxide nanoparticles against foodborne pathogens. J. Nanobiotechnol. 14: 54.

Hegde, K., S.K. Brar, M. Verma and R.Y. Surampalli. 2016. Current understandings of toxicity, risks and regulations of engineered nanoparticles with respect to environmental microorganisms. Nanotechnol. Environ. Eng. 1: 5.

Herlekar, M.B., S. Barve and R. Kumar. 2015. Biological synthesis of iron oxide nanoparticles using agro-wastes and feasibility for municipal wastewater treatment. Conference: 47th Indian Water Works Annual Convention. 1–6.

Hisey, B., P.J. Ragogna and E.R. Gillies. 2017. Phosphonium-functionalized polymer micelles with intrinsic antibacterial activity. Biomacromolecules. 18: 914–923.

Hofmann-Amtenbrink, M., D.W. Grainger and H. Hofmann. 2015. Nanoparticles in medicine: Current challenges facing inorganic nanoparticle toxicity assessments and standardizations. Nanomedicine. 11: 1689–1694.

Hong, X., J. Wen, X. Xiong and Y. Hu. 2016. Shape effect on the antibacterial activity of silver nanoparticles synthesized via a microwave-assisted method. Environ. Sci. Pollut. Res. 23: 4489–4497.

Hoseinzadeh, A., A. Habibi-Yangjeh and M. Davari. 2016. Antifungal activity of magnetically separable Fe_3O_4/ZnO/AgBr nanocomposites prepared by a facile microwave-assisted method. Prog. Nat. Sci.: Mater. Int. 26: 334–340.

Hossain, Z., G. Mustafa and S. Komatsu. 2015. Plant responses to nanoparticle stress. Int. J. Mol. Sci. 16: 26644–26653.

Hou, S., J. Li, X. Huang, X. Wang, L. Ma, W. Shen, et al. 2017. Silver nanoparticles-loaded exfoliated graphite and its anti-bacterial performance. Appl. Sci. 7: 852.

Hristozov, D., S. Gottardo, E. Semenzin, A. Oomen, P. Bos, W. Peijnenburg, et al. 2016. Frameworks and tools for risk assessment of manufactured nanomaterials. Environ. Int. 95: 36–53.

Hu, C.-M.J., W.-S. Chang, Z.-S. Fang, Y.-T. Chen, W.-L. Wang, H.-H. Tsai, et al. 2017. Nanoparticulate vacuolar ATPase blocker exhibits potent host-targeted antiviral activity against feline coronavirus. Sci. Rep. 7: 13043.

Hu, F., Z. Zhou, Q. Xu, C. Fan, L. Wang, H. Ren, et al. 2019. A novel pH-responsive quaternary ammonium chitosan-liposome nanoparticles for periodontal treatment. Int. J. Biol. Macromol. 129: 1113–1119.

Hung, W.-C., K.-H. Wu, D.-Y. Lyu, K.-F. Cheng and W.-C. Huang. 2017. Preparation and characterization of expanded graphite/metal oxides for antimicrobial application. Mater. Sci. Eng. C. 75: 1019–1025.

Huo, L., R. Chen, L. Zhao, X. Shi, R. Bai, D. Long, et al. 2015. Silver nanoparticles activate endoplasmic reticulum stress signaling pathway in cell and mouse models: The role in toxicity evaluation. Biomaterials. 61: 307–315.

Huy, T.Q., N.T.H. Thanh, N.T. Thuy, P. Van Chung, P.N. Hung, A.-T. Le, et al. 2017. Cytotoxicity and antiviral activity of electrochemical–synthesized silver nanoparticles against poliovirus. J. Virol. Methods. 241: 52–57.

Ibrahim, H.M.M. 2015. Green synthesis and characterization of silver nanoparticles using banana peel extract and their antimicrobial activity against representative microorganisms. J. Radiat. Res. Appl. Sci. 8: 265–275.

Ifuku, S., Y. Tsukiyama, T. Yukawa, M. Egusa, H. Kaminaka, H. Izawa, et al. 2015. Facile preparation of silver nanoparticles immobilized on chitin nanofiber surfaces to endow antifungal activities. Carbohydr. Polym. 117: 813–817.

Imada, K., S. Sakai, H. Kajihara, S. Tanaka and S. Ito. 2016. Magnesium oxide nanoparticles induce systemic resistance in tomato against bacterial wilt disease. Plant Pathol. 65: 551–560.

Inam, M., J.C. Foster, J. Gao, Y. Hong, J. Du, A.P. Dove, et al. 2019. Size and shape affects the antimicrobial activity of quaternized nanoparticles. J. Polym. Sci. Part A: Polym. Chem. 57: 255–259.

Irshad, R., K. Tahir, B. Li, A. Ahmad, A.R. Siddiqui and S. Nazir. 2017. Antibacterial activity of biochemically capped iron oxide nanoparticles: A view towards green chemistry. J. Photochem. Photobiol. B. 170: 241–246.

Ismail, R.A., G.M. Sulaiman, S.A. Abdulrahman and T.R. Marzoog. 2015. Antibacterial activity of magnetic iron oxide nanoparticles synthesized by laser ablation in liquid. Mater. Sci. Eng. C. 53: 286–297.

Iswarya, V., M. Bhuvaneshwari, N. Chandrasekaran and A. Mukherjee. 2016a. Individual and binary toxicity of anatase and rutile nanoparticles towards *Ceriodaphnia dubia*. Aquat. Toxicol. 178: 209–221.

Iswarya, V., J. Manivannan, A. De, S. Paul, R. Roy, J.B. Johnson, et al. 2016b. Surface capping and size-dependent toxicity of gold nanoparticles on different trophic levels. Environ. Sci. Pollut. Res. 23: 4844–4858.

Jagielski, T., Z. Bakuła, M. Pleń, M. Kamiński, J. Nowakowska, J. Bielecki, et al. 2018. The activity of silver nanoparticles against microalgae of the Prototheca genus. Nanomedicine. 13: 1025–1036.

Jahnel, J. 2015. Conceptual questions and challenges associated with the traditional risk assessment paradigm for nanomaterials. Nanoethics. 9: 261–276.

Jamdagni, P., P. Khatri and J.S. Rana. 2018. Green synthesis of zinc oxide nanoparticles using flower extract of Nyctanthes arbor-tristis and their antifungal activity. J. King Saud. Univ. Sci. 30: 168–175.

Javadi, F., M.E.T. Yazdi, M. Baghani and A. Es-haghi. 2019. Biosynthesis, characterization of cerium oxide nanoparticles using Ceratonia siliqua and evaluation of antioxidant and cytotoxicity activities. Mater. Res. Express. 6: 065408.

Jayaramudu, T., G.M. Raghavendra, K. Varaprasad, G.V.S. Reddy, A.B. Reddy, K. Sudhakar, et al. 2016. Preparation and characterization of poly (ethylene glycol) stabilized nano silver particles by a mechanochemical assisted ball mill process. J. Appl. Polym. Sci. 133: 43027.

Jeevanandam, J., Y.S. Chan and M.K. Danquah. 2016. Biosynthesis of metal and metal oxide nanoparticles. ChemBioEng Rev. 3: 55–67.

Jeevanandam, J., A. Barhoum, Y.S. Chan, A. Dufresne and M.K. Danquah. 2018a. Review on nanoparticles and nanostructured materials: history, sources, toxicity and regulations. Beilstein. J. Nanotechnol. 9: 1050–1074.

Jeevanandam, J., K. Pal and M.K. Danquah. 2018b. Virus-like nanoparticles as a novel delivery tool in gene therapy. Biochimie. 157: 38–47.

Jeevanandam, J., Y.S. Chan and M.K. Danquah. 2019a. Evaluating the antibacterial activity of MgO nanoparticles synthesized from aqueous leaf extract. Med One 4: e190011.

Jeevanandam, J., Y.S. Chan and M.K. Danquah. 2019b. Zebrafish as a model organism to study nanomaterial toxicity. Emerg. Sci. J. 3: 195–208.

Jeevanandam, J., Y.S. Chan, S. Pan and M.K. Danquah 2019c. Metal oxide nanocomposites: Cytotoxicity and targeted drug delivery applications. pp. 111–147. *In:* K. Pal [ed.]. Hybrid nanocomposites: Fundamentals, synthesis and applications. Pan Stanford Publishing.

Jeevanandam, J., Y. San Chan, M.K. Danquah and M.C. Law. 2019d. Cytotoxicity analysis of morphologically different sol-gel-synthesized MgO nanoparticles and their *in vitro* insulin resistance reversal ability in adipose cells. Appl. Biochem. Biotechnol. 1–26.

Jeevanandam, J., A. Sundaramurthy, V. Sharma, C. Murugan, K. Pal, M.H.A. Kodous, et al. 2020. Sustainability of one-dimensional nanostructures: Fabrication and industrial applications. pp. 83–113. *In:* G. Szekely and A. Livingston [eds]. Sustainable Nanoscale Engineering: From Materials Design to Chemical Processing. Elsevier.

Jesline, A., N.P. John, P.M. Narayanan, C. Vani and S. Murugan. 2015. Antimicrobial activity of zinc and titanium dioxide nanoparticles against biofilm-producing methicillin-resistant Staphylococcus aureus. Appl. Nanosci. 5: 157–162.

Ji, H., H. Sun and X. Qu. 2016. Antibacterial applications of graphene-based nanomaterials: Recent achievements and challenges. Adv. Drug Delivery Rev. 105: 176–189.

Ji, W., D. Zhu, Y. Chen, J. Hu and F. Li. 2017. *In-vitro* cytotoxicity of biosynthesized zinc oxide nanoparticles towards cardiac cell lines of Catla catla. Biomed. Res. 28: 2262–2266.

Jiang, S., Y. Qin, J. Yang, M. Li, L. Xiong and Q. Sun. 2017. Enhanced antibacterial activity of lysozyme immobilized on chitin nanowhiskers. Food Chem. 221: 1507–1513.

Kadiyala, U., E.S. Turali-Emre, J.H. Bahng, N.A. Kotov and J.S. VanEpps. 2018. Unexpected insights into antibacterial activity of zinc oxide nanoparticles against methicillin resistant Staphylococcus aureus (MRSA). Nanoscale. 10: 4927–4939.

Karimiyan, A., H. Najafzadeh, M. Ghorbanpour and S.H. Hekmati-Moghaddam. 2015. Antifungal effect of magnesium oxide, zinc oxide, silicon oxide and copper oxide nanoparticles against *Candida albicans*. Zahedan J. Res. Med. Sci. 17: e2179.

Katwal, R., H. Kaur, G. Sharma, M. Naushad and D. Pathania. 2015. Electrochemical synthesized copper oxide nanoparticles for enhanced photocatalytic and antimicrobial activity. J. Ind. Eng. Chem. 31: 173–184.

Khalil, A.T., M. Ovais, I. Ullah, M. Ali, Z.K. Shinwari and M. Maaza. 2017. Biosynthesis of iron oxide (Fe_2O_3) nanoparticles via aqueous extracts of Sageretia thea (Osbeck.) and their pharmacognostic properties. Green Chem. Lett. Rev. 10: 186–201.

Khan, B.A., V.S. Chevali, H. Na, J. Zhu, P. Warner and H. Wang. 2016. Processing and properties of antibacterial silver nanoparticle-loaded hemp hurd/poly (lactic acid) biocomposites. Compos. B. Eng. 100: 10–18.

Khan, I., K. Saeed and I. Khan. 2019. Nanoparticles: Properties, applications and toxicities. Arabian J. Chem. 12: 908–931.

Khan, S.A., F. Noreen, S. Kanwal and G. Hussain. 2017. Comparative synthesis, characterization of Cu-doped ZnO nanoparticles and their antioxidant, antibacterial, antifungal and photocatalytic dye degradation activities. Dig. J. Nanomater. Biostruct. 12: 877–889.

Khatami, M., H. Heli, P.M. Jahani, H. Azizi and M.A.L. Nobre. 2017. Copper/copper oxide nanoparticles synthesis using *Stachys lavandulifolia* and its antibacterial activity. IET Nanobiotechnol. 11: 709–713.

Khatir, N.M., Z. Abdul-Malek, A.K. Zak, A. Akbari and F. Sabbagh. 2016. Sol–gel grown Fe-doped ZnO nanoparticles: Antibacterial and structural behaviors. J. Sol-Gel Sci. Technol. 78: 91–98.

Kora, A.J. and J. Arunachalam. 2011. Assessment of antibacterial activity of silver nanoparticles on *Pseudomonas aeruginosa* and its mechanism of action. World J. Microbiol. Biotechnol. 27: 1209–1216.

Krishnaraj, C., B.J. Ji, S.L. Harper and S.I. Yun. 2016. Plant extract-mediated biogenic synthesis of silver, manganese dioxide, silver-doped manganese dioxide nanoparticles and their antibacterial activity against food- and water-borne pathogens. Bioprocess Biosyst. Eng. 39: 759–772.

Kroll, A., M.H. Pillukat, D. Hahn and J. Schnekenburger. 2009. Current *in vitro* methods in nanoparticle risk assessment: Limitations and challenges. Eur. J. Pharm. Biopharm. 72: 370–377.

Kubo, A.L., I. Capjak, I.V. Vrček, O.M. Bondarenko, I. Kurvet, H. Vija, et al. 2018. Antimicrobial potency of differently coated 10 and 50 nm silver nanoparticles against clinically relevant bacteria *Escherichia coli* and *Staphylococcus aureus*. Colloids Surf. B Biointerfaces. 170: 401–410.

Kühnel, D., W. Busch, T. Meißner, A. Springer, A. Potthoff, V. Richter, et al. 2009. Agglomeration of tungsten carbide nanoparticles in exposure medium does not prevent uptake and toxicity toward a rainbow trout gill cell line. Aquat. Toxicol. 93: 91–99.

Kumar, P.P.N.V., U. Shameem, P. Kollu, R.L. Kalyani and S.V.N. Pammi. 2015. Green synthesis of copper oxide nanoparticles using *Aloe vera* leaf extract and its antibacterial activity against fish bacterial pathogens. Bionanosci. 5: 135–139.

Kumar, R., G.C. Sahoo, M. Chawla-Sarkar, M.K. Nayak, K. Trivedi, S. Rana, et al. 2016. Antiviral effect of glycine coated iron oxide nanoparticles iron against H1N1 influenza A virus. Int. J. Infect. Dis. 45: 281–282.

Kumari, R., M. Barsainya and D.P. Singh. 2017. Biogenic synthesis of silver nanoparticle by using secondary metabolites from Pseudomonas aeruginosa DM1 and its anti-algal effect on *Chlorella vulgaris* and *Chlorella pyrenoidosa*. Environ. Sci. Pollut. Res. 24: 4645–4654.

Kwon, J.Y., P. Koedrith and Y.R. Seo. 2014. Current investigations into the genotoxicity of zinc oxide and silica nanoparticles in mammalian models *in vitro* and *in vivo*: Carcinogenic/genotoxic potential, relevant mechanisms and biomarkers, artifacts, and limitations. Int. J. Nanomed. 9: 271.

Laux, P., J. Tentschert, C. Riebeling, A. Braeuning, O. Creutzenberg, A. Epp, et al. 2018. Nanomaterials: Certain aspects of application, risk assessment and risk communication. Arch. Toxicol. 92: 121–141.

Le, N.T.T., L.G. Bach, D.C. Nguyen, T.H.X. Le, K.H. Pham, D.H. Nguyen, et al. 2019. Evaluation of factors affecting antimicrobial activity of bacteriocin from *Lactobacillus plantarum* microencapsulated in alginate-gelatin capsules and its application on pork meat as a bio-preservative. Int. J. Environ. Res. Public Health. 16: 1017.

Lefcheck, J.S., J.E.K. Byrnes, F. Isbell, L. Gamfeldt, J.N. Griffin, N. Eisenhauer, et al. 2015. Biodiversity enhances ecosystem multifunctionality across trophic levels and habitats. Nat. Commun. 6: 6936.

Levina, A.S., M.N. Repkova, E.V. Bessudnova, E.I. Filippova, N.A. Mazurkova and V.F. Zarytova. 2016. High antiviral effect of TiO_2·PL–DNA nanocomposites targeted to conservative regions of (–) RNA and (+) RNA of influenza A virus in cell culture. Beilstein J. Nanotechnol. 7: 1166–1173.

Li, J., D. Zhai, F. Lv, Q. Yu, H. Ma, J. Yin, et al. 2016. Preparation of copper-containing bioactive glass/eggshell membrane nanocomposites for improving angiogenesis, antibacterial activity and wound healing. Acta Biomater. 36: 254–266.

Li, H., M. Zhang, Y. Song, H. Wang, Chang'an Liu, Y. Fu, et al. 2018. Multifunctional carbon dot for lifetime thermal sensing, nucleolus imaging and antialgal activity. J. Mater. Chem. B. 6: 5708–5717.

Li, X., R. Huang, F.-K. Tang, W.-C. Li, S.S.W. Wong, K.C.-F. Leung, et al. 2019. Red-emissive guanylated polyene-functionalized carbon dots arm oral epithelia against invasive fungal infections. ACS Appl. Mater. Interfaces. 11: 46591–46603.

Lin, Z., N.A. Monteiro-Riviere, R. Kannan and J.E. Riviere. 2015. A computational framework for interspecies pharmacokinetics, exposure and toxicity assessment of gold nanoparticles. Nanomedicine. 11: 107–119.

Linkov, I. and F.K. Satterstrom 2008. Nanomaterial risk assessment and risk management. pp. 129–157. *In:* I. Linkov, E. Ferguson and V.S. Magar [eds]. Real-Time and Deliberative Decision Making. NATO Science for Peace and Security Series C: Environmental Security. Springer, Dordrecht.

Liu, J., S. Lu, Q. Tang, K. Zhang, W. Yu, H. Sun, et al. 2017a. One-step hydrothermal synthesis of photoluminescent carbon nanodots with selective antibacterial activity against *Porphyromonas gingivalis*. Nanoscale. 9: 7135–7142.

Liu, N., Y. Chang, Y. Feng, Y. Cheng, X. Sun, H. Jian, et al. 2017b. {101}–{001} Surface heterojunction-enhanced antibacterial activity of titanium dioxide nanocrystals under sunlight irradiation. ACS Appl. Mater. Interfaces. 9: 5907–5915.

Logeswari, P., S. Silambarasan and J. Abraham. 2015. Synthesis of silver nanoparticles using plants extract and analysis of their antimicrobial property. J. Saudi Chem. Soc. 19: 311–317.

Ma, Y., J. Liu, H. Yin, X. Xu, Y. Xie, D. Chen, et al. 2018. Remarkably improvement in antibacterial activity of carbon nanotubes by hybridizing with silver nanodots. J. Nanosci. Nanotechnol. 18: 5704–5710.

Mahmoud, N.N., A.M. Alkilany, E.A. Khalil and A.G. Al-Bakri. 2017. Antibacterial activity of gold nanorods against *Staphylococcus aureus* and *Propionibacterium acnes*: misinterpretations and artifacts. Int. J. Nanomed. 12: 7311–7322.

Marchiore, N.G., I.J. Manso, K.C. Kaufmann, G.F. Lemes, A.P. de Oliveira Pizolli, A.A. Droval, et al. 2017. Migration evaluation of silver nanoparticles from antimicrobial edible coating to sausages. LWT-Food Sci. Technol. 76: 203–208.

Mariano-Torres, J.A., A. López-Marure, M. García-Hernández, G. Basurto-Islas and M.Á. Domínguez-Sánchez. 2018. Synthesis and characterization of glycerol citrate polymer and yttrium oxide nanoparticles as a potential antibacterial material. Mater. Trans. 59: 1915–1919.

Marvin, H.J.P., Y. Bouzembrak, E.M. Janssen, M. van der Zande, F. Murphy, B. Sheehan, et al. 2017. Application of Bayesian networks for hazard ranking of nanomaterials to support human health risk assessment. Nanotoxicology. 11: 123–133.

Mashitah, M.D., Y.S. Chan and J. Jason 2016. Antimicrobial properties of nanobiomaterials and the mechanism. pp. 261–312. *In:* A.M. Grumezescu [ed.]. Nanobiomaterials in Antimicrobial Therapy: Applications of Nanobiomaterials. William Andrew Publishing.

Medda, S., A. Hajra, U. Dey, P. Bose and N.K. Mondal. 2015. Biosynthesis of silver nanoparticles from *Aloe vera* leaf extract and antifungal activity against *Rhizopus* sp. and *Aspergillus* sp. Appl. Nanosci. 5: 875–880.

Meghana, S., P. Kabra, S. Chakraborty and N. Padmavathy. 2015. Understanding the pathway of antibacterial activity of copper oxide nanoparticles. RSC Adv. 5: 12293–12299.

Melagraki, G. and A. Afantitis. 2015. A risk assessment tool for the virtual screening of metal oxide nanoparticles through Enalos InSilicoNano Platform. Curr. Top. Med. Chem. 15: 1827–1836.

Mihailescu, I.N., D. Bociaga, G. Socol, G.E. Stan, M.-C. Chifiriuc, C. Bleotu, et al. 2016. Fabrication of antimicrobial silver-doped carbon structures by combinatorial pulsed laser deposition. Int. J. Pharm. 515: 592–606.

Mikiciuk, J., E. Mikiciuk and A. Szterk. 2017. Physico-chemical properties and inhibitory effects of commercial colloidal silver nanoparticles as potential antimicrobial agent in the food industry. J. Food Process. Preserv. 41: e12793.

Mocan, T., C.T. Matea, T. Pop, O. Mosteanu, A.D. Buzoianu, S. Suciu, et al. 2017. Carbon nanotubes as anti-bacterial agents. Cell. Mol. Life Sci. 74: 3467–3479.

Mohamed, M.M., S.A. Fouad, H.A. Elshoky, G.M. Mohammed and T.A. Salaheldin. 2017. Antibacterial effect of gold nanoparticles against *Corynebacterium pseudotuberculosis*. Int. J. Vet. Sci. Med. 5: 23–29.

Mohamed, N.A. and N.A.A. El-Ghany. 2018. Novel aminohydrazide cross-linked chitosan filled with multi-walled carbon nanotubes as antimicrobial agents. Int. J. Biol. Macromol. 115: 651–662.

Mohammad, M.R., D.S. Ahmed and M.K.A. Mohammed. 2019. Synthesis of Ag-doped TiO_2 nanoparticles coated with carbon nanotubes by the sol-gel method and their antibacterial activities. J. Sol-Gel Sci. Technol. 90: 498–509.

Mohammed, M.K.A., D.S. Ahmed and M.R. Mohammad. 2019. Studying antimicrobial activity of carbon nanotubes decorated with metal-doped ZnO hybrid materials. Mater. Res. Express. 6: 055404.

Mohan, A.N., M.B and S. Panicker. 2019. Facile synthesis of graphene-tin oxide nanocomposite derived from agricultural waste for enhanced antibacterial activity against *Pseudomonas aeruginosa*. Sci. Rep. 9: 4170.

Molina, G.A., R. Esparza, J.L. López-Miranda, A.R. Hernández-Martínez, B.L. España-Sánchez, E.A. Elizalde-Peña, et al. 2019. Green synthesis of Ag nanoflowers using *Kalanchoe Daigremontiana* extract for enhanced photocatalytic and antibacterial activities. Colloids Surf. B. 180: 141–149.

Mukherjee, A., I. Mohammed Sadiq, T.C. Prathna and N. Chandrasekaran. 2011. Antimicrobial activity of aluminium oxide nanoparticles for potential clinical applications. pp. 245–251. *In*: A. Méndez-Vilas [ed.]. Science against Microbial Pathogens: Communicating Current Research and Technological Advances. Formatex Research Center.

Murugan, K., D. Dinesh, D. Nataraj, J. Subramaniam, P. Amuthavalli, J. Madhavan, et al. 2018. Iron and iron oxide nanoparticles are highly toxic to *Culex quinquefasciatus* with little non-target effects on larvivorous fishes. Environ. Sci. Pollut. Res. 25: 10504–10514.

Muthu, S., V. Raju, V.B. Gopal, A. Gunasekaran, K.S. Narayan, S. Malairaj, et al. 2019. A rapid synthesis and antibacterial property of selenium nanoparticles using egg white lysozyme as a stabilizing agent. SN. Appl. Sci. 1: 1543.

Muthukumar, T., Sudhakumari, B. Sambandam, A. Aravinthan, T.P. Sastry and J.H. Kim. 2016. Green synthesis of gold nanoparticles and their enhanced synergistic antitumor activity using HepG2 and MCF7 cells and its antibacterial effects. Process Biochem. 51: 384–391.

Naicker, P.K., P.T. Cummings, H. Zhang and J.F. Banfield. 2005. Characterization of titanium dioxide nanoparticles using molecular dynamics simulations. J. Phys. Chem. B. 109: 15243–15249.

Natarajan, S., D.S. Lakshmi, V. Thiagarajan, P. Mrudula, N. Chandrasekaran and A. Mukherjee. 2018. Antifouling and anti-algal effects of chitosan nanocomposite (TiO_2/Ag) and pristine (TiO_2 and Ag) films on marine microalgae *Dunaliella salina*. J. Environ. Chem. Eng. 6: 6870–6880.

Nayak, D., S. Ashe, P.R. Rauta, M. Kumari and B. Nayak. 2016. Bark extract mediated green synthesis of silver nanoparticles: evaluation of antimicrobial activity and antiproliferative response against osteosarcoma. Mater. Sci. Eng. C. 58: 44–52.

Nazaktabar, A., M.S. Lashkenari, A. Araghi, M. Ghorbani and H. Golshahi. 2017. *In vivo* evaluation of toxicity and antiviral activity of polyrhodanine nanoparticles by using the chicken embryo model. Int. J. Biol. Macromol. 103: 379–384.

Ncube, N.S., A.J. Afolayan and A.I. Okoh. 2008. Assessment techniques of antimicrobial properties of natural compounds of plant origin: Current methods and future trends. Afr. J. Biotechnol. 7: 1797–1806.

Nolte, T.M., N.B. Hartmann, J.M. Kleijn, J. Garnæs, D. van de Meent, A.J. Hendriks, et al. 2017. The toxicity of plastic nanoparticles to green algae as influenced by surface modification, medium hardness and cellular adsorption. Aquat. Toxicol. 183: 11–20.

Nored, A.W., M.C.G. Chalbot and I.G. Kavouras. 2018. Characterization of paint dust aerosol generated from mechanical abrasion of TiO_2-containing paints. J. Occup. Environ. Hyg. 15: 629–640.

Nowakowska, J.B., K.I. Wolska and A.M. Grudniak. 2018. The activity of silver nanoparticles against microalgae of the *Prototheca genus*. Nanomedicine (Lond). 13: 1025–1036.

Nthunya, L.N., S. Derese, L. Gutierrez, A.R. Verliefde, B.B. Mamba, T.G. Barnard, et al. 2019. Green synthesis of silver nanoparticles using one-pot and microwave-assisted methods and their subsequent embedment on PVDF nanofibre membranes for growth inhibition of mesophilic and thermophilic bacteria. New J. Chem. 43: 4168–4180.

Ojo, S.A., A. Lateef, M.A. Azeez, S.M. Oladejo, A.S. Akinwale, T.B. Asafa, et al. 2016. Biomedical and catalytic applications of gold and silver-gold alloy nanoparticles biosynthesized using cell-free extract of bacillus safensis LAU 13: Antifungal, dye degradation, anti-coagulant and thrombolytic activities. IEEE Trans. Nanobioscience. 15: 433–442.

Oomen, G.A., A.J.E. Bleeker, M.J.P. Bos, F. Van Broekhuizen, S. Gottardo, M. Groenewold, et al. 2015. Grouping and read-across approaches for risk assessment of nanomaterials. Int. J. Environ. Res. Public Health. 12: 13415–13434.

Orłowski, P., A. Kowalczyk, E. Tomaszewska, K. Ranoszek-Soliwoda, A. Węgrzyn, J. Grzesiak, et al. 2018. Antiviral activity of tannic acid modified silver nanoparticles: Potential to activate immune response in herpes genitalis. Viruses. 10: 524.

Pachfule, P., D. Shinde, M. Majumder and Q. Xu. 2016. Fabrication of carbon nanorods and graphene nanoribbons from a metal–organic framework. Nat. Chem. 8: 718–724.

Pang, C., D. Hristozov, A. Zabeo, L. Pizzol, M.P. Tsang, P. Sayre, et al. 2017. Probabilistic approach for assessing infants' health risks due to ingestion of nanoscale silver released from consumer products. Environ. Int. 99: 199–207.

Parveen, S., A.H. Wani, M.A. Shah, H.S. Devi, M.Y. Bhat and J.A. Koka. 2018. Preparation, characterization and antifungal activity of iron oxide nanoparticles. Microb. Pathog. 115: 287–292.

Pathak, T.K., R.E. Kroon and H.C. Swart. 2018. Photocatalytic and biological applications of Ag and Au doped ZnO nanomaterial synthesized by combustion. Vacuum. 157: 508–513.

Pathania, D., G. Sharma and R. Thakur. 2015. Pectin @ zirconium (IV) silicophosphate nanocomposite ion exchanger: Photo catalysis, heavy metal separation and antibacterial activity. Chem. Eng. J. 267: 235–244.

Patil, B.N. and T.C. Taranath. 2016. *Limonia acidissima* L. leaf mediated synthesis of zinc oxide nanoparticles: A potent tool against Mycobacterium tuberculosis. Int. J. Mycobact. 5: 197–204.

Payne, J.N., H.K. Waghwani, M.G. Connor, W. Hamilton, S. Tockstein, H. Moolani, et al. 2016. Novel synthesis of kanamycin conjugated gold nanoparticles with potent antibacterial activity. Front. Microbiol. 7: 607.

Peña-González, C.E., P. García-Broncano, M.F. Ottaviani, M. Cangiotti, A. Fattori, M. Hierro-Oliva, et al. 2016. Dendronized anionic gold nanoparticles: Synthesis, characterization, and antiviral activity. Chem. Eur. J. 22: 2987–2999.

Penders, J., M. Stolzoff, D.J. Hickey, M. Andersson and T.J. Webster. 2017. Shape-dependent antibacterial effects of non-cytotoxic gold nanoparticles. Int. J. Nanomed. 12: 2457–2468.

Pietroiusti, A., A. Magrini and L. Campagnolo. 2016. New frontiers in nanotoxicology: Gut microbiota/microbiome-mediated effects of engineered nanomaterials. Toxicol. Appl. Pharmacol. 299: 90–95.

Pikula, K.S., A.M. Zakharenko, V.V. Chaika, K.Y. Kirichenko, A.M. Tsatsakis and K.S. Golokhvast. 2020. Risk assessment in nanotoxicology: Bioinformatics and computational approaches. Curr. Opin. Toxicol. 19: 1–6.

Piras, A.M., G. Maisetta, S. Sandreschi, M. Gazzarri, C. Bartoli, L. Grassi, et al. 2015. Chitosan nanoparticles loaded with the antimicrobial peptide temporin B exert a long-term antibacterial activity *in vitro* against clinical isolates of *Staphylococcus epidermidis*. Front. Microbiol. 6: 372.

Ponmurugan, P., K. Manjukarunambika, V. Elango and B.M. Gnanamangai. 2016. Antifungal activity of biosynthesised copper nanoparticles evaluated against red root-rot disease in tea plants. J. Exp. Nanosci. 11: 1019–1031.

Priyadarshini, E., K. Rawat, T. Prasad and H.B. Bohidar. 2018. Antifungal efficacy of Au@ carbon dots nanoconjugates against opportunistic fungal pathogen, *Candida albicans*. Colloids Surf. B. 163: 355–361.

Rafiei, S., S.E. Rezatofighi, M.R. Ardakani and O. Madadgar. 2015. *In vitro* anti-foot-and-mouth disease virus activity of magnesium oxide nanoparticles. IET Nanobiotechnol. 9: 247–251.

Rafińska, K., P. Pomastowski and B. Buszewski. 2019. Study of *Bacillus subtilis* response to different forms of silver. Sci. Total Environ. 661: 120–129.

Rahdar, A., M. Aliahmad, M. Samani, M. HeidariMajd and M.A.B.H. Susan. 2019. Synthesis and characterization of highly efficacious Fe-doped ceria nanoparticles for cytotoxic and antifungal activity. Ceram. Int. 45: 7950–7955.

Raja, P.M.V., G. Lacroix, J.A. Sergent, F. Bois, A.R. Barron, E. Monbelli, et al. 2017. Nanotoxicology: Role of physical and chemical characterization and related *in vitro*, *in vivo*, and *in silico* methods. pp. 363–380. *In:* E. Mansfield, D.L. Kaiser, D.F. Professor and M.V. de Voorde Professor [eds]. Metrology and Standardization of Nanotechnology: Protocols and Industrial Innovations. Wiley.

Ramanathan, R., Y. Jiang, B. Read, S. Golan-Paz and K.A. Woodrow. 2016. Biophysical characterization of small molecule antiviral-loaded nanolipogels for HIV-1 chemoprophylaxis and topical mucosal application. Acta Biomater. 36: 122–131.

Ramaswamy, S.V.P., S. Narendhran and R. Sivaraj. 2016. Potentiating effect of ecofriendly synthesis of copper oxide nanoparticles using brown alga: Antimicrobial and anticancer activities. Bull. Mater. Sci. 39: 361–364.

Randazzo, W., M.J. Fabra, I. Falcó, A. López-Rubio and G. Sánchez. 2018. Polymers and biopolymers with antiviral activity: Potential applications for improving food safety. Compr. Rev. Food Sci. Food Saf. 17: 754–768.

Ravichandran, K., N. Chidhambaram, T. Arun, S. Velmathi and S. Gobalakrishnan. 2016. Realizing cost-effective ZnO: Sr nanoparticles@ graphene nanospreads for improved photocatalytic and antibacterial activities. RSC Adv. 6: 67575–67585.

Reddy, P.V.L., J.A. Hernandez-Viezcas, J.R. Peralta-Videa and J.L. Gardea-Torresdey. 2016. Lessons learned: Are engineered nanomaterials toxic to terrestrial plants? Sci. Total Environ. 568: 470–479.

Renault, S., M. Baudrimont, N. Mesmer-Dudons, P. Gonzalez, S. Mornet and A. Brisson. 2008. Impacts of gold nanoparticle exposure on two freshwater species: A phytoplanktonic alga (Scenedesmus subspicatus) and a benthic bivalve (Corbicula fluminea). Gold Bull. 41: 116–126.

Rotomskis, R., Ž. Jurgelėnė, M. Stankevičius, M. Stankevičiūtė, N. Kazlauskienė, K. Jokšas, et al. 2018. Interaction of carboxylated CdSe/ZnS quantum dots with fish embryos: Towards understanding of nanoparticles toxicity. Sci. Total Environ. 635: 1280–1291.

Sahraee, S., J.M. Milani, B. Ghanbarzadeh and H. Hamishehkar. 2017. Effect of corn oil on physical, thermal, and antifungal properties of gelatin-based nanocomposite films containing nano chitin. LWT-Food Sci. Technol. 76: 33–39.

Sardana, N., K. Singh, M. Saharan, D. Bhatnagar and R.S. Ronin. 2018. Synthesis and characterization of dendrimer modified magnetite nanoparticles and their antimicrobial activity for toxicity analysis. J. Int. Sci. Tech. 6.

Sathe, P., M.T.Z. Myint, S. Dobretsov and J. Dutta. 2016. Removal and regrowth inhibition of microalgae using visible light photocatalysis with ZnO nanorods: A green technology. Sep. Purif. Technol. 162: 61–67.

Sawangphruk, M., P. Srimuk, P. Chiochan, T. Sangsri and P. Siwayaprahm. 2012. Synthesis and antifungal activity of reduced graphene oxide nanosheets. Carbon. 50: 5156–5161.

Schlich, K. and K. Hund-Rinke. 2015. Influence of soil properties on the effect of silver nanomaterials on microbial activity in five soils. Environ. Pollut. 196: 321–330.

Seddighi, N.S., S. Salari and A.R. Izadi. 2017. Evaluation of antifungal effect of iron-oxide nanoparticles against different *Candida* species. IET Nanobiotechnology. 11: 883–888.

Seino, S., Y. Imoto, T. Kosaka, T. Nishida, T. Nakagawa and T.A. Yamamoto. 2016. Antiviral activity of silver nanoparticles immobilized onto textile fabrics synthesized by radiochemical process. MRS Adv. 1: 705–710.

Shah, S.T., W.A. Yehya, O. Saad, K. Simarani, Z. Chowdhury, A.A. Alhadi, et al. 2017. Surface functionalization of iron oxide nanoparticles with gallic acid as potential antioxidant and antimicrobial agents. Nanomaterials (Basel). 7: 306.

Shaheen, T.I. and A. Fouda. 2018. Green approach for one-pot synthesis of silver nanorod using cellulose nanocrystal and their cytotoxicity and antibacterial assessment. Int. J. Biol. Macromol. 106: 784–792.

Shanthi, S.I., S. Poovaragan, M.V. Arularasu, S. Nithya, R. Sundaram, C.M. Magdalane, et al. 2018. Optical, magnetic and photocatalytic activity studies of Li, Mg and Sr doped and undoped zinc oxide nanoparticles. J. Nanosci. Nanotechnol. 18: 5441–5447.

Shao, W., X. Liu, H. Min, G. Dong, Q. Feng and S. Zuo. 2015. Preparation, characterization, and antibacterial activity of silver nanoparticle-decorated graphene oxide nanocomposite. ACS Appl. Mater. Interfaces. 7: 6966–6973.

Shariati, M., A. Mallakin, F. Malekmohammady and F. Khosravi-Nejad. 2018a. Inhibitory effects of functionalized indium doped ZnO nanoparticles on algal growth for preservation of adobe mud and earthen-made artworks under humid conditions. Int. Biodeterior. Biodegrad. 127: 209–216.

Shariati, M.R., A. Samadi-Maybodi and A.H. Colagar. 2018b. Dual cocatalyst loaded reverse type-I core/shell quantum dots for photocatalytic antibacterial applications. J. Mater. Chem. A. 6: 20433–20443.

Sharma, G., A. Kumar, M. Naushad, D. Pathania and M. Sillanpää. 2016. Polyacrylamide@ Zr(IV) vanadophosphate nanocomposite: Ion exchange properties, antibacterial activity, and photocatalytic behavior. J. Ind. Eng. Chem. 33: 201–208.

Shende, S., A.P. Ingle, A. Gade and M. Rai. 2015. Green synthesis of copper nanoparticles by *Citrus medica* Linn. (Idilimbu) juice and its antimicrobial activity. World J. Microbiol. Biotechnol. 31: 865–873.

Shibata, T., N. Hamada, K. Kimoto, T. Sawada, T. Sawada, H. Kumada, et al. 2007. Antifungal effect of acrylic resin containing apatite-coated TiO_2 photocatalyst. Dent. Mater. J. 26: 437–444.

Sierra-Fernandez, A., S.C. De la Rosa-García, L.S. Gomez-Villalba, S. Gómez-Cornelio, M.E. Rabanal, R. Fort, et al. 2017. Synthesis, photocatalytic, and antifungal properties of MgO, ZnO and Zn/Mg oxide nanoparticles for the protection of calcareous stone heritage. ACS Appl. Mater. Interfaces. 9: 24873–24886.

Sirelkhatim, A., S. Mahmud, A. Seeni, N.H.M. Kaus, L.C. Ann, S.K.M. Bakhori, et al. 2015. Review on zinc oxide nanoparticles: Antibacterial activity and toxicity mechanism. Nano-Micro Lett. 7: 219–242.

Sobhani-Nasab, A., Z. Zahraei, M. Akbari, M. Maddahfar and S.M. Hosseinpour-Mashkani. 2017. Synthesis, characterization, and antibacterial activities of $ZnLaFe_2O_4/NiTiO_3$ nanocomposite. J. Mol. Struct. 1139: 430–435.

Sodagar, A., S. Khalil, M.Z. Kassaee, A.S. Shahroudi, B. Pourakbari and A. Bahador. 2016. Antimicrobial properties of poly (methyl methacrylate) acrylic resins incorporated with silicon dioxide and titanium dioxide nanoparticles on cariogenic bacteria. J. Orthod. Sci. 5: 7–13.

Sohn, E.K., Y.S. Chung, S.A. Johari, T.G. Kim, J.K. Kim, J.H. Lee, et al. 2015. Acute toxicity comparison of single-walled carbon nanotubes in various freshwater organisms. BioMed. Res. Int. 2015: 323090.

Su, Y., X. Zheng, A. Chen, Y. Chen, G. He and H. Chen. 2015. Hydroxyl functionalization of single-walled carbon nanotubes causes inhibition to the bacterial denitrification process. Chem. Eng. J. 279: 47–55.

Sun, X., B. Chen, Q. Li, N. Liu, B. Xia, L. Zhu, et al. 2018. Toxicities of polystyrene nano- and microplastics toward marine bacterium *Halomonas alkaliphila*. Sci. Total Environ. 642: 1378–1385.

Supraja, N., T. Prasad, T.G. Krishna and E. David. 2016. Synthesis, characterization, and evaluation of the antimicrobial efficacy of *Boswellia ovalifoliolata* stem bark-extract-mediated zinc oxide nanoparticles. Appl. Nanosci. 6: 581–590.

Surendra, T.V., S.M. Roopan, D. Devipriya, M.M.R. Khan and R. Hassanien. 2019. Multi-perspective CuO@ C nanocomposites: Synthesis using drumstick peel as carbon source and its optimization using response surface methodology. Compos. B. Eng. 172: 690–703.

Suresh, M., J. Jeevanandam, Y.S. Chan, M.K. Danquah and J.M.V. Kalaiarasi. 2019. Opportunities for metal oxide nanoparticles as a potential mosquitocide. BioNanoSci. 10, 292–310.

Swain, S., S.K. Barik, T. Behera, S.K. Nayak, S.K. Sahoo, S.S. Mishra, et al. 2016. Green synthesis of gold nanoparticles using root and leaf extracts of vetiveria zizanioides and cannabis sativa and its antifungal activities. BioNanoScience. 6: 205–213.

Tahir, K., S. Nazir, A. Ahmad, B. Li, A.U. Khan, Z.U.H. Khan, et al. 2017. Facile and green synthesis of phytochemicals capped platinum nanoparticles and *in vitro* their superior antibacterial activity. J. Photochem. Photobiol. B. 166: 246–251.

Tan, K.X., A. Barhoum, S. Pan and M.K. Danquah 2018. Risks and toxicity of nanoparticles and nanostructured materials. pp. 121–139. *In:* A. Barhoum and A.S.H. Makhlouf [eds]. Emerging Applications of Nanoparticles and Architecture Nanostructures. Elsevier.

Tavakoli, A., A. Ataei-Pirkooh, G. Mm Sadeghi, F. Bokharaei-Salim, P. Sahrapour, S.J. Kiani, et al. 2018. Polyethylene glycol-coated zinc oxide nanoparticle: An efficient nanoweapon to fight against herpes simplex virus type 1. Nanomedicine. 13: 2675–2690.

Tennant P., A. Gubba, M. Royel and G. Fermin. 2018. Viruses as pathogens: Plant viruses. pp. 135–156. *In:* P. Tennant, G. Fermin and J.E. Foster [eds]. Viruses: Molecular Biology, Host Interactions, and Applications to Biotechnology. Academic Press.

Thompson, L.B., G.L.F. Carfagno, K. Andresen, A.J. Sitton, T. Bury, L.L. Lee, et al. 2017. Differential uptake of gold nanoparticles by 2 species of tadpole, the wood frog (*Lithobates sylvaticus*) and the bullfrog (*Lithobates catesbeianus*). Environ. Toxicol. Chem. 36: 3351–3358.

Tian, X., Y. Sun, S. Fan, M.D. Boudreau, C. Chen, C. Ge, et al. 2019. Photogenerated charge carriers in molybdenum disulfide quantum dots with enhanced antibacterial activity. ACS Appl. Mater. Interfaces 11: 4858–4866.

Tiwari, N., R. Pandit, S. Gaikwad, A. Gade and M. Rai. 2016. Biosynthesis of zinc oxide nanoparticles by petals extract of *Rosa indica* L., its formulation as nail paint and evaluation of antifungal activity against fungi causing onychomycosis. IET Nanobiotechnol. 11: 205–211.

Tlili, A., J.r.m. Jabiol, R. Behra, C. Gil-Allué and M.O. Gessner. 2017. Chronic exposure effects of silver nanoparticles on stream microbial decomposer communities and ecosystem functions. Environ. Sci. Technol. 51: 2447–2455.

Varga, O.E., A.K. Hansen, P. Sandøe and I.A.S. Olsson. 2010. Validating animal models for preclinical research: A scientific and ethical discussion. Altern. Lab. Anim. 38: 245–248.

Velmurugan, P., J. Shim, K. Kim and B.-T. Oh. 2016. Prunus× yedoensis tree gum mediated synthesis of platinum nanoparticles with antifungal activity against phytopathogens. Mater. Lett. 174: 61–65.

Venkateswarlu, S., B. Viswanath, A.S. Reddy and M. Yoon. 2018. Fungus-derived photoluminescent carbon nanodots for ultrasensitive detection of Hg^{2+} ions and photo-induced bactericidal activity. Sens. Actuators B. Chem. 258: 172–183.

Verma, J., S. Nigam, S. Sinha and A. Bhattacharya. 2018a. Comparative Studies on Polyacrylic Based Anti-Algal Coating Formulation with $SiO_2@TiO_2$ Core-Shell Nano-particles. Asian J. Chem. 30: 1120–1124.

Verma, J., S. Nigam, S. Sinha and A. Bhattacharya. 2018b. Development of polyurethane based anti-scratch and anti-algal coating formulation with silica-titania core-shell nanoparticles. Vacuum. 153: 24–34.

Vidhu, V.K. and D. Philip. 2015. Biogenic synthesis of SnO_2 nanoparticles: Evaluation of antibacterial and antioxidant activities. Spectrochim. Acta. A Mol. Biomol. Spectrosc. 134: 372–379.

Viet, P.V., H.T. Nguyen, T.M. Cao and L.V. Hieu. 2016. Fusarium antifungal activities of copper nanoparticles synthesized by a chemical reduction method. J. Nanomater. 2016.

Vijayaprasath, G., R. Murugan, S. Palanisamy, N.M. Prabhu, T. Mahalingam, Y. Hayakawa, et al. 2016. Role of nickel doping on structural, optical, magnetic properties and antibacterial activity of ZnO nanoparticles. Mater. Res. Bull. 76: 48–61.

Vimbela, G.V., S.M. Ngo, C. Fraze, L. Yang and D.A. Stout. 2017. Antibacterial properties and toxicity from metallic nanomaterials. Int. J. Nanomed. 12: 3941.

Vinita, C., S.H. Sontakke and J. Khadse. 2017. Toxic effect of silver nanoparticles on bull and cattle spermatozoa. Int. J. Vet. Sci. & Anim. Husb. 2: 4–9.

Wang, X., J. Zhang, W. Sun, W. Yang, J. Cao, Q. Li, et al. 2015. Anti-algal activity of palladium oxide-modified nitrogen-doped titanium oxide photocatalyst on *Anabaena* sp. PCC 7120 and its photocatalytic degradation on Microcystin LR under visible light illumination. Chem. Eng. J. 264: 437–444.

Wang, X., Z. Zhou and F. Chen. 2017. Surface modification of carbon nanotubes with an enhanced antifungal activity for the control of plant fungal pathogen. Materials. 10: 1375.

Wen, S., J. Zhou, K. Zheng, A. Bednarkiewicz, X. Liu and D. Jin. 2018. Advances in highly doped upconversion nanoparticles. Nat. Commun. 9: 1–12.

Whitehead, K.A., M. Vaidya, C.M. Liauw, D.A.C. Brownson, P. Ramalingam, J. Kamieniak, et al. 2017. Antimicrobial activity of graphene oxide-metal hybrids. Int. Biodeterior. Biodegrad. 123: 182–190.

Wu, T., C. Wu, S. Fu, L. Wang, C. Yuan, S. Chen, et al. 2017. Integration of lysozyme into chitosan nanoparticles for improving antibacterial activity. Carbohydr. Polym. 155: 192–200.

Xia, Z.-K., Q.-H. Ma, S.-Y. Li, D.-Q. Zhang, L. Cong, Y.-L. Tian, et al. 2016. The antifungal effect of silver nanoparticles on *Trichosporon asahii*. J. Microbiol. Immunol. Infect. 49: 182–188.

Xie, C., J. Zhang, Y. Ma, Y. Ding, P. Zhang, L. Zheng, et al. 2019. *Bacillus subtilis* causes dissolution of ceria nanoparticles at the nano–bio interface. Environ. Sci. Nano. 6: 216–223.

Xu, C., C. Peng, L. Sun, S. Zhang, H. Huang, Y. Chen, et al. 2015. Distinctive effects of TiO_2 and CuO nanoparticles on soil microbes and their community structures in flooded paddy soil. Soil Biol. Biochem. 86: 24–33.

Xu, J., M. Yu, C. Peng, P. Carter, J. Tian, X. Ning, et al. 2018. Dose dependencies and biocompatibility of renal clearable gold nanoparticles: from mice to non-human primates. Angew. Chem. Int. Ed. Engl. 57: 266–271.

Yadav, H.M., S.V. Otari, V.B. Koli, S.S. Mali, C.K. Hong, S.H. Pawar, et al. 2014. Preparation and characterization of copper-doped anatase TiO_2 nanoparticles with visible light photocatalytic antibacterial activity. J. Photochem. Photobiol. A. 280: 32–38.

Yang, W., J.S. Owczarek, E. Fortunati, M. Kozanecki, A. Mazzaglia, G.M. Balestra, et al. 2016. Antioxidant and antibacterial lignin nanoparticles in polyvinyl alcohol/chitosan films for active packaging. Ind. Crops Prod. 94: 800–811.

Yang, Z., X. Hao, S. Chen, Z. Ma, W. Wang, C. Wang, et al. 2019. Long-term antibacterial stable reduced graphene oxide nanocomposites loaded with cuprous oxide nanoparticles. J. Colloid Interface Sci. 533: 13–23.

Ye, S., K. Shao, Z. Li, N. Guo, Y. Zuo, Q. Li, et al. 2015. Antiviral activity of graphene oxide: How sharp edged structure and charge matter. ACS Appl. Mater. Interfaces. 7: 21571–21579.

Yela, A.V., V.J. Jimenez, D.V. Rodriguez and G.P. Quishpe. 2016. Evaluation of the Antifungal Activity of Sulfur and Chitosan Nanocomposites with Active Ingredients of *Ruta graveolens*, Thymus vulgaris and *Eucalyptus melliodora* on the Growth of *Botrytis fabae* and *Fusarium oxysporum*. Biol. Med. (Aligarh) 8: 291.

Yien, L., N.M. Zin, A. Sarwar and H. Katas. 2012. Antifungal activity of chitosan nanoparticles and correlation with their physical properties. Int. J. Biomater. 2012.

Yuan, Y.G., Q.L. Peng and S. Gurunathan. 2017. Effects of silver nanoparticles on multiple drug-resistant strains of *Staphylococcus aureus* and *Pseudomonas aeruginosa* from mastitis-infected goats: An alternative approach for antimicrobial therapy. Int. J. Mol. Sci. 18: 569.

Yugandhar, P., T. Vasavi, Y. Jayavardhana Rao, P. Uma Maheswari Devi, G. Narasimha and N. Savithramma. 2018. Cost effective, green synthesis of copper oxide nanoparticles using fruit extract of *Syzygium alternifolium* (Wt.) Walp., characterization and evaluation of antiviral activity. J. Cluster Sci. 29: 743–755.

Zare, Y. and I. Shabani. 2016. Polymer/metal nanocomposites for biomedical applications. Mater. Sci. Eng. C. 60: 195–203.

Zhang, C.C., X. Duan, Y. Ding and C. Srinivasakannan. 2017. Influence of Ce^{3+} doping on the algal inhibiting properties of copper/sepiolite nanofibers. Environ. Prot. Eng. 43: 253–263.

Zhang, M., H. Wang, P. Liu, Y. Song, H. Huang, M. Shao, et al. 2019. Biotoxicity of degradable carbon dots towards microalgae *Chlorella vulgaris*. Environ. Sci. Nano. 6: 3316–3323.

Zheng, Y., C. Cai, F. Zhang, J. Monty, R.J. Linhardt and T.J. Simmons. 2016. Can natural fibers be a silver bullet? Antibacterial cellulose fibers through the covalent bonding of silver nanoparticles to electrospun fibers. Nanotechnology. 27: 055102.

Zheng, K., M.I. Setyawati, D.T. Leong and J. Xie. 2018. Antimicrobial silver nanomaterials. Coord. Chem. Rev. 357: 1–17.

Zhu, C., H. Li, H. Wang, B. Yao, H. Huang, Y. Liu, et al. 2019. Negatively charged carbon nanodots with bacteria resistance ability for high-performance antibiofilm formation and anticorrosion coating design. Small. 15: 1900007.

5

Nanocrystalline Cellulose (NCC) Composites with Antimicrobial Properties

John HT Luong[1*], Jeremy D. Glennon[1] and Bansi D. Malhotra[2]

[1]School of Chemistry
University College Cork, Cork, Ireland, T12 YN60
Email: j.luong@ucc.ie, luongprof@gmail.com; j.glennon@ucc.ie

[2]Department of Biotechnology
Delhi Technological University, Delhi 110042, India
Email: bansi.malhotra@gmail.com

INTRODUCTION

Nanocrystalline cellulose (NCC), also known as cellulose nanocrystals (CNCs) or cellulose nanowhiskers (CNWs) are rod-like nanoparticles and can be derived from cellulose, an abundant renewable bioresource. Cellulose is a semi-crystalline homopolymer of β-D-glucopyranose units linked by β-1, 4-linkages, each unit with three hydroxyl (–OH) groups at the C2, C3, and C6 positions. Thus, cellulose is composed of amorphous (disordered) and crystalline (ordered) regions. The reactivity of the surface −OH groups follows O2 > O6 > O3 for native cotton, but O6 is more reactive than O2, followed by O3 for *Valonia ventricosa* (bubble alga) cellulose (Wohlhauser et al. 2018). Cellulose is one of the major components of plant cell-walls, certain sea creatures, e.g., tunicates, and algae. Cellulose is also produced by bacteria, known as bacterial cellulose (BC) or microbial cellulose.

*For Correspondence: j.luong@ucc.ie, luongprof@gmail.com

In the past, NCC was obtained by acid hydrolysis of cellulose using sulfuric acid, HCl, and phosphoric acid. Of recent significance is the use of ammonium persulfate, a strong oxidant, to produce NCC with carboxyl groups (Leung et al. 2011). Acids or strong oxidants penetrate amorphous cellulose molecules to effectuate the cleavage of glycosidic bonds, to release individual crystallites or NCC. Thus, NCC exhibits high crystallinity, tunable aspect ratio, and high stiffness/strength, compared to cellulose. As a material which is physiologically benign with negligible cytotoxicity, NCC has been advocated for diversified potential applications, such as wound dressing, packaging materials, dialysis membranes, tissue engineering, high-performance biomaterials, acoustics, sensing, controlled drug delivery, etc. (Bacakova et al. 2019). NCC with reactive hydroxyl groups can be grafted or conjugated with small organic molecules, such as antibiotics, enzymes, biomolecules, polymers, etc. to form nanocomposites with desirable antimicrobial activities (Figure 5.1). NCC with carboxyls allows many types of chemical modifications to produce functional nanomaterials with a wide spectrum of properties and functions.

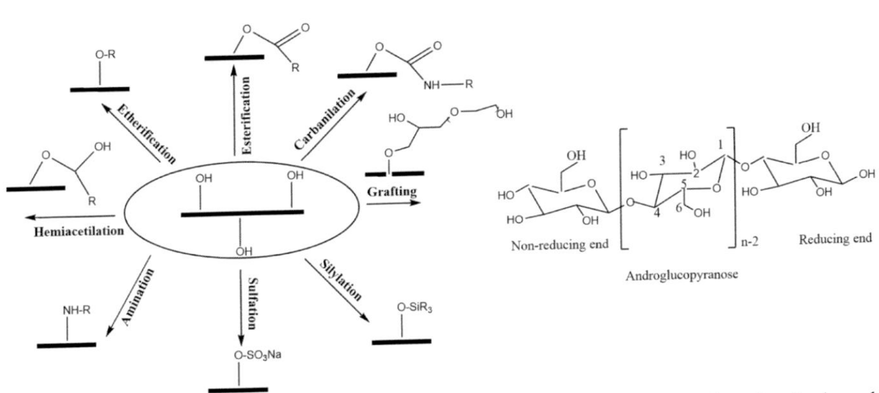

Figure 5.1 Grafting and chemical modification of cellulose (left). There are three free hydroxyl groups on each glucose unit in the cellulose chain; O6 and O2 are more reactive than O3 (right).

NCC has typical dimensions of 50 nm^{-1} μm in length and 5–50 nm in breadth, depending on the cellulose material. Therefore, the modification step must be carefully designed to preserve its dimensions, crystallinity, and other physicochemical properties. Bacterial cellulose (BC) or microbial cellulose is mainly produced by genera *Acetobacter*, *Sarcina*, and *Agrobacterium*. Besides native cellulose from plants, significant efforts focus on *Acetobacter xylinum*, because this bacterial cellulose exhibits high purity and unique mechanical properties. In addition, *A. xylinum* is capable of synthesizing high levels of BC from different sources of carbon and nitrogen (Petersen and Gatenholm 2011). Thus, bacterial cellulose could become a promising starting material for extracting NCC for some medical applications. However, a high cost associated with the production of BC is one major drawback to promote its widespread application.

Of importance is the preparation of NCC composites with polymers, antibiotics, enzymes, and metal nanoparticles with enhanced antimicrobial activities. Indeed, nanocellulose-based antimicrobial materials have been investigated as packaging materials, wound dressing materials, and drug carriers. NCC nanocomposites with very large surface areas and high antimicrobial activities should be effective against antibiotic-resistant microorganisms. NCC-polymers and NCC nanocomposites display action against yeasts, gram-positive, and gram-negative bacteria. There are considerable efforts to develop new and effective antimicrobial reagents with respect to the emerging increase of microbial organisms resistant to multiple antibiotics. NCC offers biocompatibility and serves as a versatile platform for the preparation of nanocomposites with enzymes, antibiotics, quaternary ammonium compounds, and metal nanoparticles. In brief, metal nanoparticles (NPs), such as Ag NPs and Zn NPs exhibit attractive antibacterial properties, as the reduced particle size leads to enhanced particle surface reactivity. This feature allows for their high loadings on NCC. Apparently, this new class of antimicrobials is more applicable than traditional antibiotics in wound dressing, hygienic products, packaging, medical devices, and prophylaxis in antimicrobial applications.

The chapter focuses on the preparation of NCC and its composites with antimicrobial activities together with opportunities and technical challenges. The chapter also aims to highlight potential applications of NCC nanocomposites in wound dressing, food packaging, and antimicrobial control systems.

SYNTHESIS AND CHARACTERISTICS OF NCC

Cellulose chains exhibit extensive inter- and intramolecular H-bonding into crystalline structures, consisting of crystalline and amorphous regions. Cellulose with both amorphous and crystalline domains is often referred to as cellulose nanofibrils (CNF), nanofibrillated cellulose (NFC), or microfibrillated cellulose (MFC). Cellulose chains can be subject to mechanical, enzymatic, or chemical treatments, resulting in NCC. In practice, these steps are often used in combination to shorten the hydrolysis time together with high yields. The conventional treatment with sulfuric (Habibi et al. 2010) (55–65 wt% sulfuric acid for 30–60 min at 35 to 70°C) or phosphoric acid results in the formation of sulfate or phosphate ester groups on NCC, respectively, whereas the hydrolysis by ammonium persulfate, a very strong oxidant, produces a carboxyl group (Lam et al. 2013). The presence of such functional groups on NCC controls the charge density and its dispersibility in liquid solvents or polymer matrices. The modulation of hydrophilicity or hydrophobicity of carboxylated NCC is also easier for various applications. In all cases, the presence of such groups reduces the thermal stability and may provoke the toxicological and antimicrobial of NCC. Unlike negatively charged NCC synthesized by sulfuric acid of ammonium persulfate hydrolysis, NCC hydrolyzed by hydrochloric or hydrobromic acid exhibits a very low surface charge. The NCC produced has low colloidal stability due to the lack of sufficient electrostatic repulsive forces that prevent consequent hydrogen bonding between adjacent NCC nanoparticles. Such dispersion aggregation is obviously problematic for subsequent chemical modification. NCC is also oxidized by 2,2,6,6-tetramethylpiperidin-1-yl)oxyl (TEMPO)-mediated oxidation to lead to the surface carboxyl group

(Habibi et al. 2006). However, this oxidant is costly and might not be suitable for large scale processing except for some special applications.

NCC often has a different number of individual cellulose molecules, depending significantly on the cellulose sources and the preparation procedure. Plant and bacterial cellulose microfibrils might have 36 chains (Cosgrove 2014), corresponding to a cross-sectional area of 11.6 nm^2. However, a more precise depiction of 18-chains has been suggested (Fernandes et al. 2011), considering most current estimates of 7 nm^2 for the above-mentioned cross-sectional area. The primary wall microfibril diameter of about 3 nm is too small to contain 36 chains. Solid-state NMR unravels that primary cell wall cellulose microfibrils should have at least 24 chains (Wang and Hong 2016). NCC has a rod-shape with a diameter ranging from 5 to 40 nm and a length of 100–500 nm. Like pristine cellulose, NCC with a high density of surface hydroxyl groups can interact with diversified polymers via hydrogen bonding to form polymer composites. The negative charges of carboxylated NCC interact with metal nanoparticles and facilitate covalent coupling with a myriad of polymers and biomolecules to form nanocomposites. Typical dimensions of NCC reported in the literature are shown in Table 5.1.

Table 5.1 Dimensions of NCC derived from different cellulose sources

Dimensions (Breadth × Length, **nm × nm***)*	*Raw Materials*
5–10 × 100–150	Cotton
5–15 × 70–200	Ramie[a]
3–5 × 100–300	Wood, Sisal[b]
20 × 1000–2000	*Valonia ventricosa*[c]
10–20 × 500–2000	Tunicates[d]
10–20 × 100–300	Bacterial cellulose hydrolyzed by HCl

Notes: (a) One of the oldest fiber crops for fabric production, at least 6,000 years (b) Sisal plant, a Mexican agave with large fleshy leaves, cultivated for fiber production of fibers (c) Bubble algae found in the ocean (d) A marine invertebrate animal, a member of the subphylum *Tunicata*.

PLAUSIBLE CYTOTOXICITY OF NCC COMPOSITES

Considering the potential large-scale applications, NCC has been under intense scrutiny for its plausible cytotoxicity and biocompatibility. Fibrosis, asbestosis, lung cancer, mesothelioma, and pleural plaques are attributed to asbestos fibers (Donaldson and Tran 2004), so the word "fibers" in cellulose nanofibers raises grave concern to the workers associated with the production of NCC in large quantity. They are constantly exposed to large quantities of NCC in the production facilities. The risk might be real, considering that inhalation might happen during the production of NCC and its associated products. However, the human risk associated with the exposure concentrations (Donaldson et al. 2013) or doses is still unclear and must be assessed thoroughly.

In brief, NCC with realistic doses (0.03–10 mg/mL) and exposure scenarios only exhibits a limited associated toxic potential (Kovacs et al. 2010). In contrast, noticeably inhibitory effects on human embryonic kidney cells (HEK 293) and Sf9 insect cells are observed for NCC ranging from 0.25–5 mg/mL

(Mahmoud et al. 2010). Indeed, there is some conflicting literature information pertaining to the cytotoxicity effect of NCC (Table 5.2). There is an increase in the LDH (lactate dehydrogenase) activity, an indicator of cytotoxicity after the mice have aspired with 50 μg/mouse. The finding is troubling because the result is comparable to asbestos aspiration in the context of cytotoxicity (Yanamala et al. 2014). Such discrepancies can be attributed to the different tested cell systems, sources of NCC and methods of preparation, exposure conditions, and methods used for the cytotoxicity assay. Certain forms of NCC with specific physical characteristics also affect cell behavior, encompassing mobility, growth, and replication. Similarly, there is also conflicting information pertaining to inflammation and oxidative stress caused by NCC (Endes et al. 2016).

Table 5.2 Selected information concerning the plausible cytotoxicity effect of NCC

Type of NCC	*Exposure Dose*	*Other Remarks*	*References*
200 × 10 × 5 nm	0.03–10 g/L-negligible cytotoxicity	Derived from Kraft pulp	Kovacs et al. 2010
130–200 × 10–20 nm (FITC-labeled NCC)	0.25–5 mg/mL-membrane rupture using human embryonic kidney cells (HEK 293) and Sf9 insect cells	Isolated from enzyme-treated flax fibers	Mahmoud et al. 2010
FITC labeled NCC	0.01–0.05 mg/mL	No detectable cytotoxicity in a wide range of barrier	Dong et al. 2012
Cellulose nanofibrils (33 ± 2.5 μm × 10–10 nm)	0.25–1 mg/mL	Only the exposure to high concentrations impairs metabolic activity and cell proliferation	Colic et al. 2015
NCC derived from wood pulp	50, 100 and 200 μg/mouse	Cytotoxicity comparable to asbestos aspiration (50 μg/mouse)	Yanamala et al. 2014

ANTIMICROBIAL PROPERTIES OF NCC NANOCOMPOSITES

The antimicrobial mechanism generally falls within one of four mechanisms: (i) inhibition or regulation of enzymes involved in cell wall synthesis, (ii) interfering nucleic acid metabolism and repair, (iii) inhibition of protein synthesis, and (iv) disruption of membrane structure (Kohanski et al. 2010). Antibiotics prevent the cycle of multiplying cells and inhibit some popular enzyme targets, including topoisomerases, transpeptidases, transglycosylases, peptidyltransferases, and RNA polymerase. However, the bacterial responses to antibiotic drug treatments are quite complex and involve multiple genetic and biochemical pathways.

Antibiotic resistance is the ability of a microorganism to suppress the effects of an antibiotic by several different mechanisms. This is one of the biggest threats to global health, food security, and the environment. There is a growing number of infections, which are harder to treat because of antibiotic resistance. Resistant strains have developed increased virulence and improved transmissibility, a rundown of global crisis. Thus, comprehensive efforts are needed to develop a new class of antimicrobial agents to overcome antibiotic-resistant bacteria. Several product types have been under clinical trials: bacteriophages, vaccines, immune stimulators, antimicrobial peptides, probiotics, and antibodies to target *Staphylococcus aureus*, *Pseudomonas aeruginosa*, *Staphylococcus aureus*, *Acinetobacter baumannii*, *Neisseria gonorrhoeae*, etc. These bacteria are listed by WHO as 11 priority pathogens, which are required new antibiotics (WHO 2017).

Table 5.3 NCC is conjugated with enzymes, proteins, polymers, and antibiotics

Modified NCC	*Tested Microorganisms and Antimicrobial Effects*	*References*
	Enzyme/protein conjugated NCC	
Lysozyme	*C. albicans*, *A. niger*, *S. aureus*, and *E. coli*. Lysozyme hydrolyzes peptidoglycan (polyamino acid) of the bacterial cell wall	Jebali et al. 2013
Bovine lactoferrin	*E. coli* and *S. aureus*. Bovine lactoferrin is a glycoprotein (≈80 kDa) with a wide range of antimicrobial activities	Padrao et al. 2016
	Antibiotic-NCC Complex	
NFC-polymyxin B	*P. aeruginosa, Serratia marcescens, S. typhimurium, E. coli, Enterobacter cloacae*	Campia et al. 2017
BC-ceftriaxone	*S. aureus*	Kaplan et al. 2014
BC-chloromycetin	*S. aureus, Streptococcus pneumonia, E. coli*	Lacin 2014
BC-ampicillin	*S. aureus, P. aeruginosa, E. coli, E. faecalis*	Kaplan et al. 2014
BC-gentamycin	*S. aureus, P. aeruginosa, E. coli, E. faecalis*	Kaplan et al. 2014
BC-tetracycline	*S. aureus, E. coli*	Liao et al. 2015
	Polymer-NCC Complex	
NCC/chitosan nanopaper	*B. cereus* and *Salmonella typhimurium*	Dehnad et al. 2014
NCC/chitosan/poly (vinyl pyrrolidone) composite	*S. aureus* and *P. aeruginosa*	Poonguzhali et al. 2017
NFC/chitosan-BKC biocomposite	*E. coli* and *S. aureus*	Liu et al. 2013
BC-chitosan antimicrobial films	Films are prepared in LiBr solution. The antibacterial activities of the composite films against *E. coli* and *S. aureus* increase greatly as the ratio of chitosan increases	Yang et al. 2018
BC/polylysine nanofibers	*E. coli* and *S. aureus*	Gao et al. 2014

NCC at concentrations below 0.03 g/L is a benign compound with minimal cytotoxicity, i.e., its antimicrobial activity is expected to be negligible. Thus, NCC must be grafted with antimicrobial functional groups or bioconjugated with enzymes, polymers, small molecules, e.g., allicin (an organosulfur compound obtained from garlic) (Jafary et al. 2015) and triclosan (Rodriguez-Felix et al. 2016) (inhibiting fatty acid synthesis), or even antibiotics (Table 5.3). Worth noticing is the use of metal nanoparticles, e.g., ZnO nanoparticles, to generate reactive oxygen species (ROS) to kill bacteria. These radicals damage lipids, proteins, and oxidize the deoxynucleotide pools. Bactericidal antibiotics have well-established mechanisms of action; however, this classical view has been expanded to consider the role of ROS in the antibiotic-mediated killing of bacteria.

Selected antibiotics with known mechanisms have been integrated with NCC to form new arsenal weapons for combating pathogens. Except for chloramphenicol, the three remaining antibiotics have at least 1 amino group, which can be covalently coupled to the carboxyl group of NCC by the well-known carbodiimide procedure (Nozaki 1999). Such antibiotics interfere with the synthesis of cellular membranes or proteins, eventually leading to cell lysis (Table 5.4).

Table 5.4 Six popular antibiotics and their antimicrobial mode against bacteria

Antibiotics	*Mode of Action*
Ampicillin	An irreversible inhibitor of the enzyme transpeptidase, that makes the cell wall. It inhibits the third and final stage of bacterial cell wall synthesis in binary fission, leading to cell lysis
Ceftriaxone	Electively and irreversibly binds transpeptidases or transamidases. The enzymes catalyze the cross-linking of the peptidoglycan polymers forming the bacterial cell wall
Chloromycetin Chloramphenicol	Preventing protein chain elongation by inhibiting the peptidyltransferase activity of the ribosome. It specifically binds A2451 and A2452 residues in the 23S rRNA of the 50S ribosomal subunit to prevent the peptide bond formation
Gentamycin	Binding the 30S subunit of the bacterial ribosome, adversely impacting protein synthesis
Tetracycline	Binding reversibly to the bacterial 30S ribosomal subunit and blocking incoming aminoacyl tRNA from binding to the ribosome acceptor site
Polymyxin B	A complex antibiotic, cationic, basic peptide. It exhibits bactericidal action against almost all gram-negative *Bacilli* except the *Proteus* and *Neisseria* genera. It binds the cell membrane and alters its structure, making it more permeable, leading to elevated water uptake to cause cell lysis

The grafting of NCC with quaternary ammonium organic compounds (QAC) is a logical choice, considering dequalinium chloride or 1,1'-(1,10-decanediyl)-bis(4-amino-2-methyl)quinolinium dichloride has been widely used as an antiseptic and disinfectant against various bacteria, parasites, and fungi (Tischer et al. 2012). Besides some popular QACs, such as CP (cationic porphrin), BBCTC (2,4-bi[(3-benzyl-3-bimethylammonium)propylamino]-6-choro-1,3,5-triazine chloride), BKC (benzalkonium chloride), and MDBAC ([2-(methacryloyloxy)ethyl]dimethyl benzyl ammonium chloride), newer bisquaternary ammonium compounds (BQACs) with high activities, especially against gram-positive bacteria, have been reported (Tischer et al. 2012). Other BQACs are entering clinical trials and work is in progress to decipher their antimicrobial action, albeit nonspecific interactions with membranes have been considered the main mechanism of QACs. The preparation of the NCC-

QACs is very straightforward using carboxylated NCC prepared by ammonium persulfate. For NCC obtained by acid hydrolysis, the pre-oxidation step by a strong oxidant, e.g., TEMPO as mentioned earlier, is required. QAC salts can be adsorbed on the surface of NCC via ionic interactions (Salajková et al. 2012) (Figure 5.2). NCC and selected cellulose-based materials have been conjugated with QACs and proven their antimicrobial activities against various bacteria (Table 5.5).

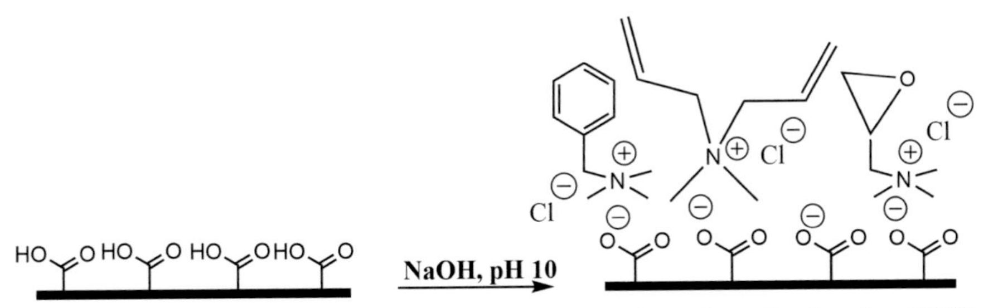

Figure 5.2 Adsorption of quaternary ammonium salts on the surface of carboxylated NCC.

Table 5.5 Cellulose-based materials have been conjugated with QACs

NCC-QACs	*Tested Microorganisms and Some Key Points*	*References*
CP (cationic porphrin)-NCC	Against *Mycobacterium smegmatis*, *S. aureus* but only a slight effect on *E. coli*	Feese et al. 2011
GTM-NFC: grinding NFC with glycidyltrimethyl ammonium chloride (GTMAC) or high-pressure homogenization of NFC and GTMAC	Antimicrobial properties against *E. coli, S. aureus,* and *P. aeruginosa*	Chaker and Boufi 2015
BC grafted with aminoalkyl groups	Lethal to *E. coli* and *S. aureus*	Favi et al. 2013
BC-BKC (benzalkonium chloride) mats	The mats release BKC to kill *E. coli* and *S. aureus*	Wei et al. 2011
Methacryloyloxyethyldimethylbenzyl ammonium chloride-coated BC	Eradicating *E. coli* in *water*	Lu et al. 2004

NCC-Metal Nanoparticles with Antimicrobial Properties

NCC is capable of forming nanocomposites with metal nanoparticles, such as Ag, ZnO, MgO, etc. These metal nanoparticles have been tested extensively due to their ease of preparation with known antimicrobial action (Table 5.6). Both *E. coli* (gram-negative, facultatively anaerobic, rod-shaped, coliform bacterium) and *S. aureus* (gram-positive, round-shaped bacterium) have been tested extensively as the model systems. However, metal nanoparticles can be used as effective growth inhibitors in various microorganisms. In general, gram-positive bacteria without an outer membrane are more sensitive than gram-negative bacteria. An antimicrobial agent must pass this layer before its interaction with intracellular components. The antimicrobial action or metal nanoparticles involves ROS, reactive oxygen species with ZnO, or silver nanoparticles as two representative models. ROS are free radicals or oxygen radicals, such as peroxides, superoxide, hydroxyl radical, singlet oxygen, and α-oxygen. They are an unstable molecule and easily react with cellular molecules, causing damage to DNA, RNA, and proteins. Similarly, the antimicrobial effect also stems from the generation of superoxide on the surface of MgO nanoparticles. There is an increase in pH from the hydration of MgO in aqueous media, as $MgO + H_2O \rightarrow Mg(OH)_2$. Thus, this combined effect impairs the integrity of the cell membrane, leading to the leakage of intracellular contents and eventual cell death.

Table 5.6 Antimicrobial activities of NCC-metal nanoparticles

NCC/ Cellulose-Metal Nanoparticles	***Tested Bacteria***	***References***
Ag NP decorated cellulose paper	Gram-negative *Proteus mirabilis*, *Vibrio parahemolyticus*, *E. faecalis*, and *Serratia marcescens*, virulent fish and shrimp pathogenic bacteria	Uddin et al. 2017
TiO_2-NCC or TiO_2-NFC	*S. aureus* and *E. coli*. ROS are released by TiO_2 nanoparticles (anatase forms) to kill bacteria	Liu et al. 2014, El-Wakil et al. 2015, Galkina et al. 2015
CuO-NFC	*S. aureus*, *E. coli*, *C. albicans*. Nanoparticles penetrate in the cell membrane to inhibit vital enzymes	Muthulakshmi et al. 2017, Natan et al. 2015, Barua et al. 2013
ZnO-NCC	*S. aureus*, *E. coli*, *B. subtilis*	Lefatshe et al. 2017, Nath et al. 2016
ZnO-BC	*E. coli*, *S. aureus*	Ul-Islam et al. 2014, Katepetch et al. 2013, Wang et al. 2014, Janpetch et al. 2016, Abdalkarim et al. 2017
MgO-NCC	*E. coli*, damaging the cell membrane and causing the leakage of intracellular materials	Rabie et al. 2016, Zhao et al. 2014

ZnO is an interesting semiconductor, consisting of a conduction band (CB) and a valence band (VB). Incident photon radiation above 3.3 eV (bandgap energy) is rapidly absorbed to move the electrons from the VB to the CB to create positive holes (h^+) in the VB. Positive hole (h^+) then serves as a direct oxidant for the formation of reactive hydroxyl radicals ($OH^\bullet$). ZnO NPs in aqueous solution under UV radiation has a phototoxic effect because of the formation of reactive species, such as $^1O_2$ (singlet oxygen), H_2O_2 (hydrogen peroxide), $^\bullet O_2^-$ (superoxide radical), and $OH^\bullet$ (hydroxyl radical) as follows:

$$ZnO + hv \rightarrow e^- + h^+$$
$$h^+ + H_2O \rightarrow H^+ + OH^\bullet$$
$$e^- + O_2 \rightarrow {}^\bullet O_2^-$$
$${}^\bullet O_2^- + H^+ \rightarrow HO_2^\bullet$$
$$HO_2^\bullet + H^+ + e^- \rightarrow H_2O_2$$

Therefore, effective ZnO antibacterial activities can be obtained from its exposure to UV (Sirelkhatim et al. 2015). The release of the ROS to damage bacterial active enzymes, DNA, and proteins that support cell growth and proliferation has been reported (Ann et al. 2014, Kirkinezos and Moraes 2001). Notice that the superoxide radical with a negative charge cannot penetrate the cell membrane, i.e., it only acts on the outer surface of bacteria. In contrast, hydrogen peroxide should be able to penetrate the bacterial cell wall to trigger cell lysis. The size of antibiotics also plays an important role, e.g., vancomycin, a big glycopeptide is known to prevent cell wall construction by interfering with transglycosylases of gram-positive bacteria. However, it cannot penetrate the outer cytoplasmic membrane of gram-negative bacteria. Indeed, gram-negative bacteria are intrinsically resistant to many antibiotics. In contrast, the exclusion by the cell wall of gram-positive bacteria is not problematic, considering antimicrobial peptides such as nisin (3354 Da) and defensins (3000–3500 Da) are able to penetrate the wall layer to interact with the cytoplasmic membrane (Friedrich et al. 2000). Indeed, lysozyme (14, 307 Da) can reach the peptidoglycan in the cell wall of gram-positive bacteria (Buckland and Wilton 2000, Foreman-Wykert et al. 1999).

ZnO NPs exhibit attractive antibacterial properties due to increased specific surface area, as the reduced particle size leads to enhanced particle surface reactivity. ZnO is a bio-safe material that possesses photo-oxidizing and photocatalytic effects on chemical and biological species. ZnO NPs and even zinc powders show antimicrobial activities against both gram-negative and gram-positive bacteria (Adams et al. 2006, Jones et al. 2008), including oral microbes known to contribute to caries (Fang et al. 2006). As bacterial activities depend on the total contact surface area, smaller particles of ZnO NPs should be more effective than their larger counterparts. ZnO NPs exhibit different morphologies, depending on the method of synthesis. Simple microwave decomposition produces spherical ZnO NPs, whereas a simple wet chemical route results in ZnO with different shapes, such as nano and micro-flowers, dumbbell-shaped, rice flakes, and rings (Sirelkhatim et al. 2015). Nanorods are synthesized by a hydrothermal technique, whereas nano-flakes are produced from a simple precipitation method. The solvothermal procedure produces nano-flowers, nanorods, and nano-spheres. More exotic shapes, such as Mulberry-like and hexagonal prismatic rods, are prepared by the microwave hydrothermal method and hydrothermal synthesis, respectively (Ma et al. 2013).

Like ZnO NPs, TiO_2-based nanocomposites subjected to light excitation are effective in eliciting microbial death. The photocatalytic action decreases the expression of a large array of genes/proteins specific for regulatory, signaling, and growth functions. The photocatalytic action also exhibits selective effects on ion homeostasis, coenzyme-independent respiration, and cell wall structure.

The cytotoxicity of Ag nanoparticles deserves a brief discussion, as the antibacterial effects of Ag salts have been well recognized. Albeit the mechanism is only partially known, Ag NPs are expected to accumulate in the bacterial membrane via ionic interactions. Ag NPs generate free radicals to induce bacterial membrane damage (Kim et al. 2007). The binding of silver(I) to the amino acids is postulated similar to that to copper(I), which binds the amino nitrogen, the carbonyl oxygen, and the oxygen and sulfur on the side chain (Lee et al. 1998).

The minimum inhibitory concentration (MIC) of metal nanoparticles decorated on nanocellulose on bacterial growth is appreciably lower than that of metal or metal oxide nanoparticles, which is another appealing feature. As an example, a MIC of 27 μg/mL of silver NPs was observed for *E. coli*, compared to 5.4 μg/mL of silver NPs decorated on cellulose. NCC with aldehyde groups was prepared by periodate oxidation, which reduced Ag^+ into Ag^0 in mild alkaline conditions. The result was attributed to the consequence of tight binding of cellulose nanocrystals to the bacterial envelope (Drogat el al. 2010). Excessive discharge of metal or metal oxide nanoparticles will lead to bacterial resistance and environmental pollution. Much lower MIC and the use of NCC to entrap metal nanoparticles could be the right approach to alleviate this environmental concern.

POTENTIAL APPLICATIONS OF NCC-NANOCOMPOSITES

Food Packaging

Food spoilage caused by food-borne pathogens is a serious global problem, therefore, the demand for antibacterial drugs in food packaging is growing to prolong the shelf life of food products. However, such antimicrobial substances must have only a slightly negative influence on food quality and taste. Besides the traditional methods for food preservation, "active packaging" or "intelligent packaging" is a new concept to provide interaction between food and packaging materials (Hirota et al. 2010, Kirkinezos and Moraes 2001). *Campylobacter* sp., *Salmonella* sp., *Yersinia enterocolitica*, *Escherichia coli*, and *Listeria monocytogenes* are the main culprits in food spoilage and deterioration. Polymeric food packaging films are most commonly used in the food package, and NCC might play an important role in food packaging. The functionalization of the hydroxyl group of NCC will circumvent its sensitivity to water. A cellulose matrix film with pediocin (cationic proteins with anti-listerial activity) inhibits the growth of pathogenic bacteria (Santiago-Silva et al. 2009). Several pediocins exhibit thermostability and retain their activity at a wide pH range with the bactericidal action against gram-positive food spoilage and pathogenic bacteria (Papagianni and Anastasiadou 2009). Essential oils and NCC are incorporated in the PLA (poly-lactic acid) film to preserve ground beef (Talebi et al. 2018). Essential oils have antimicrobial activities against *Staphylococcus aureus*, *Enterobacteriaceae*, *Pseudomonas* sp., *E. coli*, *B. subtilis*, etc. Lysozyme has also been used in "smart packaging" (Mecitoglu et al. 2006) and a cellulose film containing nisin (a polycyclic antibacterial peptide produced by *Lactococcus lactis*) is prepared for meat packaging applications to suppress the growth of *L. monocytogenes* (Saini et al. 2016) Nisin-grafted carboxylated cellulose nanofibers or a simple mixture of NCC-nisin show antimicrobial properties against various gram-positive bacteria (Nguyen et al. 2008). Among a plethora of antimicrobial molecules (Huang et al. 2019), of interest is the use of propylparaben, a colorless and odorless substance that inhibits the growth of molds and yeasts (Moir and Eyles 1992). It is an *n*-propyl ester of *p*-hydroxybenzoic acid, a natural substance found in many plants and some insects. The antibacterial action of this compound may be related to its effect on bacterial cytoplasmic membrane.

Wound Dressing

Wound dressings are designed to be effective in the treatment of chronic non-healing wounds. BC-based wound dressings have been developed to reduce pain and improve wound healing/cleaning (Piatkowski et al. 2011). BC with ultrafine fiber structures can absorb and retain a large amount of water. Nanoporous structured BC facilitates the diffusion of antimicrobial active molecules to the wound, whereas its tensile strength and flexibility serve as a good barrier for wound protection. In this context, NCC has also been developed for wound healing (Sampaio et al. 2016). NCC-composites or BC-composites with pertinent active molecules can be easily fabricated, e.g., BC/Au-DAPT composite (Zhang et al. 2016). However, both NCC and BC must compete with chitosan, a similar polymer or carboxylated chitosan nanocrystals, which is similar to carboxylated NCC (Luong et al. 2019). Chitosan exhibits antimicrobial activities with the presence of charged groups in the polymer backbone and displays ionic interactions with bacterial wall constituents. The BC-chitosan composite shows high antimicrobial activity against *S. aureus* and *E. coli* (Li et al. 2017).

Drug Delivery

Perhaps, this is one of the most difficult areas for NCC, albeit cellulose has a long history of use for the control and release of binding drugs. One of the NCC competitors, chitosan with a positive charge on amine groups of each sugar residue, interacts with negatively charged drugs (Springate et al. 2005, 2008). The amino group of chitosan has a pK_a value of 6.5; i.e., chitosan is positively charged and soluble in weakly acidic solutions. However, the charge density is dependent on the pH and the degree of deacetylation. Chitosan could have a degree of deacetylation close to 0% or 100%, and this parameter will affect many physiochemical and biological properties of chitosan and its antimicrobial activities (Yuan et al. 2011).

NCC must also compete with cellulose, cellulose derivatives, and microcrystalline cellulose in the formulation of drug-loaded tablets to form dense matrices for oral administration. Nevertheless, some work has advocated the use of acid-hydrolyzed NCC to bind significant quantities of two ionizable drugs, tetracycline and doxorubicin, which are released rapidly over a one-day period. The binding event forms a strong ionic bond between negatively charged NCC and the drugs with a positive charge at physiological pH 8. The binding drugs can be released at pH 8 over a day period. The NCC-CTAB (cetyltrimethylammonium bromide (CTAB) complex binds significant quantities of three hydrophobic anticancer drugs: docetaxel, paclitaxel, and etoposide (Jackson et al. 2011). Notice that NCC has some key properties, which are suitable for a drug delivery system. First, NCC is very stable at pH 7 and is unlikely to be vulnerable to enzymatic degradation. Thus, it should stay long enough in the circulation system without being removed by the mononuclear phagocytic system (the reticuloendothelial system or Kupffer cells) and phagocytic macrophages. Second, nanoparticles must be sufficiently large so that they will not permeate quickly into fenestrated blood vessels. In contrast, they need to be sufficiently small to elude phagocytosis by macrophages. NCC with the dimension of 5–10 nm × 100-few microns fulfils the requirement for a drug delivery system. Cellulose-based nanocarriers as platforms for cancer therapy have been discussed elsewhere (Meng et al. 2017).

CONCLUSIONS AND OUTLOOKS

Considering the diversified applications of NCC in various fields, there are some commercial attempts to produce NCC for semi-industrial applications. In the past, some research activities were noted in Canada, Sweden, China, Japan, and the USA (An et al. 2017). However, the industrial production of NCC has not materialized and the search continues for green methods with low energy and low cost. NCC must compete with other cheaper nanomaterials for similar applications, e.g., nanoclays at 2 Rs/kg in India or 2.8 cents/kg in US. The price of NCC is estimated to be ~2 US$/kg on a wet basis (Nelson 2019). In addition, NCC must compete with bacterial cellulose, microcrystalline cellulose, and chitosan, similar and abundant biopolymers with proven track records. In addition, there are still some technical issues related to the production of NCC. The effective dispersion of NCC in hydrophobic plastic resins is still far from perfection, albeit NCC with three hydroxyl groups can be functionalized to alter its surface properties. Coating NCC with black lignin might prevent intra- and intermolecular hydrogen bonding of NCC during conventional freeze-drying (Nelson 2019). However, a hydrogen bonding blocker must be transparent to be compatible with hydrophobic plastics. Another expensive step is the characterization of NCC for the process and product quality control. The characterization of NCC by SEM, TEM, AFM, etc. is time-consuming [over 30–40 min] and expensive. The cytotoxicity and biocompatibility of NCC are still not clearly known, therefore, demonstrating the safety of NCC and its related products must be thoroughly scrutinized. The fate of NCC after the environmental release is another important and pending issue. Currently, there is a lack of standard NCC materials and even methods for assessing biosafety. NCC might be more useful in environmental remediation, e.g., the use of nanocellulose-based adsorbents for the removal of organic dyes (He et al. 2013), antibiotics, chemicals, heavy metals, etc. in waste and wastewaters. NCC-based adsorbents are difficult to recover from treated wastewater and might involve several steps, such as precipitation, filtration, or high-speed centrifugation (Mahmoud et al. 2013). In this context, NCC magnetite (Fe_3O_4) materials can be prepared to facilitate the recovery of spent adsorbents by a magnetic field (Beyki et al. 2016, Zhang et al. 2017). However, the performance and economic viability of NCC-based materials for treating actual wastewater remain to be proven.

The concept of using NCC composites with antimicrobial properties is appealing for numerous applications, including food packaging and wound dressing. Various metal nanoparticles, notably ZnO NPs can be prepared by wet chemistry and incorporated in NCC as a new class of antimicrobial weapon to eradicate both gram-negative and gram-positive bacteria. ZnO NPs with different shapes and sizes are biocompatible to human cells (Zhou et al. 2006), a prerequisite for their applications related to food packaging, wound dressing, and other applications. The action mode of ZnO NPs is partially understood, which involves the release of reactive oxygen species (ROS) to impair the bacterial cell wall. In some cases, enhanced antibacterial activity stems from surface defects on the ZnO abrasive surface texture. Of note is the presence of bacterial superoxide dismutases and catalases that degrade superoxide and hydrogen peroxide, respectively. However, no cellular detoxification mechanism is known for hydroxyl radicals. Of course, there is equilibrium between the enzymes and radicals, and the amount of NCC-metal oxide nanoparticles can be adjusted to alleviate bacterial resistance, i.e., exceeding the antioxidant capacity of bacteria. It is desirable to have antimicrobial agents with low MIC, but a low dose might be beneficial to bacteria to induce resistance. Another approach is to dope ZnO NPs with metals such as Ag, Co, Ta, reduced graphene oxide, etc. to improve microbial activity (Sivakumara et al. 2018). ZnO with neutral hydroxyl groups can be readily functionalized by NCC and other biomolecules. ZnO NPs can be absorbed and degraded in the body as zinc is involved in the synthesis and degradation of lipids and proteins to regulate homeostasis (Bisht and Rayamajhi 2016). Besides iron, zinc is the second most required element in the human diet, which is about 15 mg Zn per day (King and Turnlund 1989). It is recognized as a safe (GRAS) material by the US Food and Drug Administration (FDA) (Zhou et al. 2006). However, elevated levels of administered intracellular ZnO show enhanced cytotoxicity through oxidative stress and zinc-mediated protein activity disequilibrium (Bisht and Rayamajhi 2016).

The development of novel antibacterial agents against bacteria strains, most major food pathogens, such as *Escherichia coli* O157:H, *Campylobacter jejuni*, *Staphylococcus aureus*, *Pseudomonas aeruginosa*, *Enterococcus faecalis*, *Salmonella* types, and *Clostridium perfringens*, has become of utmost importance. Doubtlessly, the application of antimicrobial materials can be extended to a variety of other applications, such as textile, coating, military, and household equipment.

References

Abdalkarim, S.Y.H., H.Y. Yu, D. Wang and J. Yao. 2017. Electrospunpoly(3-hydroxybutyrate-co-3-hydroxy-valerate)/cellulose reinforced nanofibrous membranes with ZnO nanocrystals for antibacterial wound dressings. Cellulose. 24: 2925–2938.

Adams, L.K., D.Y. Lyon, A. McIntosh and P.J. Alvarez. 2006. Comparative toxicity of nanoscale TiO_2, SiO_2 and ZnO water suspensions. Water Sci. Technol. 54: 327–334.

An, X., D. Cheng, J. Sheng, Q. Jia, Z. He, L. Zheng, et al. 2017. NCC productions nanocellulosic materials: Research/production activities and applications. J. Biores. Bioprod. 2: 45–49.

Ann, L.C., S. Mahmud, S.K.M. Bakhori, A. Sirelkhatim, D. Mohamad, H. Hasan, et al. 2014. Effect of surface modification and UVA photoactivation on antibacterial bioactivity of zinc oxide powder. Appl. Surf. Sci. 292: 405–412.

Bacakova, L., J. Pajorova, M. Bacakova, A. Skogberg, P. Kallio, K. Kolarova, et al. 2019. Versatile application of nanocellulose: From industry to skin tissue engineering and wound healing. Nanomaterials. 9: 2.

Barua, S., G. Das, L. Aidew, A.K. Buragohain and N. Karak. 2013. Copper-copper oxide coated nanofibrillar cellulose, a promising biomaterial. RSC Adv. 3: 14997–15004.

Beyki, M.H., M. Bayat and F. Shemirani. 2016. Fabrication of core-shell structured magnetic nanocellulose base polymeric ionic liquid for effective biosorption of Congo red dye. Bioresour. Technol. 218: 326–334.

Bisht, G. and S. Rayamajhi. 2016. ZnO nanoparticles: A promising anticancer agent. Nanobiomedicine. 3: 9.

Buckland, A.G. and D.C. Wilton. 2000. The antibacterial properties of secreted phospholipases A2. Biochim. Biophys. Acta. 488: 71–82.

Campia, P., E. Ponzini, B. Rossi, S. Farris, T. Silvetti, L. Merlini, et al. 2017. Aerogels of enzymatically oxidized galactomannans from leguminous plants, Versatile delivery systems of antimicrobial peptides and enzyme. Carbohydr. Polym. 158: 102–111.

Chaker, A. and S. Boufi. 2015. Cationic nanofibrillar cellulose with high antibacterial properties. Carbohydr. Polym. 131: 224–232.

Colic, M., D. Mihajlovic, A. Mathew, N. Naseri and I.V. Koko. 2015. Cytocompatibility and immunomodulatory properties of wood based nanofibrillated cellulose. Cellulose. 22: 763–778.

Cosgrove, D.J. 2014. Re-constructing our models of cellulose and primary cell wall assembly. Curr. Opin. Plant Biol. 22: 122–131.

Dehnad, D., H. Mirzaei, Z. Emam-Djomeh, S.M. Jafari and S. Dadashi. 2014. Thermal and antimicrobial properties of chitosan-nanocellulose films for extending shelf life of ground meat. Carbohydr. Polym. 109: 148–154.

Donaldson, K. and C.L. Tran. 2004. An introduction to the short-term toxicology of respirable industrial fibres. Mutation Res. Fundam. Mol. Mech. Mutagen. 553: 5–9.

Donaldson, K., A. Schinwald, F. Murphy, W.-S. Cho, R. Duffin, T. Lang, et al. 2013. The biologically effective dose in inhalation nanotoxicology. Acc. Chem. Res. 46: 723–732.

Dong, S., A.A. Hirani, K.R. Colacino, Y.W. Lee and M. Roman. 2012. Cytotoxicity and cellular uptake of cellulose nanocrystals. Nano LIFE. 02: 1241006.

Drogat, N., R. Granet, V. Sol, A. Memmi, N. Saad, C.K. Koerkamp, et al. 2010. Antimicrobial silver nanoparticles generated on cellulose nanocrystals. J. Nanopart. Res. 13:1557–1562.

El-Wakil, N.A, E.A. Hassan, R.E. Abou-Zeid and A. Dufresne. 2015. Development of wheat gluten/nanocellulose/titanium dioxide nanocomposites for active food packaging. Carbohydr. Polym. 124: 337–346.

Endes, C., S. Camarero-Espinosa, S. Mueller, E.J. Foster, A. Petri-Fink, B. Rothen-Rutishauser, et al. 2016. A critical review of the current knowledge regarding the biological impact of nanocellulose. J. Nanobiotechnol. 14: 78.

Fang, M., J.H. Chen, J.X.L. Xu, P.H. Yang and H.F. Hildebrand. 2006. Antibacterial activities of inorganic agents on six bacteria associated with oral infections by two susceptibility tests. Int. J. Antimicrob. Agents. 27: 513–517.

Favi, P.M., R.S. Benson, N.R. Neilsen, R.L. Hammonds, C.C. Bates, C.P. Stephens, et al. 2013. Cell proliferation, viability, and *in vitro* differentiation of equine mesenchymal stem cells seeded on bacterial cellulose hydrogel scaffolds. Mater. Sci. Eng. C. 33: 1935–1944.

Feese, E., H. Sadeghifar, H.S. Gracz, D.S. Argyropoulos and R.A. Ghiladi. 2011. Photobactericidal porphyrin-cellulose nanocrystals, synthesis, characterization, and antimicrobial properties. Biomacromolecules. 12: 3528–3539.

Fernandes, A.N., L.H. Thomas, C.M. Altaner, P. Callow, V.T. Forsyth, D.C. Apperley, et al. 2011. Nanostructure of cellulose microfibrils in spruce wood. Proc. Natl. Acad. Sci. USA. 108: E1195–E120.

Foreman-Wykert, A.K., Y. Weinrauch, P. Elsbach, and J. Weiss. 1999. Cell-wall determinants of the bactericidal action of group IIA phospholipase A2 against Gram-positive bacteria. J. Clin. Invest. 103: 715–721.

Friedrich, C.L., D. Moyles, T.J. Beveridge and R.E. Hancock. 2000. Antibacterial action of structurally diverse cationic peptides on gram-positive bacteria. Antimicrob. Agents Chemother. 44: 2086–2092.

Galkina, O.L., K. Onneby, P. Huang, V.K. Ivanov, A.V. Agafonov, A.G. Seisenbaeva, et al. 2015. Antibacterial and photochemical properties of cellulose nanofiber-titania nanocomposites loaded with two different types of antibiotic medicines. J. Mater. Chem. B. 3: 7125–7134.

Gao, C., T. Yan, J. Du, F. He, H. Luo and Y. Wan. 2014. Introduction of broad spectrum antibacterial properties to bacterial cellulose nanofibers via immobilising ε-polylysine nanocoatings. Food Hydrocoll. 36: 204–211.

Habibi, Y., H. Chanzy and M.R. Vignon. 2006. TEMPO-mediated surface oxidation of cellulose whiskers. Cellulose. 13: 679–687.

Habibi, Y., L.A. Lucia and O.J. Rojas. 2010. Cellulose nanocrystals, chemistry, self-assembly, and applications. Chem. Rev. 110: 3479–3500.

He, X., K.B. Male, P.N. Nesterenko, D. Brabazon, B. Paull and J.H.T. Luong. 2013. Adsorption and desorption of methylene blue on porous carbon monoliths and nanocrystalline cellulose. ACS Appl. Mater. Int. 5: 8796–8804.

Hirota, K., M. Sugimoto, M. Kato, K. Tsukagoshi, T. Tanigawa and H. Sugimoto. 2010. Preparation of zinc oxide ceramics with a sustainable antibacterial activity under dark conditions. Ceram. Int. 36: 497–506.

Huang, T., Y. Qian, J. Wei and C. Zhou. 2019. Polymeric antimicrobial food packaging and its applications. MDPI Polymers. 11: 560.

Jackson, J.K., K. Letchford, B.Z. Wasserman, L. Ye, W.Y. Hamad and H.M. Burt. 2011. The use of nanocrystalline cellulose for the binding and controlled release of drugs. Int. J. Nanomedicine. 6: 321–330.

Jebali, A., S. Hekmatimoghaddam, A. Behzadi, I. Rezapor, B.H. Mohammadi, T. Jasemizad, et al. 2013. Antimicrobial activity of nanocellulose conjugated with allicin and lysozyme. Cellulose. 20: 2897–2907.

Jafary, R., M.K. Mehrizi, S.H. Hekmatimoghaddam and A. Jebali. 2015. Antibacterial property of cellulose fabric finished by allicin conjugated nanocellulose. J. Text. Inst. 106: 683–689.

Janpetch, N., N. Saito and R. Rujiravanit. 2016. Fabrication of bacterial cellulose-ZnO composite via solution plasma process for antibacterial applications. Carbohydr. Polym. 148: 335–344.

Jones, N., B. Ray, K.D. Ranjit and A.C. Manna. 2008. Antibacterial activity of ZnO nanoparticle suspensions on a broad spectrum of microorganisms. FEMS Microbiol. Lett. 279: 71–76.

Kaplan, E., T. Ince, E. Yorulmaz, F. Yener, E. Harputlu and N.T. Lacin. 2014. Controlled delivery of ampicillin and gentamycin from cellulose hydrogels and their antibacterial efficiency. J. Biomater. Tissue Eng. 4: 543–549.

Katepetch, C., R. Rujiravanit and H. Tamura. 2013. Formation of nanocrystalline ZnO particles into bacterial cellulose pellicle by ultrasonic-assisted *in situ* synthesis. Cellulose. 20: 1275–1292.

Kim, J.S., E. Kuk, K.N. Yu, J. H. Kim, S.J. Park, H.J. Lee, et al. 2007. Antimicrobial effects of silver nanoparticles. Nanomedicine, Nanotechnology, Biology, Medicine. 3: 95–101.

King, J. and J. Turnlund. 1989. Human zinc requirements. pp. 335–350. *In*: C. Mills [ed.]. Zinc in Human Biology. ILSI Human Nutrition Reviews. Springer, London.

Kirkinezos, I.G. and C.T. Moraes. 2001. Reactive oxygen species and mitochondrial diseases. Semin. Cell Dev. Biol. 12: 449–457.

Kohanski, M.A., D.J. Dwyer and J.J. Collins. 2010. How antibiotics kill bacteria: From targets to networks. Nat. Rev. Microbiol. 8: 423–435.

Kovacs, T., V. Naish, B. O'Connor, C. Blaise, F. Gagne, L. Hall, et al. 2010. An ecotoxicological characterization of nanocrystalline cellulose (NCC). Nanotoxicology. 4: 255–270.

Lacin, N.T. 2014. Development of biodegradable antibacterial cellulose based hydrogel membranes for wound healing. Int. J. Biol. Macromol. 67: 22–27.

Lam, E., A.C.W. Leung , Y. Liu, E. Majid, S. Hrapovic, K.B. Male, et al. 2013. A green strategy guided by Raman spectroscopy for the synthesis of ammonium carboxylated nanocrystalline cellulose and the recovery of by-products. ACS Sust. Chem. Eng. 1: 278–283.

Lee, V.W.-M., H. Li, T.-C. Lau, R. Guevremont and K.W.M. Siu. 1998. Relative silver(I) ion binding energies of α-amino acids, a determination by means of the kinetic method. J. Am. Soc. Mass. Spectrom. 9: 760–766.

Lefatshe, K., C.M. Muiva and L.P. Kebaabetswe. 2017. Extraction of nanocellulose and *in-situ* casting of ZnO/cellulose nanocomposite with enhanced photocatalytic and antibacterial activity. Carbohydr. Polym. 164: 301–308.

Leung, A.C.W., S. Hrapovic, E. Lam, Y. Liu, K.B. Male, K.A. Mahmoud, et al. 2011. Characteristics and properties of carboxylated cellulose nanocrystals prepared from a novel one-step procedure. Small 7: 302–305.

Li, Y., Y. Tian, W. Zheng, Y. Feng, R. Huang, J. Shao, et al. 2017. Composites of bacterial cellulose and small molecule-decorated gold nanoparticles for treating gram-negative bacteria-infected wounds. Small. 13: 1700130.

Liao, N., A.R. Unnithan, M.K. Joshi, A.P. Tiwari, S.T. Hong, C.H. Park, et al. 2015. Electrospun bioactive poly(ε-caprolactone)–cellulose acetate–dextran antibacterial composite mats for wound dressing applications. Colloids Surf. A. 469: 194–201.

Liu, K., X. Lin, L. Chen, L. Huang, S. Cao and H. Wang. 2013. Preparation of microfibrillated cellulose/chitosan-benzalkonium chloride biocomposite for enhancing antibacterium and strength of sodium alginate films. J. Agric. Food Chem. 61: 6562–6567.

Liu, W., P. Su, S. Chen, N. Wang, Y. Ma, Y. Liu, et al. 2014. Synthesis of TiO_2 nanotubes with ZnO nanoparticles to achieve antibacterial properties and stem cell compatibility. Nanoscale. 6: 9050–9062.

Lu, D.N., X.R. Zhou, X.D. Xing, X.G. Wang and Z. Liu. 2004. Quaternary ammonium salt (QAS) grafted cellulose fiber--preparation and anti-bacterial function. Acta. Polym. Sin. 25: 107–113.

Luong, J.H.T., E. Lam, C.EW. Leung, S. Hrapovic and K.B. Male. 2019. Chitin nanocrystals and process for preparation thereof. US Patent # 20160272731.

Ma, J., J. Liu, Y. Bao and Z. Zhu. 2013. Synthesis of large-scale uniform mulberry-like ZnO particles with microwave hydrothermal method and its antibacterial property. Ceramics International. 39: 2803–2810.

Mahmoud, K.A., J.A. Mena, K.B. Male, S. Hrapovic, A. Kamen and J.H.T. Luong. 2010. Effect of surface charge on the cellular uptake and cytotoxicity of fluorescent labeled cellulose nanocrystals. ACS Appl. Mater. Interfaces. 2: 2924–2932.

Mahmoud, K.A., E. Lam, S. Hrapovic and J.H.T. Luong. 2013. Preparation of well-dispersed gold/magnetite nanoparticles embedded on cellulose nanocrystals for efficient immobilization of papain enzyme. ACS Appl. Mat. Int. 5: 4978–4985.

Mecitoglu, C., A. Yemenicioglu, A. Arslanoglu, Z.S. Elmaci, F. Korel and A.E. Cetin. 2006. Incorporation of partially purified hen egg white lysozyme into zein films for antimicrobial food packaging. Food Res. Int. 39: 12–21.

Meng, L.Y., B. Wang, M.G. Ma and J.F. Zhu. 2017. Cellulose-based nanocarriers as platforms for cancer therapy. Curr. Pharm. Des. 23: 5292–5300.

Moir, C.J. and M.J. Eyles. 1992. Inhibition, injury, and inactivation of four psychrotrophic foodborne bacteria by the preservatives methyl ρ-hydroxybenzoate and potassium sorbate. J. Food Prot. 55: 360–366.

Muthulakshmi, L., N. Rajini, H. Nellaiah, T. Kathiresan, M. Jawaid and A.V. Rajulu. 2017. Preparation and properties of cellulose nanocomposite films with *in situ* generated copper nanoparticles using *Terminalia catappa* leaf extract. Int. J. Biol. Macromol. 95: 1064–1071.

Natan, M., O. Gutman, R. Lavi, S. Margel, and E. Banin. 2015. Killing mechanism of stable N-halamine cross-linked polymethacrylamide nanoparticles that selectively target bacteria. ACS Nano. 9: 1175–1188.

Nath, B.K., C. Chaliha, E. Kalita, and M.C. Kalita. 2016. Synthesis and characterization of ZnO, CeO_2, nanocellulose, PANI bionanocomposite. A bimodal agent for arsenic adsorption and antibacterial action. Carbohydr. Polym. 148: 397–405.

Nelson, K. 2019. Challenges and opportunities in manufacturing panel. https//www.nano.gov/sites/default/files/nelson_challenges_and_opportunities_in_manufacturing_panel_api_final.pdf. Accessed Nov. 13, 2019.

Nguyen, V.T., M.J Gidley and G.A. Dykes. 2008. Potential of a nisin-containing bacterial cellulose film to inhibit *Listeria monocytogenes* on processed meats. Food Microbiol. 25: 471–474.

Nozaki, S. 1999. Effects of amounts of additives on peptide coupling mediated by a water-soluble carbodiimide in alcohols. J. Pept. Res. 54:162–167.

Padrao, J., S. Goncalves, J.P. Silva, V. Sencadas, S. Lanceros-Mendez, A.C. Pinheiro, et al. 2016. Bacterial cellulose-lactoferrin as an antimicrobial edible packaging. Food Hydrocoll. 58: 126–140.

Papagianni, M. and S. Anastasiadou. 2009. Pediocins: The bacteriocins of Pediococci. Sources, production, properties and applications. Microb. Cell Fact. 8: 3.

Petersen, N. and P. Gatenholm. 2011. Bacterial cellulose-based materials and medical devices, current state and perspectives. Appl. Microbiol. Biotechnol. 91: 1277–1286.

Piatkowski, A., N. Drummer, A. Riessen, D. Ulrich and N. Pallua. 2011. Randomized controlled single center study comparing a polyhexanide containing biocellulose dressing with silver sulfadiazine cream in partial-thickness dermal burns. Burns. 37: 800–804.

Poonguzhali, R., S.K. Basha and V.S. Kumari. 2017. Synthesis and characterization of chitosan-PVP-nanocellulose composites for *in-vitro* wound dressing application. Int. J. Biol. Macromol. 105: 111–120.

Rabie, E., J.C. Serem, H.M. Oberholzer, A.R.M. Gaspar and M.J. Bester. 2016. How methylglyoxal kills bacteria, an ultrastructural study. Ultrastruct. Pathol. 40: 107–111.

Rodriguez-Felix, D.E., M.M. Castillo-Ortega, A.L. Najera-Luna, A.G. Montano-Figueroa, I.Y. Lopez-Pena, T. Del Castillo-Castro, et al. 2016. Preparation and characterization of coaxial electrospun fibers containing triclosan for comparative study of release properties with amoxicillin and epicatechin. Curr. Drug Delivery. 13: 49–56.

Saini, S., C. Sillard, M.N. Belgacem and J. Bras. 2016. Nisin anchored cellulose nanofibers for long term antimicrobial active food packaging. RSC Adv. 6:12437–12445.

Salajková, M., L.A. Berglund and Q. Zhou. 2012. Hydrophobic cellulose nanocrystals modified with quaternary ammonium salts. J. Mater. Chem. 22: 19798–19805.

Sampaio, L.M.P., J. Padrao, J. Faria, J.P. Silva, C.J. Silva, F. Dourado, et al. 2016. Laccase immobilization on bacterial nanocellulose membranes, antimicrobial, kinetic and stability properties. Carbohydr. Polym. 145: 1–12.

Santiago-Silva, P., N.F.F. Soares, J.E. Nobrega, M.A.W. Junior, K.B.F. Barbosa, A.C.P. Volp, et al. 2009. Antimicrobial efficiency of film incorporated with pediocin (ALTA@2351) on preservation of sliced ham. Food Control. 20: 85–88.

Sirelkhatim, A., S. Mahmud, A. Seeni, N.H.M. Kaus, L.C. Ann, S.K.M. Bakhori, et al. 2015. Review on zinc oxide nanoparticles, antibacterial activity and toxicity mechanism. Nano-Micro Lett. 7: 219–242.

Sivakumara, P., M. Lee, Y.S. Kim and M.S. Shim. 2018. Photo-triggered antibacterial and anticancer activities of zinc oxide nanoparticles. J. Mater. Chem. B. 6: 4852–4871.

Springate, C.M., J.K. Jackson, M.E. Gleave and H.M. Burt. 2005. Efficacy of an intratumoral controlled release formulation of cluster in antisense oligonucleotide complexed with chitosan containing paclitaxel or docetaxel in prostate cancer xenograft models. Cancer Chemother. Pharmacol. 56: 239–247.

Springate, C.M., J.K. Jackson, M.E. Gleave and H.M. Burt. 2008. Clusterin antisense complexed with chitosan for controlled intratumoral delivery. Int. J. Pharm. 350: 53–64.

Talebi, F., A. Misaghi, A. Khanjari, A. Kamkar, H. Gandomi and M. Rezaeigolestani. 2018. Incorporation of spice essential oils into poly-lactic acid film matrix with the aim of extending microbiological and sensorial shelf life of ground beef. LWT-Food. Sci. Technol. 96: 482–490.

Tischer, M., G. Pradel, K. Ohlsen and U. Holzgrabe. 2012. Quaternary ammonium salts and their antimicrobial potential, targets or nonspecific interactions? ChemMedChem. 7: 22–31.

Uddin, K.M.A., H. Orelma, P. Mohammadi, M. Borghei, J. Laine, M. Linder, et al. 2017. Retention of lysozyme activity by physical immobilization in nanocellulose aerogels and antibacterial effects. Cellulose. 24: 2837–2848.

Ul-Islam, M., W.A. Khattak, M.W. Ullah, S. Khan and J.K. Park. 2014. Synthesis of regenerated bacterial cellulose-zinc oxide nanocomposite films for biomedical application. Cellulose. 21: 433–447.

Wang, P., J. Zhao, R. Xuan, Y. Wang, C. Zou, Z. Zhang, et al. 2014. Flexible and monolithic zinc oxide bionanocomposite foams by a bacterial cellulose mediated approach for antibacterial applications. Dalton Trans. 43: 6762–6768.

Wang, T. and M. Hong. 2016. Solid-state NMR investigations of cellulose structure and interactions with matrix polysaccharides in plant primary cell walls. J. Exp. Bot. 67: 503–514.

Wei, B., G. Yang and F. Hong. 2011. Preparation and evaluation of a kind of bacterial cellulose dry films with antibacterial properties. Carbohydr. Polym. 84: 533–538.

WHO. 2017. https://www.who.int/news-room/detail/27-02-2017-who-publishes-list-of-bacteria-for-which-new-antibiotics-are-urgently-needed.

Wohlhauser, S., G. Delepierre, M. Labet, G. Morandi, W. Thielemans, C. Weder, et al. 2018. Grafting polymers from cellulose nanocrystals, synthesis, properties, and applications. Macromolecules 51: 6157–6189.

Yanamala, N., M.T. Farcas, M.K. Hatfield, E.R. Kisin, V.E. Kagan, C.L. Geraci, et al. 2014. *In vivo* evaluation of the pulmonary toxicity of cellulose nanocrystals, a renewable and sustainable nanomaterial of the future. ACS Sustain. Chem. Eng. 2: 1691–1698.

Yang, J., G.J. Kwon, K. Hwang and D.Y. Kim. 2018. Cellulose-chitosan antibacterial composite films prepared from LiBr solution. Polymers. 10: 1058.

Yuan, Y., B.M. Chesnutt, W.O. Haggard and J.D. Bumgardner. 2011. Deacetylation of chitosan: Material characterization and *in vitro* evaluation via albumin adsorption and pre-osteoblastic cell cultures. Materials. 4: 1399–1416.

Zhang, P., L. Chen, Q. Zhang and F.F. Hong. 2016. Using *in situ* dynamic cultures to rapidly biofabricate fabric-reinforced composites of chitosan/bacterial nanocellulose for antibacterial wound dressings. Front. Microbiol. 7: 260.

Zhang, L., H. Qi, Y. Yan, Y. Gu, W. Sun and A.A. Zewde. 2017. Sonophotocatalytic inactivation of *E. coli* using ZnO nanofluids and its mechanism. Ultrason. Sonochem. 34: 232–238.

Zhao, J., X. Zhang, R. Tu, C. Lu, X. He and W. Zhang. 2014. Mechanically robust, flame-retardant and anti-bacterial nanocomposite films comprised of cellulose nanofibrils and magnesium hydroxide nanoplatelets in a regenerated cellulose matrix. Cellulose. 21: 1859–1872.

Zhou, J., N.S. Xu and Z.L. Wang. 2006. Dissolving behavior and stability of ZnO wires in biofluids: A study on biodegradability and biocompatibility of ZnO nanostructures. Adv. Mater. 18: 2432–2435.

6

TiO_2 Nanoparticles and Composite Materials: Antimicrobial Activity, Antimicrobial Mechanism and Applications

Yage Xing[1*], Xuanlin Li[1,2], Qinglian Xu[1], Xiufang Bi[1,2] and Xiaocui Liu[1]

[1]Key Laboratory of Grain and Oil Processing and Food Safety of Sichuan Province
College of Food and Bioengineering, Xihua University
Chengdu, 610039, China

[2]Key Laboratory of Food Non-Thermal Technology
Engineering Technology Research Center of Food Non-Thermal
Yibin Xihua University Research Institute, Yibin, 644004, China

Email: xingyage1@163.com (Yage Xing); xllb0519@163.com (Xuanlin Li)
775938414@qq.com (Qinglian Xu); 811043890@qq.com (Xiufang Bi)
xiaocuiliu777@126.com (Xiaocui Liu)

INTRODUCTION

Pathogenic or spoilage microorganisms, as well as ethylene contamination, are commonly responsible for the degradation in quality and losses of fresh produce during the ripening process (de Chiara et al. 2015). High decay rates pose a significant challenge to the storage of fruits and vegetables (Xing et al. 2017, Xu et al. 2017). Since consumers constantly require healthy food that is nutritious

*For Correspondence: xingyage1@163.com

and fresh, alternative processing technologies are necessary in addition to conventional methods (Xing et al. 2017). Active packaging presents an excellent alternative to employing disinfectants and chemical treatments, since it can restrict microbial growth and preclude the adverse effect of ethylene, effectively extending the shelf life of fresh fruit and vegetables (Kaewklin et al. 2018). In this context, extensive research into the combined application of nano-biocomposite coatings has been initiated (Castillo et al. 2015, Shahabi-Ghahfarrokhi et al. 2015).

The preparation and application of nano-biocomposite packaging with polymeric materials and inorganic nanoparticles for food preservation are expected to be developed in the future (Li et al. 2009a, Xing et al. 2012, Carvalho et al. 2014). Of the various types of inorganic nanoparticles, TiO_2 is always employed as an ethylene scavenger and an antimicrobial agent for different coating materials for various fruits products, since it is inexpensive, non-toxic, chemically stable, and Generally Recognized as Safe (GRAS) (Carvalho et al. 2014, Lin et al. 2015, Zhang et al. 2017). As a result, recent years have seen increasing attention given to nanoparticle biosynthesis due to the demand for renewable materials and nontoxic chemicals (Gericke and Pinches 2006).

TiO_2 nanoparticles represent an inorganic photocatalytic bacteriostatic material, in which TiO_2 semiconductor photocatalysis is employed to remove viruses and bacteria (Pal et al. 2012, Hou et al. 2013, Ramya et al. 2013). The photocatalytic behavior relies heavily on visible light irradiation, as well as ultraviolet (UV) light, which is responsible for activating its antimicrobial properties (Yemmireddy and Hung 2015a, Cano et al. 2017, Li et al. 2009b). Various attempts were made to improve the relevance and efficacy of TiO_2, and several of these studies involved incorporation of metals such as silver (Ag) (Liu et al. 2018), zinc (Wang et al. 2014), and copper (Miao et al. 2017). This is because silver displays extensive antimicrobial activity, and the incorporation of Ag could improve the photocatalytic and antibacterial activity of TiO_2 (Skorb et al. 2008, Durango-Giraldo et al. 2019). Xu et al. (2017) found that the composite coating containing graphene oxide (GO), chitosan (CS), and TiO_2 nanoparticles at a ratio of 1:20:4 exhibited excellent antibacterial activity against *Aspergillus niger* and *Bacillus subtilis*, by inducing cell membrane rupture.

Therefore, the purpose of this chapter is to discuss the antimicrobial activity, antimicrobial mechanism, and application of TiO_2 nanoparticles and composite materials during the storage of fresh produce. The first part introduces the preparation of TiO_2 nanoparticles and composite materials. Then, the antimicrobial activity and mechanisms of TiO_2 nanoparticles and the composite materials are summarized. In addition, the ion release and the application of TiO_2 nanoparticles and composite materials in the storage of fruits and vegetables are reviewed. Finally, some useful insights are provided for further research.

PREPARATION OF TIO_2 NANOPARTICLES AND COMPOSITE MATERIALS

In recently years, the reported works on the preparation of TiO_2 nanoparticles and composite materials are increasing (Karthikeyan et al. 2017, Chen et al. 2019,

Xie and Hung 2019). The synthesis of nanoparticles can be achieved by chemical, physical, or biological methods (Abu-Dalo et al. 2019). TiO_2 nanoparticles were prepared by a modified sol-gel method described in previous works (Yeung et al. 2009). In recent years, the ecofriendly processes for nanoparticles synthesis without using toxic chemicals are being developed. Biosynthetic methods using enzymes, fungi, microorganisms, and plants or plant extracts have emerged (Abu-Dalo et al. 2019). Karthikeyan et al. (2017) had prepared TiO_2 by hydrolysis method using titanium tetra isopropoxide (TTIP) as a precursor. As reported by Sundrarajan et al. (2017), TiO_2 nanoparticles were prepared by using *M. citrifolia*. Jha et al. (2009) reported the synthesis of TiO_2 nanoparticles using *B. subtilis*. Furthermore, Subhapriya and Gomathipriya (2018) and Abu-Dalo et al. (2019) have prepared TiO_2 nanoparticles by using leaves of *Trigonella foenum* and the pristine pomegranate peel extract, respectively. More importantly, TiO_2 nanoparticles as antimicrobial agents can be incorporated into coating film, and were developed by many researchers. Yemmireddy and Hung (2015b) had prepared a total of six different suspensions by mixing TiO_2 nanoparticles with shellac, polyurethane, and polyacrylate in a porcelain mortar for about 15 minutes and with further treatment in an ultrasonic water bath for 1 hour. Moreover, Zhang et al. (2016) developed a CS/whey protein isolates (WPI) film incorporated with modified TiO_2 nanoparticles. The TiO_2/plasticized CS nano-biocomposites, the GO-CS-TiO_2 (GO-CT) coating, and the carboxymethyl cellulose (CMC)-sodium montmorillonite (Na-MMT)-TiO_2 coating were prepared by Cano et al. (2017), Xu et al. (2017), and Achachlouei and Zahedi (2018), respectively. Furthermore, Goudarzi and Shahabi-Ghahfarrokhi (2018) have prepared the starch-TiO_2 nanoparticles coating film.

Modification of TiO_2 with other nanoparticles could widen the range of light responses and improve the photoactivity and antimicrobial activity, which has become an important research topic lately. Zielinska-Jurek et al. (2015) have reported the preparation of Pt/TiO_2 and Ag-Pt/TiO_2 photocatalysts. Moreover, Ag-TiO_2 nanocomposites (NC) have been developed and investigated by Dong et al. (2019) and Durango-Giraldo et al. (2019). The composite nanoparticles, such as TiO_2:Cu nanoparticles and TiO_2-doped SiO_2 hybrid material were synthesized by Rauta et al. (2016) and Chen et al. (2019), respectively. Moreover, Fe doped TiO_2 was prepared by AL-Jawad et al. (2017) and Naghibi et al. (2015). Zhao et al. (2018) have prepared M/TiO_2 (M = Li, Na, Mg, Fe, and Co) antimicrobial materials. On the other hand, the incorporation of modified TiO_2 nanoparticles in the coating film was also developed. For example, Lin et al. (2015) and Xiao et al. (2015) prepared the Ag-TiO_2-CS NC. Zein protein (ZP) NC films were prepared by Kadam et al. (2017) with the core-and-shell nanoparticles (TiO_2 as core and SiO_2 as shell).

SURFACE PROPERTIES OF TIO_2 NANOPARTICLES AND COMPOSITE MATERIALS

The surface structure of nanoparticles is critical for the application of coating films (Xing et al. 2017). As observed by Kirthi et al. (2011), TiO_2 nanoparticles synthesized with *B. subtilis* displayed a spherical shape with a few aggregates. Furthermore, the scanning electron microscopy (SEM) photographs obtained by Sathishkumar et al. (2012) showed that the TiO_2 nanoparticles were quasi-spherical in shape and uniformly distributed. Moreover, they consisted of nanoparticles devoid of aggregates, while their size and shape were well-defined, indicating the essential role of the hydrothermally acquired *M. citrifolia* leaf extract. According to Siripatrawan and Kaewklin (2018), the SEM images illustrated the even dispersion of the TiO_2 nanoparticles in the film matrix, which is particularly apparent in the cross-sectional pictures of the films. However, high TiO_2 concentrations induced TiO_2 nanoparticle agglomeration in the film matrix. On the other hand, Chung et al. (2009) indicated that metal ion doping produced more even TiO_2 crystalline particles, which were more uniformly shaped following ball milling. Rao et al. (2019) reported that the synthesized Ag/TiO_2 were circular in shape with an average size of 11.25 nm, with a rough surface, and distributed irregularly. As reported by Durango-Giraldo et al. (2019), both the transmission electron microscopy (TEM) images and SEM pictures reveal that the synthesized TiO_2 nanoparticles and silver-modified TiO_2 nanoparticles display a spherical morphology without large agglomerates. Moreover, Achachlouei and Zahedi (2018) found that the SEM micrographs exhibited well-distributed TiO_2 nanoparticles and Na-MMT throughout the film surfaces, particularly at low concentrations. Furthermore, in a study by Subhapriya and Gomathipriya (2018), SEM indicated that all prepared TiO_2/SiO_2/Au and TiO_2/SiO_2 films were porous with a homogeneous microstructure. Meanwhile, Chen et al. (2019) observed various inter-structural gaps on the surface of SiO_2 sample, consisting of multiple evenly dispersed, loosely structured TiO_2nanoparticles. The TiO_2@SiO_2 hybrid material displayed an irregular surface region, promoting microbial cell adhesion to the prepared composite.

Similarly, the surface properties of TiO_2 nanoparticles composite materials have been analyzed by others. Cano et al. (2017) indicated the sufficient distribution of the nanofiller into the polymer matrix, some TiO_2 nanoparticles and aggregates appeared to be adequately dispersed in the polymer matrix with elevated TiO_2 nanoparticle content (Salarbashi et al. 2018, Alizadeh-Sani et al. 2018). Xu et al. (2017) observed that the GO-CS biopolymer acquired a rough texture following the addition of TiO_2 nanoparticles. Moreover, Roilo et al. (2018) indicated that the surface of the TEMPO-oxidized cellulose nanofiber (TO-CNF) coating was devoid of revealing pinholes and cracks, while the presence of TiO_2 nanoparticles was observed. However, Montaser et al. (2019) indicated that adding TiO_2 nanoparticles to the CS matrix caused the surface of the composite membrane to become irregular, displaying the presence of various tiny granules and encouraging pore formation. Furthermore, Xie and Hung (2018) found that all cellulose acetate and polylactic aid films containing TiO_2 nanoparticles exhibited an even surface and uniform TiO_2 nanoparticles dispersion, while the integration of TiO_2 nanoparticles in polycaprolactone film resulted in a porous structure. Rauta et al. (2016) indicated that the CS-TiO_2:Cu particles displayed free dispersion, which provided an irregular surface area for the adhesion of microbial cells. Additionally, the CS-TiO_2:Cu NC presented irregular clusters with coarse surfaces and primary particles that were spherical in shape.

ANTIMICROBIAL ACTIVITY OF TIO_2 NANOPARTICLES AND COMPOSITE MATERIALS

Since fresh post-harvest produce is vulnerable to a variety of microorganisms responsible for degradation, its quality can be significantly affected, leading to a reduced shelf-life. The antimicrobial activity of TiO_2 nanoparticles and composite materials are influenced by the synergistic effect of coating materials and nanoparticles, the type and structure of the test microorganism, and the testing conditions, as shown in Figure 6.1 (Xu et al. 2013, Soni et al. 2013, Wei et al. 2014, Magesan et al. 2016, Xing et al. 2019). As indicated in Figure 6.1, the antibacterial activity of metals is influenced by the physicochemical and morphological properties of nanoparticles (Mohammadi et al. 2010, Seil and Webster 2012), and depends on the size, shape, and crystal structure in the case of TiO_2 nanoparticles (Haghighi et al. 2013), while small nanoparticles have been proven to exhibit stronger bactericidal characteristics (Buzea et al. 2007, Fellahi et al. 2013, Besinis et al. 2014). Furthermore, the enhanced antibacterial effect of metal nanoparticles may be facilitated by their positive surface charge, allowing them to bind to bacteria with a negative surface charge (Seil and Webster 2012, Bera et al. 2014).

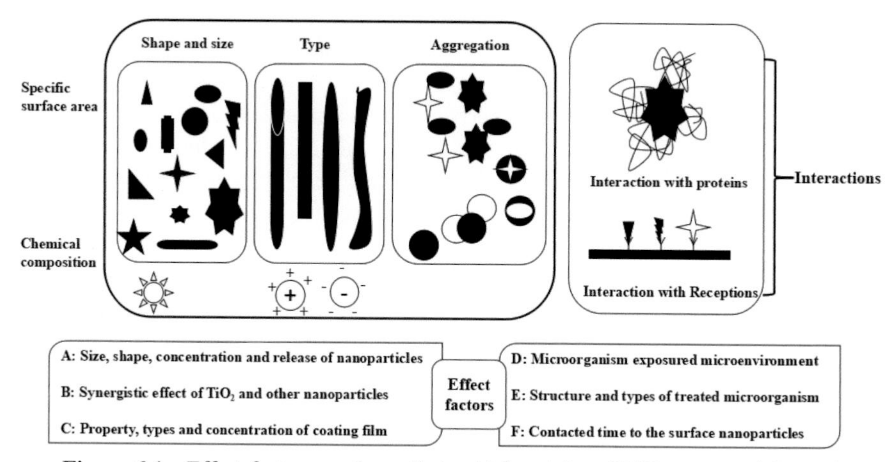

Figure 6.1 Effect factors on the antimicrobial activity of TiO_2 nanoparticles and composite materials.

Antimicrobial Activity of TiO_2 Nanoparticles

The important factors affecting the antibacterial efficiency and effectiveness of TiO_2 nanoparticles are the preparation materials and particle size (Seil and Webster 2012, Adibkia et al. 2012). According to Yeung et al. (2009), nano-TiO_2 with higher photoactivity is more effective against *B. subtilis* regarding bactericidal activity, while that of rutile TiO_2 exceeds both the mixed-phase TiO_2 and the photoactive anatase. Sani et al. (2017) established that TiO_2 nanoparticles substantially restricted the growth of test bacteria, particularly with reference to gram-positive bacteria. Moreover, as indicated by Haghighi et al. (2013), the antifungal influence of biofilms containing *C. albicans* on strains immune to fluconazole was enhanced by synthesized TiO_2 nanoparticles. According to a report by Subhapriya and Gomathipriya (2018), the TF-TiO_2 nanoparticles (biosynthesized from *Trigonella foenum-graecum)* displayed considerable antibacterial activity against *S. faecalis*, *Y. enterocolitica*, *S. aureus*, *E. coli*, and *E. faecalis*, while recording a zone of inhibitions (ZOI) measurement of about 8.5–11.5 mm with TF-TiO_2 nanoparticles. Xie and Hung (2019) indicated that direct contact between the bacterial suspension and the film, induced higher bacterial decline compared to the test environment devoid of direct contact. Verdier et al. (2014) discovered that bacterial levels were reduced more significantly when the transparent film was used to cover the solution, increasing the likelihood of interaction between the TiO_2 nanoparticle and the bacterial cells. Antimicrobial activity of TiO_2 nanoparticle on more microorganisms needs further evaluation. Overall, TiO_2 nanoparticles have diverse antimicrobial activities with different potencies and range of activities.

Antimicrobial Activity of TiO_2 Nanoparticles Composite Materials

The antimicrobial activities of various TiO_2 nanoparticles composite materials were also investigated by several researchers. According to Ye et al. (2017), the antimicrobial activity of all TiO_2/Ag composites surpassed that of Ag nanoparticles and commercial TiO_2. As reported by Rao et al. (2019), Ag/TiO_2 nanoparticles displayed the highest inhibition towards *E. coli,* then *C. albicans*, followed by methicillin-resistant *S. aureus*. Furthermore, Takai and Kamat (2011) indicated that the antibacterial activity of the TiO_2 layer containing Ag was stronger against *Bacillus* sp. than *Pseudomonas* sp. The results of a study conducted by André et al. (2015) discovered that the minimum inhibitory concentration (MIC) measurements required for the activity of TiO_2:Ag against the planktonic *C. tropicalis, C. glabrata,* and *C. albicans* strains demonstrated that the most susceptible species was *C. glabrata*, while the most resistant was *C. albicans*. In addition, Abdel-Fatah et al. (2016) demonstrated that the synthesized Ag/TiO_2 nanoparticles displayed stronger antimicrobial activity against gram-negative as well as gram-positive bacteria, increasing with elevated Ag/TiO_2 concentrations. Chen and Yu (2017) indicated that according to the ZOI and critical (minimum) concentration results, the sensitivity level of the five evaluated strains to the antibacterial effect of Ag nanoparticles/TiO_2 occurred in the following order: *E. coli* > *S. epidermidis* > *P. spinulosum* > *S. chartarum* > *A. niger*. Durango-Giraldo et al. (2019) indicated enhanced bacterial activity of the modified Ag/TiO_2-Ex (Ag modified TiO_2 particles were prepared by wet impregnation method) when compared with the altered Ag/ TiO_2-In (Ag modified TiO_2 particles were prepared by *in-situ* method), which related to higher interaction with the medium. Meanwhile, Zielinska-Jurek et al. (2015) noted that the Ag/TiO_2 particles restricted the bacterial and yeast growth, while the Ag-Pt/TiO_2 and Ag/TiO_2 samples exhibited more significant antimicrobial action against *S. aureus* and *E. coli*. Additionally, Wei et al. (2014) indicated that the antibacterial properties of Cu/TiO_2 coatings are superior to those without TiO_2, which can be ascribed to the combined antimicrobial effect provided by Cu and TiO_2. According to AL-Jawad et al. (2017), the dark reaction results suggested exceptional antimicrobial action (88%) against *S. aureus* in the presence of 3% Fe-TiO_2, while the eradication of *E. coli* did not exceed 51 percent. Furthermore, Chen et al. (2019) reported that the antibacterial ratio of TiO_2@SiO_2 hybrid materials on *E. coli* significantly increased in conjunction with elevated TiO_2 doping levels from 1.5–4.4%, while the antimicrobial effects with visible-light irradiation were lower than those exposed to UVA irradiation.

Alizadeh-Sani et al. (2018) indicated that combining TiO_2 and rosemary essential oil to form WPI/CNFs films resulted in higher antimicrobial activity against gram-positive bacteria, including *S. aureus* and *L. monocytogenes* compared to gram-negative bacteria, such as *E. coli* O_{157}:H_7, *S. enteritidis,* and *P. fluorescens*. As stated by Huang et al. (2017), it was possible to increase the log reduction for *L. monocytogenes* to >4 log colony forming units using TiO_2-polylactide composites that were illuminated with UV-A. Moreover, Teymourpour et al. (2015) and Salarbashi et al. (2018) found that soybean polysaccharide (SSPS)-TiO_2 bio-NC films displayed excellent antibacterial action against *S. aureus* and *E. coli.* Abu-Dalo et al. (2019) discovered that TiO_2 nanoparticle, pristine pomegranate peel extract (PPP), and PPP-TiO_2 had an excellent inhibitory effect on *E. coli* and *S. aureus*, where PPP-TiO_2 was responsible for a substantial number of black spaces and dead cells, illustrating the NC distribution. Furthermore, Qu et al. (2019) indicated that CS and TiO_2 nanoparticles exerted a significant synergistic antimicrobial effect compared to the single CS element, showing more substantial antibacterial activity against *S. aureus, S. enteritidis,* and *E. coli* when exposed to conditions presenting dark and ultraviolet (UV) light irradiation. Moreover, Xiao et al. (2015) illustrated that the Ag-nanoparticles@CS-TiO_2 organic-inorganic composite (CTA) displayed significantly stronger antibacterial action against *E. coli* than in the case of *C. albicans* and *S. aureus*. Rauta et al. (2016) confirmed the association between the restrictive activity of the CS-TiO_2:CuNC and nanoparticles by using SEM to examine *E. coli* bacterial cells. In general, TiO_2 nanoparticles show more antimicrobial activity after being compounded with other materials.

ANTIMICROBIAL MECHANISM OF TIO_2 NANOPARTICLES AND COMPOSITE MATERIALS

Antimicrobial Mechanism of TiO_2 Nanoparticles

The exact mechanisms for the antimicrobial effects of TiO_2 nanoparticles are still being studied. Sundrarajan et al. (2017) suggested several TiO_2 nanoparticles mechanisms, which included the following: (i) extensive active surface region, (ii) structure of the bacterial cell wall, (iii) membrane cell-wall thickness, (iv) Ti^{4+}ion release by TiO_2, (v) hydrogen peroxide production, and (vi) reactive oxygen species (ROS) on the TiO_2 nanoparticles surface.

Due to the presence of hydroxyl groups, titanium nanoparticles can disintegrate the external bacterial membranes, causing organism death (Rajakumar et al. 2012). Moreover, UV light irradiation induce TiO_2 nanoparticles to generate hydroxyl radicals (–OH), as well as other reactive oxygen species (ROS) (El-Shahawy et al. 2018). Highly active free radicals can annihilate the unsaturated bonds on the bacterial surface, causing the depolymerization of sugar, the degradation of lipids, as well as the denaturation of proteins, after which they can permeate the bacterial cell to damage its internal structure (Zhu et al. 2018, Wang et al. 2017). Furthermore, direct interaction with nanoparticles can prompt these free radicals to discharge toxic metal ions, inducing cell membrane damage and bacterial destruction (El-Shahawy et al. 2018).

Furthermore, due to their thicker cell walls containing chitin and glucan, fungi offer higher resistance than bacteria (Wang et al. 2017). Research indicated that gram-negative bacteria could resist photocatalytic deactivation more effectively, since its cell wall structure consisted of three layers, which included an external membrane, a thin peptidoglycan layer in the middle, and an internal cytoplasmic layer, therefore, restricting the penetration of the cell membrane by various molecules. Contrarily, gram-positive bacteria exhibited higher susceptibility to the destructive influence of ROS due to the absence of an external membrane (Zhu et al. 2018). TiO_2 nanoparticles can effortlessly penetrate the bacterial cell wall and rapidly damage microorganisms (Sundrarajan et al. 2017).

Antimicrobial Mechanism of the TiO_2 Nanoparticles and Composite Materials

Several popular mechanisms have been reported by researchers. As shown in Figure 6.2, first, the electrostatic interaction between the positively charged cationic polymer or free metal ions, and the negatively charged bacterial membrane could destroy the cell integrity. This effect might induce an increase in both hole formation and permeability of the membrane (Davoodbasha et al. 2016, Cano et al. 2016, Li et al. 2010, Wang et al. 2015, Li et al. 2016, Qin et al. 2006). Second, the photocatalytic reaction of nanoparticles exposed to UV and visible lights induced the generation of ROS and hydrogen peroxide on the surfaces of the particles (Carvalho et al. 2014, Yemmireddy and Hung 2015a). This oxidative damage to proteins and deoxyribonucleic acid (DNA) in the cell primarily contributed to the antimicrobial activity of nanoparticles (Yemmireddy and Hung 2015a, Banerjee et al. 2010, Khezerlou et al. 2018). Finally, the detachment of free metal ions disturbed DNA replication (Carvalho et al. 2014, Wei et al. 2014, Khezerlou et al. 2018). The above proposed antimicrobial mechanism illustrated several crucial transport processes, such as the cellular contacted oxidation, the obstructed respiratory chain, and metal ion binding (Carvalho et al. 2014).

Variations in the antimicrobial activity might be due to the size, shape, concentration, clarity, capping agent, and surface chemistry of the nanoparticles, as well as their capacity to discharge free biocidal metal ions (Khezerlou et al. 2018, Song et al. 2010). Yemmireddy and Hung (2015b) reported that the variations in the surface properties of the separate TiO_2 nanocoatings were due to three different binders being used to create them, and it is possible that the type of binders used in the TiO_2 coating could significantly affect the photocatalytic bactericidal characteristics (Carvalho et al. 2014, Heinlaan et al. 2008). On the other hand, Carré et al. (2014) indicated that the antimicrobial photocatalytic action was accompanied by lipid peroxidation, which leads to increase in the membrane fluidity and disturbing cell integrity, while doping TiO_2 nanoparticles containing metal ions can be used as a possible solution to this problem. However, Kamaraj et al. (2015) indicated that the TiO_2 layer exhibited minimal antibacterial qualities in visible light that could be associated with the surface topography and surface chemistry of the layer. Moreover, by producing reductive species, such as superoxide anions, metal ions might also be responsible for biological damage (Samuni et al. 1984). The chelation of antibiotics to metal cations, such as Cu-penicillin (Cressman et al. 2011), Zn-β-lactam (Guo et al. 2017, Si et al. 2008), Cu-aminoglycosides (Kozłowski et al. 2005, Poole 2017), and Zn-tetracycline (Pulicharla et al. 2016) can occur to increase or decrease antibiotic activity (Zawisza et al. 2016, Zarkan et al. 2016, Mohammed and Al-Amery 2016). The contact between metal cations and antibiotics minimizes the action of antibiotics while restricting the bioavailability of metal cations against microorganisms, possibly altering membrane penetration and exhibiting antibacterial activity (Wu et al. 2015, Tan et al. 2015). Furthermore, the collaborative effect of nanoparticles cations is associated with fluoroquinolones, carbapenems, tobramycin, and ceftazidime against *P. aeruginosa*, *Acinetobacter baumannii*, as well as pathogens represented by gram-negative bacteria (Bayroodi and Jalal 2016, Elkhatib and Noreddin 2014).

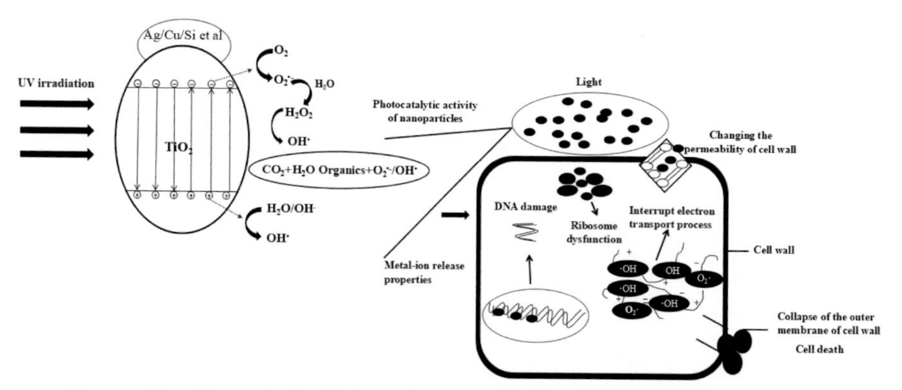

Figure 6.2 The antimicrobial mechanism of TiO_2 nanoparticles and composite nanoparticles during the storage of fruits and vegetables.

Some antimicrobial mechanisms of the TiO_2 nanoparticles composite materials have been discussed. Ye et al. (2017) suggested the following collaborative antimicrobial mechanism consisting of TiO_2/Ag compounds: (i) Ag nanoparticles modified on TiO_2 promotes the production of detrimental ROS; (ii) the continuous release of Ag^+ ions from compounds induces bacterial death in the dark, as well as in the light; (iii) the antibacterial characteristics are improved by the nanoneedle structure and a higher Brunauer-Emmett-Teller (BET) surface area. Typically, an Ag nanoparticle displays significant antimicrobial activity, since it possesses an extensive surface area for the release of Ag^+, while the TiO_2 layer presents a permeable structure suitable for the Ag nanoparticle, promoting consistent Ag^+ release. The antibacterial action of Ag relies on the Ag^+ to breach the membrane of the cell by binding with several of its donor groups, referring to active functional groups, including thiol, hydroxyl, and amino groups, which interact with Ag nanoparticle via chelation (Yuranova et al. 2003). Furthermore, the synthesized nanoparticles caused the disorganization of the bacterial membrane and leakage of cytoplasmic contents (Mei et al. 2013). The embedded Ag nanoparticles accumulated electrons, which were photogenerated in the conduction band of the TiO_2. This process produced holes in the valence of the TiO_2 and prevented the electrons from reconnecting to the holes, inducing the increased quantum efficiency of the photochemical reaction (Wei et al. 2014, Arellano et al. 2011, Manzl et al. 2004).

Arellano et al. (2011) demonstrated that the photocatalytic and textural properties of F Fe-TiO_2 5** (Fe-TiO_2 sol-gel thin film, which the precursory Fe-TiO_2 films, containing 5 wt% Fe, were thermally treated at 800°C) system are responsible for the photoactivity of the thin film, which presents a complicated texture consisting of holes created by microcrystalline material. Moreover, it is noteworthy to mention that the superimposition between the conduction bands and valence of this system is created by the imperfections, oxygen deficiencies, Fe levels, and annealing of the Fe-TiO_2 samples. Higher temperatures induce a decline in the energy band gap (Eg) value, directed at a lower energy level, and metallic conduction is initiated when the previous bands overlap. The formation of •OH radicals results from $e^- - h^+$ pairs, which are generated by visible light irradiation of the photocatalysts, and can attack the bacterial membranes adsorbed on the thin film. The F Fe-TiO_2 3** (Fe-TiO_2 sol-gel thin film, which the precursory Fe-TiO_2 films, containing 3 wt% Fe, were thermally treated at 800°C) material displays lower activity than the F Fe-TiO_2 5**, which is related to surface texture variations, exhibiting lower bacterial agglutination area per unit.

Chen et al. (2019) compared pure TiO_2 and the anatase-TiO_2@SiO_2 hybrid material, and found the latter to be more useful in increasing the number of the surface hydroxyl groups, encouraging the formation of ROS, such as –OH (Nilchi et al. 2010), as well as further enhancement of the antimicrobial performance (Torralvo et al. 2018). Furthermore, excellent adsorption characteristics were displayed by the TiO_2@SiO_2 samples. Therefore, an extensive number of bacteria were adsorbed on the specimens, as indicated by SEM images of the microstructures of the bacteria. The extensive specified surface area of the TiO_2@SiO_2 hybrid material accommodated the adsorption of large bacteria, leading to interaction with the generated ROS, causing sterilization.

Devi et al. (2010) illustrated that photocatalytic activity was improved by selectively doping the TiO_2 semiconductor with metal ions, leading to suppressed recombination of the charge carrier while modifying the interfacial rate of the charge transfer. As indicated by Manzl et al. (2004), Cu ions caused further oxidative damage to bacterial cells via SOD anions and hydrogen peroxide free radicals, which were formed by TiO_2 following exposure to sunlight. Wang et al. (2016) indicated that the boron and cerium co-doped (B/Ce)-TiO_2 nanomaterial displayed excellent antibacterial characteristics, which could be primarily attributed to the narrow band gap, a more extensive specific surface area, an increase in the active surface oxygen species, as well as the antibacterial influence of boric acid. Moreover, according to Wanag et al. (2018), the high antimicrobial efficacy of reduced graphene oxide nanomaterials could be ascribed to their particular surface properties, as well as the efficient segregation of the charge during the electron transfer. Furthermore, Zhang et al. (2017) observed alkaline phosphatase leakage during the antibacterial treatment *E. coli* cells using the CT film. This enzyme leakage indicated that the bactericidal activity of the CT film impacted the cell membrane structure, altering its permeability and inducing the escape of intracellular material, such as nucleotides.

The efficacy of the photocatalytic qualities and antimicrobial action of TiO_2 relies on (i) creating nanostructured TiO_2 particles or compounds that are stable; (ii) extending the excitation wavelength into the visible light area to facilitate the formation of electron-hole pairs, and (iii) reducing the recombination rate on these newly established electron-hole carriers. Various techniques can be employed to enhance photocatalytic productivity, such as extending the TiO_2 surface area by customizing the size of the particles pore-size dispersions, creating faulty structures to facilitate space-charge segregation using metal dopants, as well as modifying the TiO_2 surface with a metal or another semiconductor (Dawson and Kamat 2001, Yu et al. 2003).

METAL-ION RELEASE PROPERTIES OF TIO_2 NANOPARTICLES AND COMPOSITE MATERIALS

Free metal-ions released from the surface of carriers played an antimicrobial role in eradicating microorganisms on the exterior of agricultural products (Wei et al. 2014). As shown in Figure 6.3, nanoparticles enter cells by energy-dependent processes, and rapidly confine to vesicular structures endosomes and lysosomes. Acidic lysosomal pH triggers a lysosome-enhanced Trojan horse effect that causes the abundant cellular internalization of the nanoparticles via active processes, which enhances the release of relatively toxic ions (e.g., Ti, Ag, Au ions). Significant amounts of intracellularly leaked ions may then exert ion-specific toxicity (e.g., enzyme depletion/inactivation, protein denaturation, etc.) against particular cellular targets (e.g., mitochondria, RER) and/or lysosomal damage/dysfunction, resulting in ROS level increase, apoptosis, DNA and membrane damage. According to Salarbashi et al. (2018), no TiO_2 was detected with inductively coupled plasma optical emission spectroscopy (ICP-OES) in bread samples covered with a SSPS/TiO_2 film and stored for six months. A minuscule amount of TiO_2 released from the NC films was observed in water. More importantly, long-term exposure of cells to uncoated nanoparticles elicited that the presence of TiO_2 nanoparticles were in the plasma membrane of the epithelial cell line.

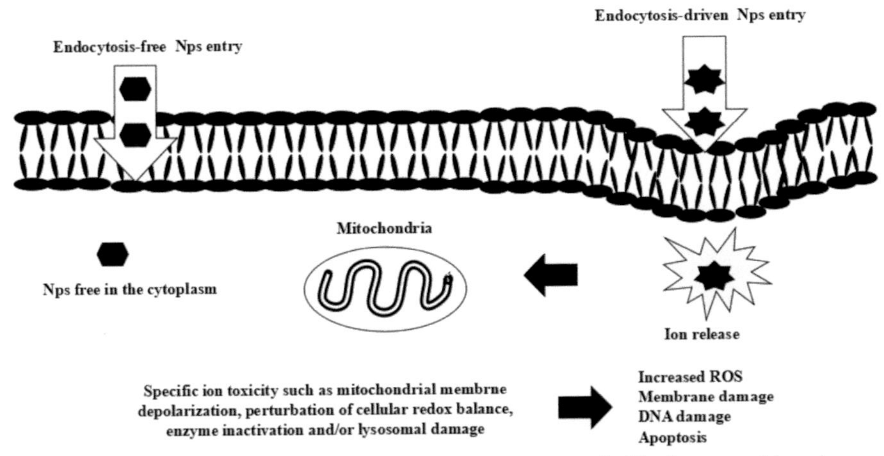

Figure 6.3 Ion release and nanoparticles induced general effectiveness with active internalization.

The release of nanoparticles or metal-ions into a system is affected by various factors. These factors include, but are not limited to the microstructure of the coating carrier, the particle, and ion diffusion to the medium via the polymer, medium migration to the polymer matrix and its expansion, as well as the polymer solubility in the medium phase (Cano et al. 2016). The film compositions and initial nanoparticles concentrations also affect the release of the nanoparticles (Huang et al. 2017). This may further modify the differences in antimicrobial activity due to variations on the surfaces of free ions or nanoparticles that are thought to be responsible for inhibiting the microorganisms.

EFFECTIVENESS OF TIO_2 NANOPARTICLES AND COMPOSITE MATERIALS ON FRUITS AND VEGETABLES

TiO_2 nanoparticles and composite nanoparticles could be used as antimicrobial agents in the coating film for increasing the storage quality of food, especially fruits and vegetables. The preservation mechanism and effectiveness, including gas modification, delayed ripening and decay controlled properties of TiO_2 nanoparticles, and composite nanoparticles under UV irradiation in the package of fruits and vegetables were indicated in Figure 6.4. As indicated in Figure 6.4, under the irradiation of UV light, ethylene (C_2H_4) gas in the package could decompose to CO_2 and H_2O. Then, the concentrations of CO_2 and C_2H_4 in the package could increase and decrease, respectively. This gas modification function could retard the respiration rate, delay ripening, and control water loss of packed fruits and vegetables (Xing et al. 2019). Roilo et al. (2018) established that the incorporation of TiO_2 nanoparticles decreased the film permeability, which might further affect the concentration of gases in the package. The respiration rate of fruits in the packaging system is related to the concentrations of CO_2 and O_2, and the consumed speed of nutrients, such as vitamin C, sugar, and organic acid in pulps. Moreover, Maneerat and Hayata (2006) evaluated the antimicrobial efficacy of TiO_2-coated films and indicated a decline in *Penicillium* rot and brown lesions in lemons. Chawengkijwanich and Hayata (2008) reported similar results, discovering that the film coated with TiO_2 reduced the microbial contamination on the surfaces of the fresh food items, therefore, minimizing the possibility of bacterial growth on freshly harvested fruits and vegetables. The TiO_2 nanoparticles and composite nanoparticles in coating materials on the produce surface might exhibit excellent antimicrobial activities, especially induced by UV light, against pathogens in the packages and grown on the surface of fresh products. Moreover, the preservation effectiveness might be due to the active properties of some coating materials as carriers for nanoparticles, such as CS. Incorporation of TiO_2 nanoparticles had displayed some effectiveness in the gas permeability of carrier coating film (Xing et al. 2019). This combined effectiveness might keep the firmness, inside color, and thus maintain a good quality and an increased shelf life of treated fruit products (Xu et al. 2013).

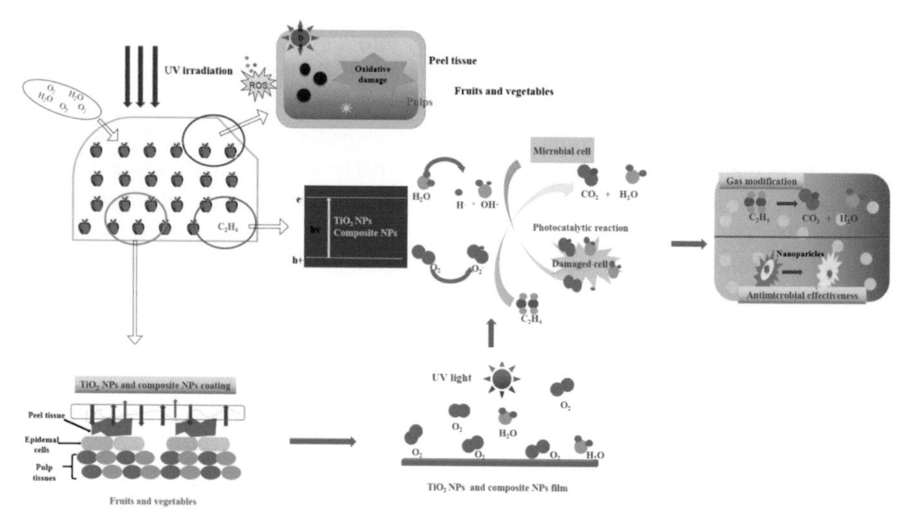

Figure 6.4 The preservation mechanism and effectiveness of TiO_2 nanoparticles and composite nanoparticles under UV irradiation in the package of fruits and vegetables.

Yang et al. (2010) obtained improved physicochemical, sensory, and physiological characteristics in strawberries using low density polyethylene (LDPE) nano-packaging (nano-Ag, kaolin, anatase TiO_2, rutile TiO_2) instead of regular LDPE packaging. These results could primarily be ascribed to the low transmission rate of oxygen, low relative humidity, and exceptional longitudinal strength of the nano-packaging. Bodaghi et al. (2013) showed that significant decrease of mesophilic bacteria and yeast cells in the pear samples packaged by TiO_2-LDPE film under fluorescent light irradiation was observed in comparison to the samples in LDPE film during storage. Furthermore, Li et al. (2017) demonstrated that nano-TiO_2-LDPE packaging reduced weight loss, which maintained the firmness and titratable acid content of strawberries, therefore, minimizing the degradation of the fruit. Zhang et al. (2017) utilized TiO_2 nano-powder during the development of a new compound CS matrix film displaying satisfactory bactericidal action, employing it to preserve red grapes, protecting the fruit from bacterial assault, therefore, achieving shelf-life prolongation in ambient conditions.

Xu et al. (2017) found that the NC coated samples exhibited a lower weight loss and maintained a better appearance, while polyphenol oxidase (PPO) activity was inhibited by the NC coating containing GO and CS-loaded TiO_2. Superoxide dismutase activity in the fruits coated with three different TiO_2 ratios of GO@CS@TiO_2 films was higher than the untreated samples and, consequently, presented considerable potential for use in the food preservation industry. However, Kaewklin et al. (2018) suggested that the ripening process of the tomatoes was delayed by exposure to the CT NC film, which degraded the ethylene. Furthermore, the inherent antimicrobial characteristics of CS (Xing et al. 2019) and the photocatalytic antimicrobial activity of TiO_2 (Bodaghi et al. 2013, Lin et al. 2015) were possibly responsible for the lack of fungal infection in the CT and CS samples. Moreover, de Chiara et al. (2015) discovered that the TiO_2 in materials containing composite titania/silica materials was responsible for the photocatalytic destruction of ethylene, inhibiting the ripening of mature green tomatoes. More recently, Lourenco et al. (2017) examined papaya, with the results indicating that higher ethylene deterioration (100% decomposition) was obtained with glass supported TiO_2 and exposure to UVA (365 nm) irradiation with 6 $\mu W/cm^2$. However, the skin of the papaya displayed a disagreeable darker coloration than the fruit in the control sample, due to the application of UVA. Consequently, since TiO_2 photocatalysis (TPC) is responsible for ethylene deterioration, it can possibly be utilized during the commercial storage of ethylene-sensitive commodities, as well as long-distance transportation. De Chiara et al. (2015) examined the impact of ethylene elimination in the ripening process of mature green tomatoes. The study revealed that the highest level of ethylene elimination (74%) was achieved with modified-TiO_2 (containing 80% TiO_2 and 20% SiO_2), along with a 3 hour exposure to UV irradiation. Furthermore, Park et al. (2016) found that combining TPC and UVC obtained the most substantial bacterial reduction in both external and internal murine norovirus 1. Ye et al. (2009) simulated the process involving the photocatalytic deterioration of ethylene in post-harvest produce in a cold storage environment using a photoreactor containing activated carbon fiber-supported TiO_2, verifying that ethylene could be eliminated with TPC treatment.

Several recent studies highlighted the possibility of utilizing a combination of high hydrostatic pressure (HHP) and TPC for the deactivation of microorganisms found in fruit juice (Chai et al. 2014, Shahbaz et al. 2016, Yoo et al. 2015), such as Angelica keiskei juice, orange juice, and apple juice. Yoo et al. (2015) examined the presence of microbial pathogens and the deactivation of yeast in commercial apple juice by employing a combination of HHP (500 MPa for 1 min) and TPC (8.45 J/cm^2 for 20 min) to accomplish the complete decontamination of *S. aureus* and *L. monocytogenes*. Shahbaz et al. (2016) succeeded in totally eliminating *E. coli O157: H7* from orange juice by applying TPC (17.2 mW/cm^2) for 20 minutes followed by a 1 minute HHP (400 MPa) treatment. Furthermore, Chai et al. (2014) established that a 20 minute combined treatment with TPC and UV irradiation (254 nm; 25 mW/cm^2) substantially reduced the *Pseudomonas, B. cereus*, and *Coliform* bacteria counts, as well as yeasts and molds in freshly squeezed Angelica keiskei juice.

FUTURE TRENDS OF TIO_2 NANOPARTICLES AND COMPOSITE NANOPARTICLES

The antibacterial activity and mechanism of TiO_2 nanoparticles and composite nanoparticles are popular research topics, while fewer articles on the antifungal activity of TiO_2 nanoparticles and composite nanoparticles in the presence of UV or visible light have been published. The modification of TiO_2 with other nanoparticles can improve the photo activity and antimicrobial activity. However, interest in the toxicity of TiO_2 nanoparticles and composite nanoparticles with the potential to induce allergic reactions, especially in freshly cut products due to the release and migration of metal ions is increasing recently. Therefore, further researches on these subjects are essential.

ACKNOWLEDGMENTS

This work is supported by the Science and technology support program of Sichuan (2019YFN0174, 2018NZ0090, 2019NZZJ0028 and 2017NFP0030), Science and technology support program of Yibin [2018ZSF002], Chengdu Science and Technology Project-key research and development program (2019-YF05-00628-SN and 2019-YF05-00190-SN), Innovation Team Construction Program of Sichuan Education Department (15TD0017).

References

Abdel-Fatah, W.I., M.M. Gobara, S.F.M. Mustafa, G.W. Ali and O.W. Guirguis. 2016. Role of silver nanoparticles in imparting antimicrobial activity of titanium dioxide. Mater. Lett. 179: 190–193.

Abu-Dalo, M., A. Jaradat, B.A. Albiss and N.A.F. Al-Rawashdeh. 2019. Green synthesis of TiO_2 nanoparticles/pristine pomegranate peel extract nanocomposite and its antimicrobial activity for water disinfection. J. Environ. Chem. Eng. 7: 103370.

Achachlouei, B.F. and Y. Zahedi. 2018. Fabrication and characterization of CMC-based nanocomposites reinforced with sodium montmorillonite and TiO_2 nanomaterials. Carbohyd. Polym. 199: 415–425.

Adibkia, K., M. Alaei-Beirami, M. Barzegar-Jalali, G. Mohammadi and M.S. Ardestani. 2012. Evaluation and optimization of factors affecting novel diclofenac sodium-eudragit RS100 nanoparticles. Afr. J. Pharm. Pharmacol. 6: 941–947.

Alizadeh-Sani, M., A. Khezerlou and A. Ehsani. 2018. Fabrication and characterization of the bionanocomposite film based on whey protein biopolymer loaded with TiO_2 nanoparticles, cellulose nanofibers and rosemary essential oil. Ind. Crop. Prod. 124: 300–315.

AL-Jawad, S.M.H., A.A. Taha and M.M. Salim. 2017. Synthesis and characterization of pure and Fe doped TiO_2 thin films for antimicrobial activity. Optik. 142: 42–53.

André, R.S., C.A. Zamperini, E.G. Mima, V.M. Longo, A.R. Albuquerque, J.R. Sambrano, et al. 2015. Antimicrobial activity of TiO_2:Ag nanocrystalline heterostructures: Experimental and theoretical insights. Chem. Phys. 459: 87–95.

Arellano, U., M. Asomoza and F. Ramirez. 2011. Antimicrobial activity of Fe-TiO_2 thin film photocatalysts. J. Photoch. Photobio. A. 222: 159–165.

Banerjee, M., S. Mallick, A. Paul, A. Chattopadhyay and S.S. Ghosh. 2010. Heightened reactive oxygen species generation in the antimicrobial activity of a three component iodinated chitosan-silver nanoparticle composite. Langmuir. 26: 5901–5908.

Bayroodi, E. and R. Jalal. 2016. Modulation of antibiotic resistance in *Pseudomonas aeruginosa* by ZnO nanoparticles. Iran. J. Microbiol. 8: 85.

Bera, R., S. Mandal and C.R. Raj. 2014. Antimicrobial activity of flfluorescent Ag nanoparticles. Lett. Appl. Microbiol. 58: 520–526.

Besinis, A., T. De Peralta and R.D. Handy. 2014. The antibacterial effects of silver, titanium dioxide and silica dioxide nanoparticles compared to the dental disinfectant chlorhexidine on *Streptococcus mutans* using a suite of bioassays. Nanotoxicology. 8: 1–16.

Bodaghi, H., Y. Mostofi, A. Oromiehie, Z. Zamani, B. Ghanbarzadeh, C. Costa, et al. 2013. Evaluation of the photocatalytic antimicrobial effects of a TiO_2 nanocomposite food packaging film by *in vitro* and *in vivo* tests. LWT-Food Sci. Technol. 50: 702–706.

Buzea, C., I.I. Pacheco and K. Robbie. 2007. Nanomaterials and nanoparticles: Sources and toxicity. Biointerphases. 2: MR17–MR71.

Cano, A., M, Cháfer, A. Chiralt and C. González-Martínez. 2016. Development and characterization of active films based on starch-PVA, containing silver nanoparticles. Food Packag. Shelf Life. 10: 16–24.

Cano, L., E. Pollet, L. Avérous and A. Tercjak. 2017. Effect of TiO_2 nanoparticles on the properties of thermoplastic chitosan-based nano-biocomposites obtained by mechanical kneading. Composites: Part A.-Appl. S.93: 33–40.

Carré, G., E. Hamon, S. Ennahar, M. Estner, M.C. Lett, P. Horvatovich, et al. 2014. TiO_2 photocatalysis damages lipids and proteins in *Escherichia coli*. Appl. Environ. Microbiol. 80: 2573–2581.

Carvalho, P., P. Sampaio, S. Azevedo, C. Vaz, J.P. Espinós, V. Teixeira, et al. 2014. Influence of thickness and coatings morphology in the antimicrobial performance of zinc oxide coatings. Appl. Surf. Sci. 307: 548–557.

Castillo, L.A., O.V. López, J. Ghilardi, M.A. Villar, S.E. Barbosa and M.A. García. 2015. Thermoplastic starch/talc bionanocomposites. Influence of particle morphology on final properties. Food Hydrocoll. 51: 432–440.

Chai, C., J. Lee, Y. Lee, S. Na and J. Park. 2014. A combination of TiO_2-UV photocatalysis and high hydrostatic pressure to inactivate *Bacillus cereus* in freshly squeezed *Angelica keiskei* juice. LWT-Food Sci. Technol. 55: 104–109.

Chawengkijwanich, C. and Y. Hayata. 2008. Development of TiO_2 powder-coated food packaging film and its ability to inactivate *Escherichia coli in vitro* and in actual tests. Int. J. Food Microbiol. 123: 288–292.

Chen, Y.C. and K.P. Yu. 2017. Enhanced antimicrobial efficacy of thermal-reduced silver nanoparticles supported by titanium dioxide. Colloid. Surface. B. 154: 195–202.

Chen, Y., X. Tang, X. Gao, B. Zhang, Y. Luo and X. Yao. 2019. Antimicrobial property and photocatalytic antibacterial mechanism of theTiO_2-doped SiO_2 hybrid materials under ultraviolet-light irradiation and visible-light irradiation. Ceram. Int. 45: 15505–15513.

Chung, C.J., H.I. Lin, C.M. Chou, P.Y. Hsieh, C.H. Hsiao, Z.Y. Shi, et al. 2009. Inactivation of *Staphylococcus aureus* and *Escherichia coli* under various light sources on photocatalytic titanium dioxide thin film. Surf. Coat. Tech. 203: 1081–1085.

Cressman, W., E. Sugita, J. Doluisio and P. Niebergall. 2011. Complexation of penicillins and penicilloic acids by cupric ion. J. Pharm. Pharmacol. 19: 774.

Davoodbasha, M., S.C. Kim, S.Y. Lee and J.W. Kim. 2016. The facile synthesis of chitosan-based silver nano-biocomposites via a solution plasma process and their potential antimicrobial efficacy. Arch. Biochem. Biophys. 605: 49–58.

Dawson, A. and P.V. Kamat. 2001. Semiconductor-metal nanocomposites. Photoinduced fusion and photocatalysis of gold-capped TiO_2 (TiO_2/gold) nanoparticles. J. Phys. Chem. B. 105: 960–966.

de Chiara, M.L.V., S. Pal, A. Licciulli, M.L. Amodio and G. Colelli. 2015. Photocatalytic degradation of ethylene on mesoporous TiO_2/SiO_2 nanocomposites: Effects on the ripening of mature green tomatoes. Biosyst. Eng. 132: 61–70.

Devi, L.G., N. Kottam, B.N. Murthy, and S.G. Kumar. 2010. Enhanced photocatalytic activity of transition metal ions Mn^{2+}, Ni^{2+} and Zn^{2+} doped polycrystalline titania for the degradation of aniline blue under UV/solar light. J. Mol. Catal. A-Chem. 328: 44–52.

Dong, P., F. Yang, X. Cheng, Z. Huang, X. Niea, Y. Xiao, et al. 2019. Plasmon enhanced photocatalytic and antimicrobial activities of Ag-TiO_2 nanocomposites under visible light irradiation prepared by DBD cold plasma treatment. Mat. Sci. Eng. C-Mater. 96: 197–204.

Durango-Giraldo, G., A. Cardona, J.F. Zapata, J.F. Santa and R. Buitrago-Sierra. 2019. Titanium dioxide modified with silver by two methods for bactericidal applications. Heliyon. 5: 01608.

Elkhatib, W. and A. Noreddin. 2014. *In vitro* antibiofilm efficacies of different antibiotic combinations with zinc sulfate against *Pseudomonas aeruginosa* recovered from hospitalized patients with urinary tract infection. Antibiotics. 3: 64–84.

El-Shahawy, A.A.G., F.I. Abo El-Ela, N.A. Mohamed, Z.E. Eldine and W.M.A. El Rouby. 2018. Synthesis and evaluation of layered double hydroxide/doxycycline and cobalt ferrite/chitosan nanohybrid efficacy on gram positive and gram-negative bacteria. Mater. Sci. Eng. C Mater. Biol. Appl. 91: 361–371.

Fellahi, O., R.K. Sarma, M.R. Das, R. Saikia, L. Marcon, Y. Coffinier, et al. 2013. The antimicrobial effect of silicon nanowires decorated with silver and copper nanoparticles. Nanotechnology. 24: 495101.

Gericke, M. and A. Pinches. 2006. Biological synthesis of metal nanoparticles. Hydrometallurgy. 83: 132–140.

Goudarzi, V. and I. Shahabi-Ghahfarrokhi. 2018. Photo-producible and photo-degradable starch/TiO_2 bionanocomposite as a food packaging material: Development and characterization. Int. J. Biol. Macromol. 106: 661–669.

Guo, Y., D.C.W. Tsang, X. Zhang and X. Yang. 2017. Cu (II)-catalyzed degradation of ampicillin: Effect of pH and dissolved oxygen. Environ. Sci. Pollut. R. 25: 4279–4288.

Haghighi, F., S. Roudbar Mohammadi, P. Mohammadi, S. Hosseinkhani and R. Shipour. 2013. Antifungal activity of TiO_2 nanoparticles and EDTA on *Candida albicans* biofilms. Infect. Epidemiol. Immunobiol. 1: 33–38.

Heinlaan, M., A. Ivask, I. Blinova, H.C. Dubourguier and A. Kahru. 2008. Toxicity ofnanosized and bulk ZnO, CuO and TiO_2 to bacteria vibrio fischeri and crus-taceans daphnia magna and thamnocephalus platyurus. Chemosphere. 71: 1308–1316.

Hou, X., J. Ma, D. Li, X. Wang, Y. Shen and T. Yu. 2013. Influence of V+-implantation on structural, chemical, optical and nanomechanical properties of TiO_2 films. Vacuum. 89: 147–152.

Huang, S., B. Guild, S. Neethirajan, P. Therrien, L.T. Lim and K. Warriner. 2017. Antimicrobial coatings for controlling *Listeria monocytogenes* based on polylactide modified with titanium dioxide and illuminated with UV-A. Food Control 73: 421–425.

Jha, A.K., K. Prasad and A.R. Kulkarni. 2009. Synthesis of TiO_2 nanoparticles using microorganisms. Colloids Surf. B Biointerfaces. 71: 226–229.

Kadam, D.M., M. Thuna, G. Srinivasan, S. Wang, M.R. Kessler, D. Grewell, et al. 2017. Effect of TiO_2 nanoparticles on thermo-mechanical properties of cast zein protein films. Food Packag. Shelf Life. 13: 35–43.

Kaewklin, P., U. Siripatrawan, A. Suwanagul and Y.S. Lee. 2018. Active packaging from chitosan-titanium dioxide nanocomposite film for prolonging storage life of tomato fruit. Int. J. Biol. Macromol. 112: 523–529.

Kamaraj, K., R.P. George, B. Anandkumar, N. Parvathavarthini and U.K. Mudali. 2015. A silver nanoparticle loaded TiO_2 nanoporous layer for visible light induced antimicrobial applications. Bioelectrochemistry. 106: 290–297.

Karthikeyan, K.T., A. Nithya and K. Jothivenkatachalam. 2017. Photocatalytic and antimicrobial activities of chitosan-TiO_2 nanocomposite. Int. J. Biol. Macromol. 104: 1762–1773.

Khezerlou, A., M. Alizadeh-Sani, M. Azizi-Lalabadi and A. Ehsani. 2018. Nanoparticles and their antimicrobial properties against pathogens including bacteria, fungi, parasites and viruses. Microb. Pathogenesis 123: 505–526.

Kirthi, A.V., A.A. Rahuman, G. Rajakumar, S. Marimuthu, T. Santhoshkumar, C. Jayaseelan, et al. 2011. Biosynthesis of titanium dioxide nanoparticles using bacterium *Bacillus subtilis*. Mater. Lett. 65: 2745–2747.

Kozłowski, H., T. Kowalik-Jankowska and M. Jeżowska-Bojczuk. 2005. Chemical and biological aspects of Cu^{2+} interactions with peptides and aminoglycosides. Coord. Chem. Rev. 249: 2323–2334.

Li, J.H., R.Y. Hong, M.Y. Li, H.Z. Li, Y. Zheng and J. Ding. 2009a. Effects of ZnO nanoparticles on the mechanical and antibacterial properties of polyurethane coatings. Prog. Org. Coat. 64: 504–509.

Li, H., F. Li, L. Wang, J. Sheng, Z. Xin, L. Zhao, et al. 2009b. Effect of nano-packing on preservation quality of Chinese jujube (*Ziziphus jujuba* Mill. var. *inermis* (*Bunge*) *Rehd*). Food Chem. 114: 547–552.

Li, W.R., X.B. Xie, Q.S. Shi, H.Y. Zeng, Y.S. Ou-Yang and Y.B. Chen. 2010. Antibacterial activity and mechanism of silver nanoparticles on *Escherichia coli*. Appl. Microbiol. Biotechnol. 85: 1115–1122.

Li, J., Y. Wu and L. Zhao. 2016. Antibacterial activity and mechanism of chitosan with ultra high molecular weight. Carbohydr. Polym. 148: 200–205.

Li, D., Q. Ye, L. Jiang and Z. Luo. 2017. Effects of nano-TiO_2-LDPE packaging on postharvest quality and antioxidant capacity of strawberry (*Fragaria ananassa* Duch.) stored at refrigeration temperature. J. Sci. Food Agr. 97: 1116–1123.

Lin, B., Y. Luo, Z. Teng, B. Zhang, B. Zhou and Q. Wang. 2015. Development of silver/titanium dioxide/chitosan adipate nanocomposite as an antibacterial coating for fruit storage. LWT-Food Sci. Technol. 63: 1206–1213.

Liu, C., L. Geng, Y.F. Yu, Y. Zhang, B. Zhao and Q. Zhao. 2018. Mechanisms of the enhanced antibacterial effect of Ag-TiO_2 coatings. Biofouling 34: 190–199.

Lourenco, R., A.A.N. Linhares, A.V. de Oliveira, M.G. da Silva, J.G. de Oliveira and M.C. Canela. 2017. Photodegradation of ethylene by use of TiO_2 sol-gel on polypropylene and on glass for application in the postharvest of papaya fruit. Environ. Sci. Pollut. R. 24: 6047–6054.

Magesan, P., P. Ganesan and M.J. Umapathy. 2016. Ultrasonic-assisted synthesis of doped TiO_2 nanocomposites: Characterization and evaluation of photocatalytic and antimicrobial activity. Optik. 127: 5171–5180.

Maneerat, C. and Y. Hayata. 2006. Antifungal activity of TiO_2 photocatalysis against Penicillium expansum *in vitro* and in fruit tests. Int. J. Food Microbiol. 107: 99–103.

Manzl, C., J. Enrich, H. Ebner, R. Dallinger and G. Krumschnabel. 2004. Copper-induced formation of reactive oxygen species causes cell death and disruption of calcium homeostasis in trout hepatocytes. Toxicology. 196: 57–64.

Mei, L., Z. Lu, W. Zhang, Z. Wu, X. Zhang, Y. Wang, et al. 2013. Bioconjugated nanoparticles for attachment and penetration into pathogenic bacteria. Biomaterials. 34: 10328–10337.

Miao, Y., X. Xu, K. Liu and N. Wang. 2017. Preparation of novel Cu/TiO_2 mischcrystal composites and antibacterial activities for *Escherichia coli* under visible light. Ceram. Int. 43: 9658–9663.

Mohammadi, G., H. Valizadeh, M. Barzegar-Jalali, F. Lotfifipour, K. Adibkia, M. Milani, et al. 2010. Development of azithromycin-PLGA nanoparticles: Physicochemical characterization and antibacterial effect against Salmonella typhi. Colloid. Surface. B. 80: 34–39.

Mohammed, S.K. and M.H. Al-Amery. 2016. Metal complexes of mixed ligands (quinolone antibiotics and α-aminonitrile derivatives) their applications: An update with Mn(II), Cu(II) and Cr(III) ions and study the biological activity. Org. Chem.: Ind. J. 12: 29–45.

Montaser, A.S., A.R. Wassel and O.N. Al-Shaye'a. 2019. Synthesis, characterization and antimicrobial activity of Schiff bases from chitosan and salicylaldehyde/TiO_2 nanocomposite membrane. Int. J. Biol. Macromol. 124: 802–809.

Naghibi, S., S. Vahed, O. Torabi, A. Jamshidi and M.H. Golabgir. 2015. Exploring a new phenomenon in the bactericidal response of TiO_2 thin films by Fe doping: Exerting the antimicrobial activity even after stoppage of illumination. Appl. Surf. Sci. 327: 371–378.

Nilchi, A., S. Janitabar-Darzi, A.R. Mahjoub and S. Rasouli-Garmarodi. 2010. New TiO_2/SiO_2 nanocomposites-phase transformations and photocatalytic studies, Colloid. Surf. Physicochem. Eng. Asp. 361: 25–30.

Pal, B., I. Singh, K. Angrish, R. Aminedi and N. Das. 2012. Rapid photokilling of gram-negative *Escherichia coli* bacteria by platinum dispersed titania nanocomposite films. Mater. Chem. Phys. 136: 21–27.

Park, D., H.M. Shahbaz, S.H. Kim, M. Lee, W. Lee, J.W. Oh, et al. 2016. Inactivation efficiency and mechanism of UV-TiO_2 photocatalysis against murine norovirus using a solidified agar matrix. Int. J. Food Microbiol. 238: 256–264.

Poole, K. 2017. At the nexus of antibiotics and metals: The impact of Cu and Zn on antibiotic activity and resistance. Trends Microbiol. 25: 820–832.

Pulicharla, R., K. Hegde, S.K. Brar and R.Y. Surampalli. 2016. Tetracyclines metal complexation: Significance and fate of mutual existence in the environment. Environ. Pollut. 221: 1–14.

Qin, C., H. Li, Q. Xiao, Y. Liu, J. Zhu and Y. Du. 2006. Water-solubility of chitosan and its antimicrobial activity. Carbohyd. Polym. 63: 367–374.

Qu, L., G. Chen, S. Dong, Y. Huo, Z. Yin, S. Li, et al. 2019. Improved mechanical and antimicrobial properties of zein/chitosan films by adding highly dispersed nano-TiO_2. Ind. Crop. Prod. 130: 450–458.

Rajakumar, G., A.A. Rahuman, S.M. Roopan, V.G. Khanna, G. Elango, C. Kamaraj, et al. 2012. Fungus-mediated biosynthesis and characterization of TiO_2 nanoparticles and their activity against pathogenic bacteria. Spectrochim. Acta Mol. Biomol. Spectrosc. 91: 23–29.

Ramya, S., S.D. Ruth Nithila, R.P. George, D.N.G. Krishna, C. Thinaharan and U. Kamachi Mudali. 2013. Antibacterial studies on Eu-Ag codoped TiO_2 surfaces. Ceram. Int. 39: 1695–1705.

Rao, T.N., Riyazuddin, P. Babji, N. Ahmad, R.A. Khan, I. Hassan, et al. 2019. Green synthesis and structural classification of *Acacia nilotica* mediated-silver doped titanium oxide (Ag/TiO_2) spherical nanoparticles: Assessment of its antimicrobial and anticancer activity. Saudi J. Biol. Sci. 26: 1385–1391.

Rauta, A.V., H.M. Yadavb, A. Gnanamanic, S. Pushpavanamd and S.H. Pawara. 2016. Synthesis and characterization of chitosan-TiO_2:Cu nanocomposite and their enhanced antimicrobial activity with visible light. Colloid. Surface. B. 148: 566–575.

Roilo, D., C.A. Maestri, M. Scarpa, P. Bettotti and R. Checchetto. 2018. Gas barrier and optical properties of cellulose nanofiber coatings with dispersed TiO_2 nanoparticles. Surf. Coat. Tech. 343: 131–137.

Salarbashi, D., M. Tafaghodi and B.S.F. Bazzaz. 2018. Soluble soybean polysaccharide/TiO_2 bionanocomposite film for food application. Carbohyd. Polym. 186: 384–393.

Samuni, A., M. Chevion and G. Czapski. 1984. Roles of copper and O_2 in the radiation induced inactivation of T7 bacteriophage. Radiat. Res. 99: 562–572.

Sani, M.A., A. Ehsani and M. Hashemi. 2017. Whey protein isolate/cellulose nanofibre/ TiO_2 nanoparticle/rosemary essential oil nanocomposite film: its effect on microbial and sensory quality of lamb meat and growth of common foodborne pathogenic bacteria during refrigeration. Int. J. Food Microbiol. 251: 8–14.

Sathishkumar, G., C. Gobinath, K. Karpaga, V. Hemamalini, K. Premkumar and S. Sivaramakrishnan. 2012. Phytosynthesis of silver nanoscale particles using *Morinda citrifolia* L. and its inhibitory activity against human pathogens. Colloid. Surf. B. 95: 235–240.

Seil, J.T. and T.J. Webster. 2012. Antimicrobial applications of nanotechnology: Methods and literature. Int. J. Nanomed. 7: 2767.

Shahabi-Ghahfarrokhi, I., F. Khodaiyan, M. Mousavi and H. Yousefi. 2015. Green bionanocomposite based on kefiran and cellulose nanocrystals produced from beer industrial residues. Int. J. Biol. Macromol. 77: 85–91.

Shahbaz, H.M., S. Yoo, B. Seo, K. Ghafoor, J.U. Kim, D.U. Lee, et al. 2016. Combination of TiO_2-UV photocatalysis and high hydrostatic pressure to inactivate bacterial pathogens and yeast in commercial apple juice. Food Bioprocess Tech. 9: 182–190.

Si, H., J. Hu, Z. Liu and Z. Zeng. 2008. Antibacterial effect of oregano essential oil alone and in combination with antibiotics against extended-spectrum β-lactamase-producing *Escherichia coli*. Pathog. Dis. 53: 190–194.

Siripatrawan, U. and P. Kaewklin. 2018. Fabrication and characterization of chitosan-titanium dioxide nanocomposite film as ethylene scavenging and antimicrobial active food packaging. Food Hydrocolloid. 84: 125–134.

Skorb, E.V., L.I. Antonouskaya, N.A. Belyasova, D.G. Shchukin, H. Mohwald and D.V. Sviridov. 2008. Antibacterial activity of thin-film photocatalysts based on metal modified TiO_2 and TiO_2:In_2O_3 nanocomposite. Appl. Catal. B-Environ. 84: 94–99.

Song, W., J. Zhang, J. Guo, J. Zhang, F. Ding, L. Li, et al. 2010. Role of the dissolved zinc ion and reactive oxygen species in cytotoxicity of ZnO nanoparticles. Toxicol. Lett. 199: 389–397.

Soni, S.S., G.S. Dave, M.J. Henderson and A. Gibaud. 2013. Visible light induced cell damage of Gram positive bacteria by N-doped TiO_2 mesoporous thin films. Thin Solid Films. 531: 559–565.

Subhapriya, S. and P. Gomathipriya. 2018. Green synthesis of titanium dioxide (TiO_2) nanoparticles by *Trigonella foenum-graecum* extract and its antimicrobial properties. Microb. Pathog. 116: 215–220.

Sundrarajan, M., K. Bama, M. Bhavani, S. Jegatheeswaran, S. Ambika, A. Sangili, et al. 2017. Obtaining titanium dioxide nanoparticles with spherical shape and antimicrobial properties using *M. citrifolia* leaves extract by hydrothermal method. J. Photoch. Photobio. B. 171: 117–124.

Takai, A. and P.V. Kamat. 2011. Capture, store and discharge. Shuttling photogenerated electrons across TiO_2-silver interface. ACS Nano. 5: 7369–7376.

Tan, Y., Y. Guo, X. Gu and C. Gu. 2015. Effects of metal cations and fulvic acid on the adsorption of ciprofloxacin onto goethite. Environ. Sci. Pollut. Control Ser. 22: 609–617.

Teymourpour, S., A.M. Nafchi and F. Nahidi. 2015. Functional, thermal, and antimicrobial properties of soluble soybean polysaccharide biocomposites reinforced by nano TiO_2. Carbohydr. Polym. 134: 726–731.

Torralvo, M., J. Sanz, I. Sobrados, J. Soria, C. Garlisi, G. Palmisano, et al. 2018. Anatase photocatalyst with supported low crystalline TiO_2: The influence of amorphous phase on the activity. Appl. Catal. B-Environ. 221: 140–151.

Verdier, T., M. Coutand, A. Bertron and C. Roques. 2014. Antibacterial activity of TiO_2 photocatalyst alone or in coatings on *E. coli*: The influence of methodological aspects. Coatings. 4: 670–686.

Wanag, A., P. Rokicka, E. Kusiak-Nejman, J. Kapica-Kozar, R.J. Wrobel, A. Markowska-Szczupak, et al. 2018. Antibacterial properties of TiO_2 modified with reduced graphene oxide. Ecotox. Environ. Safe. 147: 788–793.

Wang, Y., X. Xue and H. Yang. 2014. Modification of the antibacterial activity of Zn/TiO_2 nano-materials through different anions doped. Vacuum 101: 193–199.

Wang, Q., J.H. Zuo, Q. Wang, Y. Na and L.P. Gao. 2015. Inhibitory effect of chitosan on growth of the fungal phytopathogen, *Sclerotinia sclerotiorum*, and sclerotinia rot of carrot. J. Integr. Agr. 14: 691–697.

Wang, Y., Y. Wu, H. Yang, X. Xue and Z. Liu. 2016. Doping TiO_2 with boron or/and ceriumelements: Effects on photocatalytic antimicrobial activity. Vacuum 131: 58–64.

Wang, L., C. Hu and L. Shao. 2017. The antimicrobial activity of nanoparticles: Present situation and prospects for the future. Int. J. Nanomed. 12: 1227–1249.

Wei, X., Z. Yang, S.L. Tay and W. Gao. 2014. Photocatalytic TiO_2 nanoparticles enhanced polymer antimicrobial coating. Appl. Surf. Sci. 290: 274–279.

Wu, D., Z. Huang, K. Yang, D. Graham and B. Xie. 2015. Relationships between antibiotics and antibiotic resistance gene levels in municipal solid waste leachates in Shanghai, China. Environ. Sci. Technol. 49: 4122–4128.

Xiao, G., X. Zhang, W. Zhang, S. Zhang, H. Su and T. Tan. 2015. Visible-light-mediated synergistic photocatalytic antimicrobial effects and mechanism of Ag-nanoparticles@

chitosan-TiO_2 organic-inorganic composites for water disinfection. Appl. Catal. B-Environ. 170: 255–262.

Xie, J. and Y.C. Hung. 2018. UV-A activated TiO_2 embedded biodegradable polymer film for antimicrobial food packaging application. LWT-Food Sci. Technol. 96: 307–314.

Xie, J. and Y.C. Hung. 2019. Methodology to evaluate the antimicrobial effectiveness of UV-activated TiO_2 nanoparticle-embedded cellulose acetate film. Food Control. 106: 106690.

Xing, Y., X. Li, L. Zhang, Q. Xu, Z. Che, W. Li, et al. 2012. Effect of TiO_2 nanoparticles on the antibacterial and physical properties of polyethylene-based film. Prog. Org. Coat. 73: 219–224.

Xing, Y., Q. Xu, S.X. Yang, C. Chen, Y. Tang, S. Sun, et al. 2017. Preservation mechanism of chitosan-based coating with cinnamon oil for fruits storage based on sensor data. Sensors. 16: 1111.

Xing, Y., W. Li, Q. Wang, X. Li, Q. Xu, X. Guo, et al. 2019. Antimicrobial nanoparticles incorporated in edible coatings and films for the preservation of fruits and vegetables. Molecules 24: E1695.

Xu, Q., Y. Xing, Z. Che, T. Guan, L. Zhang, Y. Bai, et al. 2013. Effect of chitosan coating and oil fumigation on the microbiological and quality safety of fresh-cut pear. J. Food Safety. 33: 179–189.

Xu, W., W. Xie, X. Huang, X. Chen, N. Huang, X. Wang, et al. 2017. The graphene oxide and chitosan biopolymer loads TiO_2 for antibacterial and preservative research. Food Chem. 221: 267–277.

Yang, F.M., H.M. Li, F. Li, Z.H. Xin, L.Y. Zhao, Y.H. Zheng, et al. 2010. Effect of nano-packing on preservation quality of fresh strawberry (*Fragaria ananassa* Duch. cv Fengxiang) during storage at 4°C. J. Food Sci. 75: 236–240.

Ye, S., Q. Tian, X. Song and S. Luo. 2009. Photoelectrocatalytic degradation of ethylene by a combination of TiO_2 and activated carbon felts. J. Photoch. Photobiol. A. 208: 27–35.

Ye, J., H. Cheng, H. Li, Y. Yang, S. Zhang, A. Rauf, et al. 2017. Highly synergistic antimicrobial activity of spherical and flower-likehierarchical titanium dioxide/silver composites. J. Colloid Interf. Sci. 504: 448–456.

Yemmireddy, V.K. and Y.-C. Hung. 2015a. Selection of photocatalytic bactericidal titanium dioxide (TiO_2) nanoparticles for food safety applications. Lebensm–WissTechnol. 61: 1–6.

Yemmireddy, V.K. and Y.-C.Hung. 2015b. Effect of binder on the physical stability and bactericidal property of titanium dioxide (TiO_2) nanocoatings on food contact surfaces. Food Control. 57: 82–88.

Yeung, K.L., W.K. Leung, N. Yao and S. Cao. 2009. Reactivity and antimicrobial properties of nanostructured titanium dioxide. Catal. Today. 143: 218–224.

Yoo, S., K. Ghafoor, J.U. Kim, S. Kim, B. Jung, D.U. Lee, et al. 2015. Inactivation of *Escherichia coli* O157:H7 on orange fruit surfaces and in juice using photocatalysis and high hydrostatic pressure. J. Food Protect. 78: 1098–1105.

Yu, J.C., L. Zhang, Z. Zheng and J. Zhao. 2003. Synthesis and characterization of phosphated mesoporous titanium dioxide with high photocatalytic activity. Chem. Mater. 15: 2280–2286.

Yuranova, T., A.G. Rincon, A. Bozzi and S. Parra. 2003. Antibacterial textiles prepared by RF-plasma and vacuum-UV mediated deposition of silver. J. Photochem. Photobiol. A. 161: 27–34.

Zarkan, A., H.R. Macklyne, A.W. Truman, A.R. Hesketh and H.J. Hong. 2016. The frontline antibiotic vancomycin induces a zinc starvation response in bacteria by binding to Zn(II). Sci. Rep. 6: 19602.

Zawisza, B., A. Baranik, E. Malicka, E. Talik and R. Sitko. 2016. Preconcentration of Fe(III), Co(II), Ni(II), Cu(II), Zn(II) and Pb(II) with ethylenediamine-modified graphene oxide. Microchim. Acta 183: 231–240.

Zhang, W., J. Chen, Y. Chen, W. Xia, Y.L. Xiong and H. Wang. 2016. Enhanced physicochemical properties of chitosan/whey protein isolate composite film by sodium laurate-modified TiO_2 nanoparticles. Carbohyd. Polym. 138: 59–65.

Zhang, X., G. Xiao, Y. Wang, Y. Zhao, H. Su and T. Tan. 2017. Preparation of chitosan-TiO_2 compositefilm with efficient antimicrobial activities under visible light for food packaging applications. Carbohyd. Polym. 169: 101–107.

Zhao, Q., M. Wanga, H. Yang, D. Shi and Y. Wang. 2018. Preparation, characterization and the antimicrobial properties of metal ion-doped TiO_2 nano-powders. Ceram. Int. 44: 5145–5154.

Zhu, Z., H. Cai and D.W. Sun. 2018. Titanium dioxide (TiO_2) photocatalysis technology for nonthermal inactivation of microorganisms in foods. Trends Food Sci. Tech. 75: 23–35.

Zielinska-Jurek, A., Z. Wei, I. Wysocka, P. Szweda and E. Kowalska. 2015. The effect of nanoparticles size on photocatalytic and antimicrobial properties of Ag-Pt/TiO_2 photocatalysts. Appl. Surf. Sci. 353: 317–325.

7

Appraisal of Organic and Inorganic Nanomaterials in Cellular Microenvironment

Shivani Tank[1], Ragini Raghav[2], Madhyastha Radha[3], Maruyama Masugi[3], Shivani Patel[1] and Madhyastha Harishkumar[3*]

[1]Department of Biotechnology
Atmiya University, Kalawad Road, Rajkot 360005, Gujarat, India
Tel: +91-281-2563445; Email: shivanitank7@gmail.com; admin@bt.vsc.edu.in

[2]Department of Biotechnology
M & N Virani Science College, Kalawad Road, Rajkot 360005, Gujarat, India
Tel: +91-281-2563445; Email: ragini.raghav@gmail.com

[3]Department of Applied Physiology
Faculty of Medicine, University of Miyazaki, Miyazaki, 8891692, Japan
Tel: +81-985851785; Fax: +81-985857932
Email: radharao@med.miyazaki-u.ac.jp; masugi@med miyazaki-u.ac.jp;
hkumar@med.miyazaki-u.ac.jp

INTRODUCTION

Bulk materials, on sizing down to a much smaller scale approaching molecular level, display novel and varied properties than the parent bulk material. The surface properties undergo drastic changes due to increased surface free energy and chemical reactivity. This enhanced surface to volume ratio often changes

*For Correspondence: hkumar@med.miyazaki-u.ac.jp

the efficiency and toxicity aspects of these nanostructures. The control over the size of these nanomaterials can aid in achieving specificity and selectivity, as well as modifying physical, chemical, and biological properties. The controlled nanostructures have been exploited in advanced applications, such as drug delivery, diagnosis, imaging, bioengineering, cosmetics, medical devices, and theranostics (Khalid et al. 2020). In history, the employment of different inorganic metals/metal ions for the treatment of various diseases and other biomedical applications has been reported (Mukherjee et al. 2020). The innate resource of Indian medicine of Ayurveda uses gold as "Swarna Bhasma" for old age rejuvenation and revitalization (Kashani et al. 2018). Metals such as gold and silver were used in the form of "bhasma", which resemble nanoparticles, for curing certain conditions, such as asthma, anemia, chronic fever, cough, sleeplessness, weak digestion, and muscle weakness. Silver is another widely employed metal used extensively for numerous biological applications by ancient Greeks, while early Americans practiced use of silver as a sterilizing agent to store liquids (Yang et al. 2013).

The field of nanomedicine is a vast developing area of science offering alternative treatment approaches to conventional methods for disease therapy and diagnosis, and aids in understanding complex biological systems for better prospects. The cellular machinery, such as proteins (~10 nm), nucleic acids (~2 nm), and viruses (~50 nm), occur in the nanoscale range. The very existence of nanomaterials was spotted in 1850s, when gold nanoparticles (of submicroscopic size) were developed by Faraday for studying dispersion effects. Further, the idea behind nanotechnology surfaced during the famous talk by Dr Feynman at Caltech. After 1979, Prof. Taniguchi coined the term "nanotechnology", followed by K. Eric Drexler's intensive contribution to molecular nanotechnology (Taniguchi et al. 1979). Over the last three decades, nanosystems have evolved in diverse ways. The early phases of researches related to nanostructures, and nanodevices were mostly about structural manipulation and modulation of bulk to atomic-level structures while addressing their stability issues. The approach of nanosizing aided in tailoring the conventional systems for different functionalities and properties. The following phases of "nanomachines" and "nanosystems" highlighted the functional aspects of established nanoentities. One of the most notable developments were the top-down and bottom-up approaches in synthesizing and size control of nanomaterials. The current phase targets the nano-convergence level that involves commercial, regulatory, and toxicity aspects of nanosystems (Madhyastha et al. 2019). In the future, smart nanodevices will represent the biosensing attributes with subsequent targeted treatment. The exploitation of silica nanoparticles with fluorescent tagging, carbon nanotubes, metallic nanocarriers, quantum dots, magnetic nanostructures, and nanocomposites are explored as biosensors with therapeutic applications (Wang et al. 2017). Methods of drug delivery have been based upon physical and chemical properties of nanomaterials, such as high surface to volume ratio, as well as the size of nanoparticles, which play an important role in defining various attributes of the nanocarriers (Sikora et al. 2020). The sole purpose in drug delivery systems is transferring the therapeutic agents into the desired cells, offering least harm to the normal tissues by designing nanoparticles with definite and specific size, surface charge, and release of pharmacologically active agents. The use of nanoparticles for drug delivery and targeting is clinically important for cancer therapy as well as an *in vitro* testing tool (Sutariya et al. 2019, Zhao et al. 2019). To improve the efficacy and toxicity profiles of chemotherapeutic agents, nanosystems can be used as delivery vehicles, because these agents can be encapsulated, covalently attached, or adsorbed onto nanoparticles. It is also being used to overcome drug solubility problems, because more than 40% of active substances being identified through combinatorial screening programs are poorly soluble in water. The conventional and most current formulations of these drugs are frequently plagued with issues, such as poor and inconsistent bioavailability.

Paclitaxel (Taxol™) is one of the most widely used anti-cancer drugs in the clinic. The function of this drug is stabilization of microtubules that promote polymerization of tubulin, resulting in disruption of cell division and leading to cell death. It exhibits effectiveness against primary epithelial ovarian carcinoma, breast, colon, and lung cancers. However, because of the poor solubility of the drug in aqueous solution, the commercial formulations, such as Chremophor EL (polyethoxylated castor oil) and ethanol-based neem gum nano formulations (NGNF) are becoming attractive alternatives (Kamaraj et al. 2018). Protein-based nanoparticles possess advantages for drug delivery, such as biocompatibility, biodegradability, and ability to be modified. The criteria for selection of nanomaterials is dependent on various factors, including the size of nanoparticles, innate properties of the drug, such as aqueous solubility and biostability, desired drug release profile, surface charge and hydrophobicity, biocompatibility and biodegradability, and low antigenicity and toxicity profiles of the product (Rezaei et al. 2019). The reduced levels of toxicity and increased biodegradability of proteins are key features to consider them as promising nanocarriers. Generally, protein nanocarriers are ideal in the preparation of nanoparticles owing to their amphiphilicity, which enables them to interact efficiently with both the drug and solvent systems. Also, the green synthesis often involves manufacturing of nanoparticles derived from natural proteins that remain biodegradable, metabolizable, and easily amenable to surface modulations for docking of drugs and targeting ligands at the molecular level (Tanhaian et al. 2020).

Majorly, there are passive and active targeting strategies using different ligands and different conjugation methods with multiple advantages, such as:

(1) Higher level of improvization of drug stability, thus making them suitable for administration
(2) Biodistribution aspects and pharmacokinetics eventually yielding into improved efficacy of the drug
(3) Considering the early phase response effect, highly selective targeting can be employed
(4) Reduced adverse effects as a consequence of favored accumulation at target sites
(5) Decreased toxicity by employing biocompatible nanomaterials

Figure 7.1 shows the classification of the nanomaterials into organic and inorganic nanomaterials, which have been elaborated in the following section.

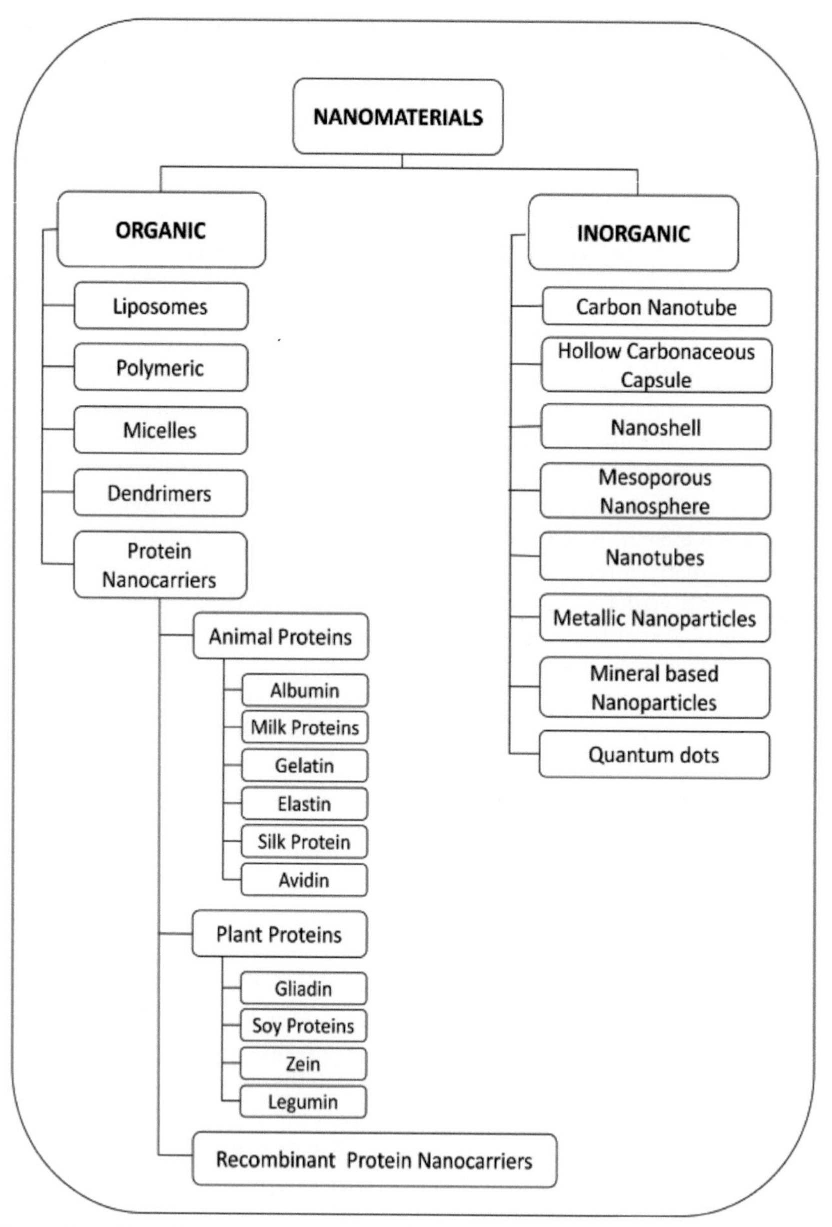

Figure 7.1 Classification of nanomaterials as organic and inorganic nanomaterials.

ORGANIC NANOTECHNOLOGY

The exploitation of synthetic chemistry for manipulation of individual molecules to produce assembled nanostructures, such as liposomes, micelles, polymeric nanocarriers, dendrimers, protein nanocarriers, etc. have been used as potent nanodelivery platforms (Schiek et al. 2008). The following section discuses about the employment of various organic nanomaterials for various biomedical applications.

Liposomes

Liposomes are spherical self-assembled artificial vesicles composed of a lipid bilayer, enclosing an aqueous core, which are able to deliver several types of biomolecules (Mirab et al. 2019). They belong to the class of most clinically established nanosystems for drug delivery, considering the biocompatibility, biodegradability, and possibility of entrapping of both hydrophilic and hydrophobic molecules. Extensive studies on Liposome-mediated delivery of siRNAs have been undertaken. siRNAs are emerging therapeutic agents that suppress the expression of targeted genes, for example, in tumor genes. The targeted approach of siRNA involves interference with the expression of specific genes with complementary nucleotide sequences through the cleavage of the target mRNA by an RNA-protein complex, viz. RNA-induced silencing complex (RISC). This results in mRNA degradation after transcription and, ultimately, no translation occurs (Fire 2007). A newly engineered liposomal-siRNA delivery platform toward triple-negative breast cancer (TNBC), leading to significant reduction of angiogenesis *in vitro* and *in vivo* has been described. The liposomes exhibit excellent features that are adapted into intelligent and switchable nanoplatforms that enable a wide range of stimuli-responsive functionalities, such as pH, temperature, ultrasound, light, magnetic field, and enzymatic response, highlighting the effectiveness of liposomes for drug delivery purposes (Mirab et al. 2019).

Polymeric Nanocarriers

In recent era, polymeric nanocarriers are drug delivery systems and are known to occur as nanospheres (matrix based of structure) or nanocapsules (vesicular system). These nanocarrier systems are prepared by binding a co-polymer onto a polymer matrix. For testing of drug delivery applications, the most outstanding candidates include polyethylene glycol (PEG), poly-lactic acid-co-glycolic acid (PLGA), polyvinyl alcohol (PVA), polyvinyl pyrrolidone (PVP), cyclodextrins (CDs), polyethylene (PE), polyanhydrides, and polyorthoesters.

Micelles

Amphiphilic molecules self-assemble into nanosized colloidal carriers with a hydrophilic shell and hydrophobic core to form micelles. Due to the hydrophobic core of the micellar structure, amphiphilic and poorly water-soluble drugs can be loaded and protected by the hydrophilic shell while being targeted to the target site (Biswas et al. 2013). The typical diameter of micelles is less than 100 nm, limiting their uptake by the Reticuloendothelial system that becomes a governing factor for choosing micelles as exquisite candidates. Shielding by hydrophilic surface enables micelles for immediate recognition and consequently increases bio-circulation times. Moreover, multifunctional micelles can be exploited for containing therapeutic and imaging agents and stimuli-responsive drug-loaded micelles are being actively investigated (Guchhait et al. 2013).

Dendrimers

The name refers to the synthetic three-dimensional polymeric macromolecules that have a well-organized structure. They have tiny sizes (up to 10 nm) and a hydrophobic interior, enabling the desired delivery of certain hydrophobic drugs (Fukushima 2015). The drug incorporation onto the external surface can be achieved by dendrimers, owing to their dendritic and branching nature. The advantages include enhanced circulation time in the blood, great penetration ability, increased bioavailability, lack of immunogenicity, and well-programmed release of drug molecules. Their electron paramagnetic resonance (EPR) effect provides for easy uptake of the nanomaterial by cancer tissues, and finds other prominent drug delivery applications (Tanbour et al. 2016). The synthesis of protein carriers at nanoscale from various protein sources has been possible using animal-based proteins (e.g., bovine and human serum albumin) and plant-based proteins (e.g., zein and gliadin). Additionally, hybrid approach and using recombinant protein as carrier in the delivery system have been considered.

Protein Nanocarriers

Animal proteins can be obtained from various sources, such as meat, fish, dairy, eggs, and other animal-derived tissues. Having the advantages of being abundantly available, biocompatible, and nontoxic, they have been used widely as protein nanocarriers. Plant proteins, such as zein, gliadin, soy proteins, legumin possess favorable properties, such as low toxicity, high bioavailability, easily modifiable structures with free functional groups, and good stability in many cases. In comparison with animal proteins, plant proteins contain higher proportions of hydrophobic amino acids, and thus are usually more hydrophobic in nature. Plant proteins are more stable in aqueous conditions without external cross-linking, have higher loading capacities for hydrophobic therapeutics, provide better protection to payloads, and show more sustainable release of payloads (Metwally et al. 2016).

Animal Proteins

Albumin. In recent years, albumin nanoparticles are one of the most important drug delivery systems studied, especially in the field of treatment of cancer. Belonging to the most abundant plasma protein, albumin acts as a transporter of nutrients and macromolecules to the cells. Looking into their secondary structure in nanoparticle preparation, they are able to deliver therapeutic agents to the tumor cells, as they deliver nutrients via the albumin-specific receptors that remain overexpressed (Arnedo et al. 2002). Tracking the recent articles on albumin and in the search for improvement of albumin-based nanoparticulate systems, the dual drug delivery development in Nanoparticle Albumin-Bound technology, self-assembly, irradiation for encapsulation of other water-insoluble drugs, investigation of methods to protect nanoparticles from degradation, enhancement of drug absorption, modification of pharmacokinetics and drug tissue distribution profiles are some of the areas of current applications for albumin nanoparticles (Arnedo et al. 2002).

Milk Proteins. Milk proteins have many applications in drug delivery owing to their safety aspects and reduced side effects. Caseins and whey proteins are two major proteins in milk. The components of casein are as1-casein, as2-casein, β-casein, and κ-casein. The caseins have proline-rich residues and possess majorly distinct hydrophobic and hydrophilic domains. Casein remains cost effective, readily available, nontoxic, exhibits biodegradability, shows powerful dispersing ability, and the capability of immediate reconstitution into physiological media (Alizadeh Sani et al. 2017). Structurally, whey proteins are globular proteins and are involved in various biological functions. The major site for the synthesis of whey proteins in bovine milk, such as β-lactoglobulin, and alpha-lactalbumin are mammary glands. The coating of casein onto the nanoparticles could help in improvisation in drug delivery applications. The study by Izadi et al. reported a pH-stable and enzymatic-responsive drug delivery vehicle composed of layer-by-layer milk protein casein coated iron oxide nanoparticles in oral cancer (Izadi et al. 2016).

Gelatin. Gelatin shows special features, such as biodegradable nature, biocompatible attributes, non-antigenicity, low cost, and easy availability. Gelatin nanoparticle synthesis, characterization, and application in biomedical arena were reported in many studies (Lee et al. 2011). The usage involves conjugation of gelatin with other materials for drug delivery systems. A study carried out by Moran et al. (2015) reported the fabrication of gelatin nanoparticles by mixing solutions of gelatin type B (either high or low gel strength) with the solutions of protamine sulfate based upon the interactions among the oppositely charged species. Through many studies, it was concluded that gelatin-based nanoparticles show excellent properties, such as being highly potent and nontoxic intracellular delivery systems, which could also be used in nonviral gene delivery systems.

Elastin. The protein elastin, mostly found within connective tissues, helps them to resume their shape soon after stretching or contracting. Some of the needed properties involve biodegradability, biocompatibility, and non-antigenicity. These properties turn it to be a potent drug delivery vehicle. In recent years, the focus on the use of elastin with other proteins, such as silk proteins and the fabrication of recombinant elastin-like (SELP) materials has been growing in the area of dendrimer biology (Fukushima et al. 2015).

Silk Proteins. The silk protein fiber is mainly composed of the compound fibroin, synthesized mainly by an insect's larvae to form their cocoons. Silk fibroin (SF) can be enlisted for the delivery of therapeutic agents, as it has shown biocompatibility, limited biodegradability, and low toxicity/immunogenicity. Parker et al. (2019) reported the use of silk fibroin (SF) nanoparticles, and the drug loaded into it was curcumin (model drug), which is a potent anticancer molecule (Parker et al. 2019). The combination of silk with magnetic nanoparticles was also reported as a model drug-loaded magnetic SF core-shell nanoparticle for sustained release of curcumin into breast cancer cells (Mickoleit et al. 2018). The particles revealed superior properties for biomedical applications because of their unique sizes. Furthermore, they had particularly desired characteristics for cell internalization, as well as the magnetic cores inside the particles that provide the possibility of using an external magnet for cancer targeting.

Avidin. It is extracted from the eggs of reptiles and amphibians. The interaction of Avidin-biotin is considered one of the most specific and stable noncovalent interactions, which is about 103 to 106 times higher than an antigen-antibody interaction. Sato et al. (2004) reported the preparation methods, biomedical applications, and diagnostic ability of neutravidin conjugated polystyrene aggregation pattern nanoparticles in the detection of a single mutation in the cell.

Plant Protein Nanocarriers

Gliadin. Gliadin is a wheat gluten associated single chain polypeptide having a molecular weight of 25–100 kDa with strong disulfide bonds. Gliadin is used in the form of water-soluble polypeptide–gelatin. Further, their composite nanoparticle for the delivery and controlled release of cyclophosphamide has been studied. The results showed that gliadin/gelatin had a better impact on the tumors than gliadin alone (Ezpeleta et al. 1999).

Soy Protein. Soy protein isolates (SPI) are one of the promising candidates for preparing nanoparticle delivery systems, owing to their high-nutritive and excellent gelation properties, followed by desired characteristics, such as biodegradability and biocompatibility nature (Liu and Tang 2013). The improved creaming stability, pickering nature, and emulsion coefficient help soy protein in being a good stabilizing agent in the nano form. The increased pickering nature is largely due to the formation of gel-like structure in the nano formulation.

Zein. Belonging to the class of water-insoluble plant proteins, zein is mostly derived from maize, and owing to its ability to maintain humidity levels in prepared goods, it has found certain outstanding applications in the packaging and food industries. It is composed majorly of nonpolar amino acids, such as alanine, phenylalanine, leucine, and proline, but considering protein structure and its hydrophobic properties, it can be phagocytosed by macrophages and can, therefore, stimulate the immune system. Perez Herrera and Vasanthan (2018) discussed the evaluations of zein hydrophobic protein nanoparticles. Zein nanoparticles were prepared by liquid-liquid phase separation; sodium caseinate (CAS)-stabilized zein nanoparticles were prepared using different zein/CAS mass ratios. The effects of CAS on cytotoxicity, cell uptake, and epithelial transport of zein nanoparticles were investigated. Looking at the near future, the research trends on zein nanoparticles as drug delivery systems are likely to be focused in three major directions:

(1) The fabrication of complex delivery systems composed of zein and other biopolymers for different delivery applications, including colon targeting and transdermal delivery (Liu and Tang 2013)

(2) The development of zein-based nanocomposites with inorganic nanocrystal materials, such as quantum dots (QDs), that show better capability to encapsulate and control the release of drugs for a wide variety of applications, and

(3) The *in vivo* evaluations of zein-based delivery systems for elaborating their biological fate after oral consumption and their efficacy and toxicity in physiological conditions. Finally, it is assumed that zein-based delivery systems will continuously attract increasing interest, and that the clinical applications and commercialization will be realized in the near future (Liu and Tang 2013).

Legumin. Legumin is most popularly known as vegetable casein, as it is analogous to the casein of milk, obtained from beans, peas, lentils, vetches, hemp (specifically edestin), and other leguminous seeds. Perez Herrera and Vasanthan (2018) used legumin nanoparticles as a drug carrier and discussed their advantages. Methylene blue (MB) was applied as the hydrophilic drug. The results indicated that legumin nanoparticles with a size of about 250 nm were prepared using coacervation method and chemical cross-linking. This method avoided the use of organic solvents with a yield of 27% of protein, and also helped in the improvement of soil bacterial communities (Ge et al. 2014).

Recombinant Protein Nanocarriers

Recombinant proteins have been used to generate nanoparticles as carriers for drugs that are proteinaceous in nature. One such example is Recombinant human gelatin (rHG), which was used to prepare nanoparticles as an alternative to natural gelatin-based nanoparticles. A new strategy of recombinant protein tetra-H2A (TH) derived from histone H2A, an alternative of protamine as a conditionally reversible, nucleic acid condensing agent was developed (Bowerman et al. 2019). The new recombinant protein overcame the disadvantage of insufficient release of nucleic acid therapeutics, which get captured by protamine during siRNA delivery. The TH/siRNA condensates were designed into core-membrane liposomal nanoparticles that effectively displayed a higher silencing efficiency of target genes in both *in vitro* and *in vivo* systems. Not only that, the nanoparticles assembled with protamine as a nucleic acid condensing agent (Wang et al. 2013).

INORGANIC NANOTECHNOLOGY

The fundamental advantages of inorganic nanomaterials over bioorganic nanomaterials, such as liposome, dendrimer, and biodegradable polymer in terms of their size and shape control and surface functionalization have been widely discussed (Briggs and Knecht 2012). In addition, since inorganic nanomaterials permit a wide range of functionality arising from their unique optical, electrical and physical properties, they may provide a solution for many physical barriers of the cell that limit biomedical applications. Nanotubes, nanoshells, and mesoporous nanoparticles are attractive vehicles for drug/gene delivery because of their hollow and porous structures and facile surface functionalization. The inner void can take up a large amount of drug, and the open ends of pores serve as gates that can control the release of drug/gene (Cattani-Scholz 2017).

Carbon Nanotube

Carbon materials are one of the leading inorganic nanomaterials for biomedical applications, because of their cell penetrating nature and osteogenic potential, which were extensively investigated in recent times. Carbon nanotubes (CNTs) have grabbed the attention as new vectors for the delivery of therapeutic molecules, because of the ease of translocation across cell membranes and reduced cell stress during the osteogenic differentiation. However, the burning issues about the ultimate biocompatibility of CNTs have limited their widespread use in biomedical applications (Son et al. 2007). Carbon nanotubes clearance was also studied. Water-soluble functionalized carbon nanotubes (f-CNTs) were rapidly cleared from the systemic blood circulation through renal excretion and respiratory tract clearance route, indicating the rate of body clearance (Sturm 2017). The modifications included labelling of nanotubes (SWNTs) with the diethylenetriamine pentaacetate (DTPA)/indium (111In) complex. SWNTs have strong optical absorbance in the near infrared (NIR) range. This enabled triggering of release of drug/gene from the SWNT upon illumination with NIR light. The absorbance of NIR light produces localized heat that eventually stimulates the release of drugs or genes from the nanotube surface. The designed folate-conjugated SWNTs have been described by many researchers and the role of macrophage in biodistribution has been investigated (Septiadi et al. 2018).

Hollow Carbonaceous Capsule

Preparations of hollow carbonaceous capsules with reactive surface layers were created by hydrothermal methods, with the anionic surfactant sodium dodecyl sulfate (SDS) and glucose as starting materials. The void size within the capsule and the shell thickness can be tuned by adjusting the amount of SDS and the hydrothermal parameters, such as time, temperature, and glucose concentrations (Vandivort et al. 2017). These capsules have potential biomedical applications, such as drug delivery and clinical diagnostics.

Calcium Phosphate Nanoshell

Nanoshells derived of calcium phosphate, the main component of the skeletons and teeth of vertebrate animals, is biocompatible and biodegradable. Also, the solid calcium phosphate [$Ca_2(PO_4)_yOH_z$] particles have been widely investigated for their applications in biotechnology. Calcium phosphate nanospheres of around 80 nm diameter were used to encapsulate DNAs for a targeted gene delivery. Further studies displayed promising transfection efficiency, both *in vitro* and *in vivo*. Nano-shells, unlike nanospheres where the drug is dispersed throughout the particles, are vesicular systems where the drug can be stored and protected from a premature inactivation. As a concluding remark, calcium phosphate nanoshells have potential as sustained drug delivery vehicles (Thakkar et al. 2012).

Mesoporous Nanosphere

The synthesis of the mesoporous material MCM-41 in ordered fashion along with its application as a drug delivery system was studied. The materials were synthesized by a self-assembly of a 0020 silica surfactant (Anjugam et al. 2018), C16-trimethylammonium bromide and C12-trimethylammonium bromide (TAB), in water. The pore size distribution is centered at 2.5 and 1.8 nm for the samples prepared from C16-TAB and C12-TAB, respectively. Ibuprofen was introduced into these two MCM-41 materials. The weight percentage ratio of drug/MCM-41 was 30 percent. The drug release profiles in a simulated body fluid show that the drug as a whole gets incorporated into the MCM-41 matrix, and can be released into a solution over a period of three days through *in vitro* studies. Several studies have reported magnetic nanotubes (MNTs), and silica nanotubes embedded with magnetite nanoparticles. The main advantage of the drug carrier having a magnetic property is that it allows the use of a powerful imaging technique, magnetic resonance imaging (MRI), to track drug delivery. In addition, controlled movement of the drug carrier can be obtained with the generation of an external magnetic field, and consequently be directed to specific anatomical sites *in vivo*. The effective combination of attractive magnetic properties with a tubular structure makes the MNT an ideal candidate as a multifunctional nanomaterial used for biomedical applications. The inner void of MNTs can be used for controlled release of desired drug molecules into a solution. So as to achieve the desire result, the interior of the MNTs was designed differentially by incorporating the functionalized groups, such as amino-silane (aminopropyltriethoxysilane, APTS). 5-Fluorouracil (5-FU), 4-nitrophenol, and ibuprofen (Ibu) molecules were also examined as model drug molecules for controlled drug release experiments. The results suggested that the drug release rate can be controlled by regulating the modification. It may be due to the increased capacity arising from the protonation of free amine groups within the multilayers and a pH-sensitive swelling of the polyelectrolytes (Anjugam et al. 2018).

Nanotubes

Nanotubes are used for controlled drug release; surface functionalization has been used. However, this approach does not offer a full range of drug release control. An alternative payload-release strategy was explored by corking nanotubes with a chemically labile cap. Approach to cap NTs with gold nanoparticles was attempted using controlled Au nanoparticle diffusion in nanotubes. In this method, negatively charged 2 nm Au nanoparticles were selectively immobilized on the positively charged inside of the nanotube through attractive electrostatic interaction

(Bayda et al. 2018). Since the non-functionalized outer surface bears partial negative charges and the repulsive forces produced by Au nanoparticles occupying the open ends block further diffusion of Au nanoparticles into the channel of the NTs, the Au nanoparticles were localized at the open end of the nanotube and were found neither on the outer surface of NT nor on the deep inner surface. The group proposed that their capping procedure can be potentially adopted for a general *in situ* encapsulation of biomolecules (e.g., DNA or enzyme).

The extensively studied model is the skin drug delivery, which is an attractive alternative for the oral route, evident for local and/or systemic drug delivery. Considering the complex and well-organized structure, most of the drugs show difficulties in penetrating the human skin. Therefore, enormous efforts have been invested to develop intelligent drug delivery systems overcoming the skin barrier, with particular emphasis on increasing therapeutic activity and minimizing undesirable side-effects. Major strategies involve the use of singular materials with novel properties. In particular, on the basis of their inherent properties, including biocompatibility, biodegradability, and relative low-cost, inorganic nanoparticles are ideal candidates for the development of skin drug delivery systems. Metallic nanoparticles, mineral-based nanoparticles, and quantum dots have been introduced for skin drug delivery platforms (Carazo et al. 2018).

Metallic Nanoparticles

A schematic representation of different metallic particles is depicted in Figure 7.2.

Gold Nanoparticles (Au nanoparticles)

Gold nanoparticles, among the inorganic nanocarriers that have been developed, have many positive features that make them ideal candidates to be used in the biomedical field. Recently, synthesis of Au nanoparticles-based nanofiller for a polyelectrolyte membrane for transdermal drug delivery of the drug diltiazem hydrochloride has been described (Mieszawska 2013). This novel design showed improved thermo mechanical properties and transparency in the visible range, which makes it attractive for cosmetic applications. The Au nanoparticles made the film resistant to microbial growth, so it could be used for long term skin care approaches. The delivery of Au nanoparticles in cosmetics justifies the role of diverse growth factors which played a part in skin reparation process.

Silver Nanoparticles (Ag nanoparticles)

Regarding Ag nanoparticles, there are several scientific works dealing with the use of this kind of inorganic nanoparticles, not only on the basis on their inherent antimicrobial properties, but also due to various biomedical purposes (Umapathi et al. 2019). The use of Ag nanoparticles against UV-B radiation aimed at preventing skin carcinogenesis showed higher efficiency than titanium dioxide nanoparticles and zinc oxide nanoparticles. Another work has studied the possible interactions established between Ag nanoparticles and keratinocytes and fibroblasts during wound healing (Liu et al. 2010). Results obtained stated that Ag nanoparticles increase the wound closure process via two different ways: via an enhanced proliferation and migration of keratinocytes and by promoting differentiation of fibroblasts into myofibroblasts.

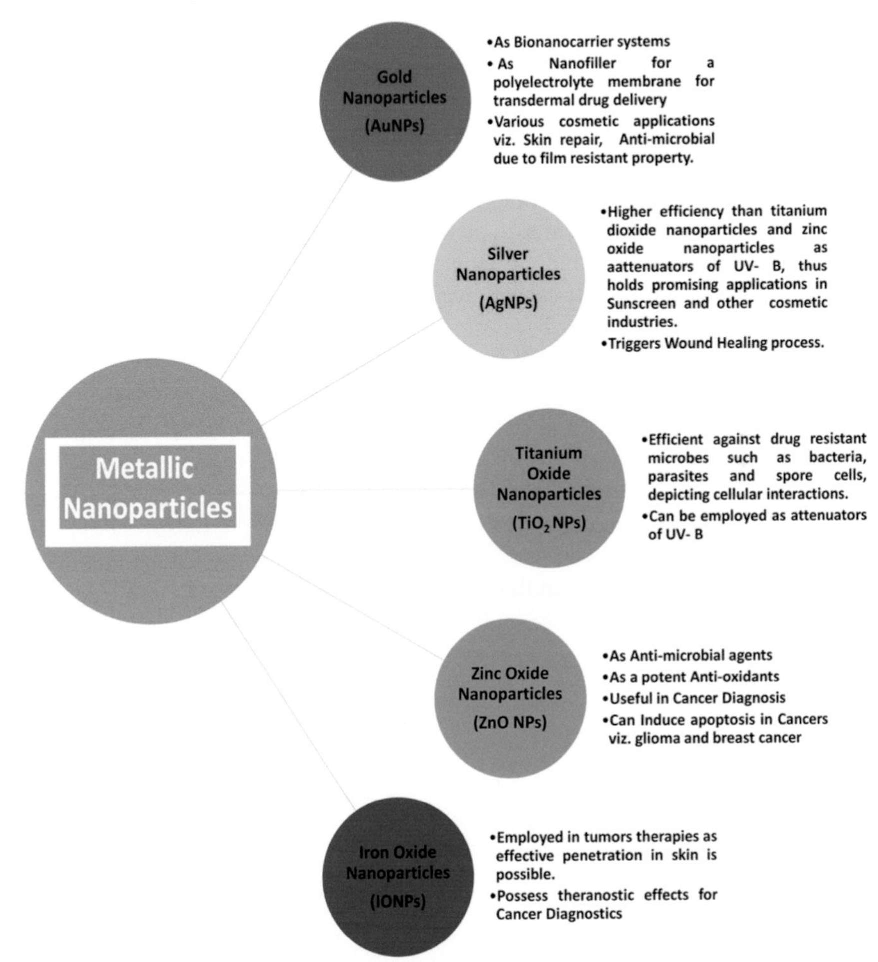

Figure 7.2 Applications of various metallic nanoparticles.

Titanium Dioxide Nanoparticles (TiO_2 nanoparticles)

By applying z-Potential measurements and confocal microscopy, the efficiency of TiO_2 nanoparticles against drug-resistant bacteria, parasites such as *Leishmania* spp., and spore cells have been fully explained, where TiO_2 nanoparticles-cell membrane interactions lead to the death of the cells. Experimental and molecular modeling techniques have been used to determine the most appropriate size of TiO_2 nanoparticles, assuming that they are spheres. TiO_2 nanoparticles can be exploited as an attenuator of the UV-B radiation in sunscreen lotions as key ingredients (Mohamed et al. 2017).

Zinc Oxide Nanoparticles (ZnO nanoparticles)

ZnO nanoparticles synthesized on aloe vera leaf extract were found to inhibit the growth of bacteria, as *E. coli*, *P. aeruginosa*, and *S. aureus* strains are resilient to conventional antibiotics. The applications of ZnO nanoparticles in the field of biomedicine are completely based on their effective antimicrobial potential and their environment-friendly, low-cost synthesis (Mirzaei and Darroudi 2017). The green and easy synthesis of colloidal ZnO nanoparticles with antimicrobial and antioxidant actions have been reported, resulting in aimed interest in justifying the activity against *Candida* sp. and their powerful antioxidant action preventing cellular damages produced by oxidative stress. Therefore, it is important to notice the role of ZnO nanoparticles in skin penetration and allergy (Raphael et al. 2013). They do show the ability to induce apoptosis in cancer cells, which is particularly effective in glioma and breast cancer, even possessing a diagnostic role.

Iron Oxide Nanoparticles (IO nanoparticles)

Magnetic nanoparticles have been used for many decades as a research tool. The use of iron oxide nanoparticles (IO nanoparticles) as drug carriers has been extensively investigated. Remarkable findings have been reported regarding the use of IO nanoparticles targeted to different tumors therapies based on several positive features. Facilitated skin penetration is obtained when drugs are loaded onto them, providing theranostic effects. Additionally, several contemporary cancer diagnostic and treatments by gene editing methodology also involve IO nanoparticles (Rohiwal et al. 2020).

Mineral-based Nanoparticles

Minerals are complex materials which are composed of diverse particle size. The diversity of surface atomic structure, crystal shape, and surface topography give them wider prospective in the field of mineral-based nanomaterials (Michael et al. 2008). The following section mentions the different types of mineral-based nanomaterials.

Natural Clay Minerals

The vast applications offered by clay minerals in pharmaceutics is the main topic studied by many research groups (Carazo et al. 2018). Particularly, the uses of clay minerals as drug nanocarriers alone and in combination with organic compounds have been carried out. Additionally, they also find applications in pelotherapy, based on the topical administration of hot-muds (peloids) consisting of inorganic gels with optimal rheological and thermal properties displayed by clay minerals and mineral-medicinal water. Besides, clay minerals have also been chosen as carriers for antibiotics in acne therapy in conjugation with essential oils, which have antimicrobial properties for treating dermatitis (Carazo et al. 2018).

Layered Double Hydroxides (LDHs)

Easy availability, biocompatibility, colloidal stability, targeted and stimuli-responsive release of drugs, and theranostic properties, apart from other desirable characteristics, have made LDHs an up-to-date desired candidate to carry drugs, genes, and other active compounds. They can be used in co-carrier, alone, or via hybrid systems with polymers, and some studies focused on the assessment of the activity of LDHs in skin repair by promoting both collagen deposition and renewal of the extracellular matrix, as well as on their safety to be used in drug delivery in implantable systems (Carazo et al. 2017b).

Mesoporous Silica Nanoparticles (MSNs)

Silica nanomaterials can be well designed by using thermo-responsive MSNs-quercetin system providing controlled release of this antioxidant, depending on the temperature of the application place as well as a protective effect of the antioxidant molecules (Carazo et al. 2017a). Besides, it was proved that MSNs act not only as efficient topical nanocarriers, but also show interesting properties to increase the stability of delicate antioxidant molecules, such as derivatives of flavonoids such as quercetin and of the vitamin E, Trolox (Navya et al. 2019). Regarding skin penetration, results found by Carazo et al. (2018) and others gave light to the relationship established between positively charged surfaces, and increased cellular uptake and skin penetration of MSNs with a size greater than 75 nm. Furthermore, the *in vitro* penetration and *in vivo* anti-inflammatory and analgesic effects of curcumin-loaded MSNs were determined. Regarding skin penetration, results found by Carazo et al. (2017b) gave light to the relationship established between positively surface charges, and increased cellular uptake and the block of skin penetration of MSNs with a size greater than 75 nm.

NANOTECHNOLOGY IN DISEASE DIAGNOSIS

Internalization into the cells becomes the most important criteria for nanoparticles to easily interact with cellular components, nucleic acids, and proteins, and thus be used for therapeutic and diagnostic purposes. The blood circulation within a living system allows the flow of 20 nm-sized particles, which could distribute into the different organs to monitor biological systems. Nanomedicine is categorized into three areas of application:

(1) Therapeutic nanomedicine
(2) Diagnostic nanomedicine, and
(3) Theranostic nanomedicine

In therapeutic nanomedicine, nanoparticles are employed for disease therapy without any need of drugs. Diagnostic nanomedicine, on the other hand, aids in early or in time identification of diseases either by non-invasive means or by measurement of biomarkers displayed in disease conditions (Mukherjee et al. 2020, Patra et al. 2019). An exquisite performance by advanced therapeutic nanomedicine involves therapy and diagnosis at the same time, and is called theranostic nanomedicine. Biomedical nanotechnology refers to the branch of nanoscience that covers cross-disciplinary areas of research in science, engineering, and medicine, aiming at broad applications for molecular imaging, molecular diagnosis, and targeted therapy. Nanometer-sized particles, such as semiconductor quantum dots and iron oxide nanocrystals, possess exploitable optical, magnetic, or structural characteristics that are not available to either molecules or bulk solids. The step of linking with ligands, such as monoclonal antibodies, peptides, or small molecules becomes inevitable, so as to bring about usefulness in targeting diseased cells and organs (such as malignant tumors and cardiovascular plaques) with high affinity and specificity. In the "mesoscopic" size range of 5–100 nm diameter, nanoparticles offer larger surface areas and functional groups for conjugating to multiple diagnostic (e.g., optical, radioisotopic, or magnetic) and therapeutic (e.g., anticancer) agents.

Recent advances have led to multifunctional nanoparticle probes for molecular and cellular imaging, nanoparticle drugs for targeted therapy, and integrated nano-devices for early disease detection and screening. Traditionally, *in vivo* imaging probes or contrast agents include radioactive small molecules in positron emission tomography (PET) and single photo emission computed tomography (SPECT), gadolinium compounds in magnetic resonance imaging (MRI), and isotope-tagged antibodies. Recent advances, such as bioconjugated QDs and targeted nanoparticles provide a number of unique features and capabilities that could significantly improve the criteria, such as sensitivity and specificity of disease imaging and diagnosis. Firstly, the size-dependent optical and electronic properties of QDs can be tuned continuously by changing the particle size. This "size effect" permits the use of a broad range of nanoparticles for simultaneous detection of multiple cancer biomarkers. Second, nanoparticles have more surface area to accommodate a large number or different types of functional groups that can be linked with multiple diagnostic (e.g., radioisotopic or magnetic) and therapeutic (e.g., anticancer) agents. This opens the opportunity to design multifunctional "smart" nanoparticles for multi-modality imaging as well as for integrated imaging and therapy. Third, extensive research has shown that nanoparticles in the size range of 10–100 nm are accumulated preferentially at tumor sites through an effect called enhanced permeability and retention (Mukherjee et al. 2020). Au nanoparticles have numerous other applications in imaging, therapy, and diagnostic systems. The advanced state of synthetic chemistry of gold nanoparticles offers precise control over physicochemical and optical properties. Furthermore, gold cores are inert and are considered to be biocompatible and nontoxic. Au nanoparticles can be incorporated into larger structures, such as polymeric nanoparticles or liposomes that deliver large payloads for enhanced diagnostic applications, efficiently encapsulate drugs for concurrent therapy, or add additional imaging labels. Intelligent "chemical nose" sensor based on the poly (*p*-phenyleneethynylene) (PPE) polymer and Au nanoparticles, was capable of distinguishing 7 different proteins (You et al. 2007). The nanosensor was composed of an array of six Au nanoparticles with different cationic coatings, each complexed with negatively charged PPE-CO_2 polymer. The PPE polymer is highly fluorescent, but the fluorescence is quenched when bound to the Au nanoparticles. The differing capping ligands used to coat the

Au nanoparticle surface provide weaker or stronger interactions with a polymer and protein analytes. Addition of protein analytes disrupts the assembly between the Au nanoparticles and PPE-CO_2 polymer, resulting in fluorescence from the polymer. The protein analytes were chosen to have different sizes and charges, and thus had differential binding to the Au nanoparticles. Therefore, each protein resulted in a unique fluorescent pattern from the array, enabling their distinction in a mixture. The array was tested against 52 protein samples and correctly identified the protein with 94.2% accuracy. This approach is an excellent example of exploitation of the tunability of Au nanoparticle surface chemistry to optimize performance.

NANOTECHNOLOGY IN IMAGING

X-ray based Imaging

Allijn et al. (2013) were able to image the 3D distribution of luminescent gold nanoshells (AuNSs) in murine tumors using two-photon induced photoluminescence. The article showed that AuNSs were coated with polyethylene glycol and accumulated in the tumor via the EPR effect. In addition to intrinsic fluorescence, Au nanoparticles can be made fluorescent by the addition of organic dyes. Recent approach on Spectral CT imaging has been done with gold nanoclusters targeted with antibodies via an avidin–biotin linkage to fibrin (Nagakura et al. 2005, Schirra et al. 2012). This brought about the specific detection of clots created *in vitro* and ultimately helps in blood diagnosis. Another model of imaging technique in nanobiology is using the tagged fluorescent probes. Au nanorods (AuNRs) exhibit transverse and longitudinal plasmon resonance bands that originate from the oscillation of the surface electrons along the x- and y-axis, respectively. For a given diameter, when the aspect ratio of the rod increases, the transverse band remains unchanged, while the longitudinal band shifts to the red. For example, Mohamed et al. (2000) found that AuNRs with an aspect ratio of 2 had an emission maximum of 540 nm, while increasing the aspect ratio to 5.4 led to an emission maximum of 740 nm. These fluorescence properties have found use in DNA biosensing. Li and coworkers have functionalized AuNRs with DNA sequences (Li et al. 2005). When complementary sequences were added, the AuNRs aggregated and the fluorescence signal decreased. This process was reversed upon heating the AuNRs and disaggregation. Surface Enhanced Raman Spectroscopy (SERS) imaging is the most accurate signature identification instrumentation technique. Raman spectroscopy is a technique used to study molecular vibrations, rotations, and other processes. Raman spectra are highly complex and can be used as "fingerprints" of molecules, thus giving the chemical composition of a sample. Although highly specific, Raman spectroscopy is limited by its low sensitivity, since only one photon in 108 is Raman scattered. Enhancement of the intensity of the vibrational spectra of Raman active molecules by several orders of magnitude can be possible by absorption upon Au nanoparticles or other metal surfaces (Allijn et al. 2013). This emergence led to a new technique called surface enhanced Raman spectroscopy (SERS). The strongest electromagnetic field enhancement occurs at sharp nanostructure edges, such as Au NR tips and between aggregated colloids.

Also, the surface plasmon absorption of Au nanostructures can be tuned into the NIR, thus avoiding absorption of excitation light by biological samples and limiting the interference for the SERS signal. Therefore, Au nanostructures are attractive as labels for flow cytometry and as contrast agents for biological SERS imaging.

Photoacoustic imaging bears advantages upon other imaging techniques, as the radiation involved is nonionizing and the implication of ultrasound as the output results in higher spatial resolution when compared with other optical methods that involve lower scattering of ultrasound in tissue. The main advantage of this technique is deeper penetration in the tissue and recording the optics from the deeper tissues. Tanaka et al. (2012), from this technology, successfully implanted the tissue engineered cartilage and its post implanted efficacy was successfully recorded.

Extraordinary light scattering properties are observed in Au nanoparticles, which are not observed in non-plasmonic nanoparticles. Any changes in size and shape of Au nanoparticles can greatly influence scattering, providing an opportunity for an agent to possess optimal light scattering performance. The strong scattering signals enabled visualization of Au nanoparticle entry to the cell nucleus (when functionalized with RGD peptides and a nuclear location sequences) or the cytoplasm (when coupled to RGD only), and their localization was traceable after cell division. The light scattering of Au nanoparticles was used by El-Sayed et al. (2005) to discriminate cancerous cells from healthy cells. Nanoparticulate systems such as quantum dots, gold nanoparticles, and dye-doped silica nanoparticles offer numerous advantages, such as better photo stability, higher quantum yield, and *in vitro* and *in vivo* stability, as compared to conventional contrast dyes. They are used in many *in vivo* imaging platforms with optical, magnetic resonance imaging (MRI), computed tomography (CT), ultrasound, positron emission tomography (PET), and single photon emission computed tomography (SPECT). Nanoparticle-based NIR nanoprobes, including NIR-emitting semi-conductor QDs, lanthanide doped up-converting nanoparticles, and NIR dye-containing nanoparticles are applied for *in vivo* imaging. Most commercially available QDs have a semiconductor core, which is often a mixture of cadmium and selenium, measuring about 2–10 nm in diameter. This core is surrounded by a shell, usually of another semiconductor material, and an outer polymer or inorganic layer. In semiconductor quantum dots, the electronic wave functions are squeezed into small areas. Stretching them in a controllable yet simple way profoundly affects their properties and can give them characteristics important for practical applications. The motion of single molecules labeled with fluorescent QDs can be monitored in live cells over time scales ranging between milliseconds to hours in order to determine individual trajectories of molecules with an accuracy at the nm level. Single quantum dot tracking (SQT) thus constitutes an extremely sensitive tool to explore how molecules in cells form assemblies that ensure cellular structures and functions (Meles et al. 2017). Single molecule spectroscopy and imaging (SMS) for measuring the signal of individual fluorescent labels is of growing interest because it provides information on a single cell instead of the population average of many cells, exposing normally hidden heterogeneities. Most of the super-resolution imaging applications involve protein-based labeling of proteins with a fluorescent tag or a dye-labeled antibody. In this way it becomes possible to follow its identity, and modifications *in vivo*. Eventually, such techniques could be useful in materials science and the life sciences, as well as for food safety control and the detection of drugs, explosives, and environmental pollutants.

NANOTECHNOLOGY IN BIOSENSING

Beginning with the easiest introduction of a biosensor, which is a device constituting a biological sensing element either intimately connected to or integrated within a transducer. Schematic is represented in Figure 7.3. A highly specific molecular recognition is a fundamental prerequisite, majorly based on affinity between pairs, such as enzyme-substrate, antibody-antigen, and receptor-hormone, and this innate property can be used for the production of concentration–proportional signals. The major properties, such as selectivity and specificity highly depend on biological recognition systems connected to a specific transducer element (Jackman et al. 2017). Gold, carbon nanotubes (CNTs), magnetic nanoparticles, and quantum dots are being gradually applied to biosensors due to their unique physical, chemical, mechanical, magnetic, and optical properties, and reportedly enhance the sensitivity and specificity of detection (Figure 7.3).

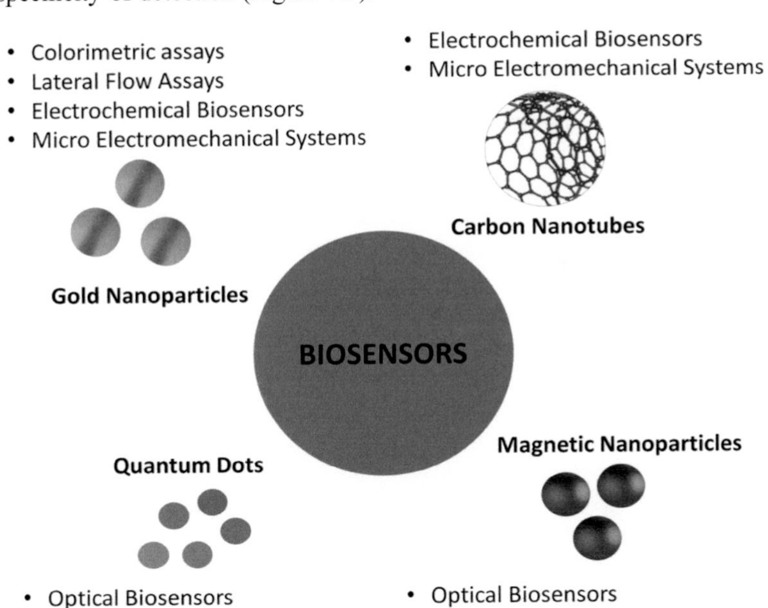

Figure 7.3 Schematic representation of various nanoparticles employed in biosensors.

Gold Nanoparticles based Biosensors

Gold nanoparticles are the most extensively investigated nanomaterials, due to their distinct physical and chemical attributes. Popularly, gold nanoparticles display a red color, while aggregated gold nanoparticles appear a blue color. Based on this phenomenon, Jackman et al. (2017) established a gold nanoparticle-based biosensor to quantitatively detect polyionic drugs, such as protamine and heparin. Gold nanoparticles in biosensors provide a biocompatible microenvironment for biomolecules, greatly increasing the amount of immobilized biomolecules on the electrode surface, and thus improving the sensitivity of the biosensor (Raghav and Srivastava 2016). The biosensor is designed to overcome the defects inherent to polymerase chain reactions (PCR). Thus, PCR-based biosensor technology is faster and reaches zeptomol concentrations, which is greatly superior to traditional fluorescence-based DNA detection systems, which have only a pico level detection limit. Au nanoparticles have become excellent scaffolds for the fabrication of novel sensors, and lots of innovative approaches have been developed for sensing in a rapid, efficient manner (Duhachek et al. 2000). A stronger binding is obtained by bio-adhesion between silica surfaces and biomolecules using a chemical linker rather than through physical adsorption. The approaches, such as Nanopatterning and Chemical linker method can be employed for improving the bioadhesion of biomolecules onto silicon surfaces. The emerging field is DNA origami that deals with nanoscale folding of DNA, generating arbitrary two and 3D shapes (Park et al. 2016). Connecting a patterned silicon substrate with biomolecules opens door to new opportunities for assembling nanostructures. Organic and biological molecules can be attached to silicon via an intermediary gold layer and a carbon-sulfur-gold bond or via siloxane chemistry directly on oxidized silicon. The exquisite building up of a scaffolding of multiple molecular layers makes it possible to imitate the surface of a living object, especially in the positive therapeutic effect on keloids, hypertrophic scars (Yang and Guy 2015). The real impact of nanotechnology appears when the simple structures discussed above are integrated in more complex entities, called Micro Electromechanical Systems (MEMS). MEMS are structures obtained through the integration of mechanical elements, sensors, and actuators on one silicon substrate through microfabrication technology. When MEMS are reduced in size by scaling down, they arrive in the size domain of nanotechnology and are named Nano Electro Mechanical Systems (NEMS).

CNTs have great potential in applications, such as nanoelectronics, biomedical engineering, and biosensing (Zhang et al. 2009). For example, polymer-CNTs composites can achieve high electrical conductivity and better mechanical properties, which offer exciting possibilities for developing ultrasensitive, electrochemical biosensors. A highly effective strategy involves carbon nanofibers as biosensor platform (Son et al. 2007). Synergistic effects of MWNTs and ZnO improved the performance of the biosensors formed (Bai 2007, Zhang et al. 2009). The report on an amperometric biosensor for hydrogen peroxide, was developed based on adsorption of horseradish peroxidase at the GCE modified with ZnO nanoflowers produced by electrodeposition onto MWNTs film. Another example by Zhang and others described a controllable layer-by-layer self-assembly modification technique of GCE with MWNTs, and introduced a controllable direct immobilization of acetylcholinesterase (AChE) on the modified electrode. By the decreasing activity of immobilized AChE caused by pesticides, the composition of pesticides can be determined.

Magnetic Nanoparticles based Biosensors

Magnetic nanoparticles, because of their special magnetic properties, have been widely explored in applications, such as hyperthermia, magnetic resonance imaging (MRI) contrast agents, tissue repair, immunoassay, drug/gene delivery, cell separation, Giant Magneto Resistance (GMR)-sensor (Bergin et al. 2016, Raphael et al. 2013). Another group prepared a new kind of magnetic dextran microsphere (MDMS) by suspension cross-linking using iron nanoparticles and dextran (Xia et al. 2005). HRP was then immobilized on a MDMS-modified glass carbon electrode (GCE). On the basis of the immobilized HRP-modified electrode with hydroquinone (HQ) as mediator, an amperometric H_2O_2 biosensor was fabricated. Hemoglobin (Hb) was successfully immobilized on the surface of MCMS modified GCE with the cross-linking of glutaraldehyde. Recently, a highly sensitive, giant magnetoresistance-spin valve (GMR-SV) biosensing chip device with high linearity and very low hysteresis was fabricated by photolithography (Rohiwal et al. 2020). The signal from even one drop of human blood and nanoparticles in distilled water was sufficient for their detection and analysis.

To study the interaction between protein molecules or detect the dynamic course of signal transduction in live cells, Fluorescence Resonance Energy Transfer (FRET) with synthesized quantum dots has been employed. These synthesized quantum dots have significant advantages over traditional fluorescent dyes, including better stability, stronger fluorescent intensity, and different colors, which are adjusted by controlling the size of the dots (Pandey et al. 2020). Therefore, quantum dots offer a new functional platform for bioanalytical sciences and biomedical engineering. For example, CdTe quantum dots led to an increased effective surface area for immobilization of enzyme, and their electrocatalytic activity promoted electron transfer reactions and catalyzed the electro-oxidation of thiocholine, thus amplifying the detection sensitivity.

Metal nanoparticles, such as copper nanoparticles with greater surface area and higher surface energy, are used as electron-conductors and displayed good catalytic ability to the reduction of H_2O_2 and enzymatic conversion mechanism (Breger et al. 2020). Owing to the biosensor applications, platinum nanoparticles are used in the immobilization of proteins and enzymes for bioanalytical applications as well. For example, metal-oxide-based semiconducting nanowires or nanotubes play an important role in electric, optical, electrochemical, and magnetic transducers. The important detection of lactate dehydrogenase (LDH) was reported by Pandey et al. (2020), and found its usefulness in cell stress determination.

POTENTIAL APPLICATION OF NANOMATERIALS-BASED BIOSENSORS

The biosensors could be used to determine the glucose levels of serum samples with higher sensitivity. The verities of analytes such as glucose, uric acid, and creatine can be detected by reacting the test samples by corresponding oxidase enzymes, which results in the formation of stoichiometric amount of H_2O_2. Oxidation of H_2O_2

can be detected by peroxidase enzymes, such as horseradish peroxidase or HRP, which is tagged with fluorescent dyes. Here electron transfer mediated mechanism plays a major role in the detection method. Classic method of detection of electron transfer is tagging with Prussian blue or $Fe_4[Fe(CN)6]_3$ nanoparticles. The first evidence of nanowire field effect transistor-based biosensor was found in a study by Maki et al. (2008). It is centered to achieve simple and ultra-sensitive electronic DNA methylation detection. Additionally, the complicated bisulfite treatment and PCR amplification can be avoided. Similarly, employing protein–ligand (antigen) interaction properties, protein-nanoparticles-based biosensors can realize the ultra-sensitive detection of special protein molecules. Liposome-mediated biosensors have successfully monitored organophosphorus pesticides, such as dicchlorvos and paraoxon at very low levels. The nanosized liposomes provide a suitable environment for the effective stabilization of acethylcholinesterase (AChE), and they can be utilized as fluorescent biosensors. Porins embedded into the lipid membrane allow for the free substrate and pesticide transport into the liposomes. Pesticide concentrations down to 10^{-10} M can be monitored (Mahajan et al. 2020). The analytical technology is dominated by mass spectrometry that involves the identification of species by molecular mass measurement that forms the basis of chemical and biological research. Nanometer scale analytical systems are based upon nanosized materials, and are limiting in size and weight. Nanotechnological analytical aspects exploit physicochemical properties of nanomaterials. This exceptionally accounts for most current uses of analytical nanoscience. Metallic nanoparticles can be used to increase the sensitivity of fluorescent detection because they generate a phenomenon known as metal-enhanced fluorescence, increasing fluorescence lifetime and quantum yields (Wang et al. 2017, Yang et al. 2013). This may result in localized surface plasmon resonance (LSPR). Quantum dots rest many of the problems related with organic fluorophores in near infrared spectroscopy. The designing of quantum dots comprises an inorganic core and shell of metal and an outer organic coating. The inorganic core and shell enable tuning of fluorescence with a narrow bandwidth enabling multiplexed detection of molecular targets. The amalgamation of novel concepts into analytical nanosystems is at a developing stage. By attaching molecules (individual retinal chromophores found in receptor cells) to C60 molecules, it is likely to trap them inside single-walled carbon nanotubes and imaging them using HR-TEM, thus ensuring that individual molecules are well separated and are protected from electron-beam damage during the observation. Multiple molecularly resolved images would be obtained with sub-second time resolution. Nanoscale carbon-based materials, including single-walled nanotubes (SWNTs), multiple-walled nanotubes (MWNTs), and carbon nanotubes, and fibers can be used in sensing functional groups, electrospun nanofibers are a good sorbent substrate for solid phase extraction-based techniques.

Nanoelectromechanical systems (NEMS) resonators can detect mass with exceptional sensitivity. Previously, mass spectra from several hundred adsorption events were assembled in NEMS-based mass spectrometry using statistical analysis. As each molecule in the sample adsorbs on the resonator, its mass and position of adsorption are determined by continuously tracking two driven vibrational modes of the device. The development of new, delicate sample-handling methods for molecular ionization/injection, enabling so-called 'native' mass spectrometry has been essential in permitting large molecules or molecular assemblies to be transported, intact, from the fluid phase to the vacuum phase for subsequent analysis (Caballero-Quintana et al. 2020). Nanoelectromechanical systems (NEMS)-based mass spectrometry (NEMS–MS) is sensitive to the inertial mass of neutral particles that accrete on the resonators, which makes it particularly well-suited to studies that require minimal ionization to avoid structural changes in the protein under investigation (Fan et al. 2012). It was the first NEMS-based mass spectrometer which overcame this problem by delivering the analytes in a way that ensured they were adsorbed uniformly across the surface of a doubly-clamped NEMS resonator (for which the relation between adsorption position and frequency shift is well known). This allowed the constituents of simple mixtures to be determined after the collection of only several hundred single-molecule adsorption events. For comparison, conventional mass spectrometry measurements typically involve the measurement of approximately 10^8 molecules. The real potential of NEMS-based mass spectrometry, by measuring the mass of individual protein macromolecules has been realized in real time. In particular, NEMS–MS systems are used to access masses greater than 500 kDa, while the performance of conventional mass spectrometry systems degrades at high masses. Improvization of the mass resolution of top-down fabricated nanomechanical devices by recent developments are attainable in the near future, offering exciting prospects for real world implications in bacterial identification, native mass spectrometry, and structural identification of large macromolecules. Recent work has significantly focused and achieved the mass resolution of bottom-up fabricated NEMS devices (Fan et al. 2012), and now presents realistic potential for ultimately generating NEMS–MS spectrometers with resolution down to a few daltons.

However, certain aspects under research include: bottom-up NEMS devices and approaches have yet to demonstrate mass measurements of individual molecules, and questions remain with large-scale integration and their compatibility. Attaining the capability of using very-large-scale-integration- (VLSI-) and CMOS-compatible NEMS with devices providing single-Dalton sensitivity enables the measurement of numerous proteins–in real time–from a small discrete sample (like a single cell), while retaining precision over the single-protein to a full range of biological interest.

CONCLUSION

Nanosized particles have emerged as state-of-the-art smart molecules in the area of regenerative medicine with harnessed and focused close links with the various cell types. By continuing the various innovative ways of synthesis techniques and formulation excipients, further understanding of the exact mechanism of nanomaterials in the cellular milieu will be elucidated. This provides in depth understanding of action to manipulate the size and shape of the nanoparticles through intelligent design for their use in drug delivery in order to fight various diseases.

ACKNOWLEDGMENTS

No funding in form of sponsorship was received for the publication of this article. All named authors meet the criteria for authorship for this manuscript, take equal responsibility for the integrity of the manuscript as a whole, and have given final approval for the version to be published.

References

Alizadeh Sani, M., A. Ehsani and M. Hashemi. 2017. Whey protein isolate/cellulose nanofibre/TiO. Int. J. Food Microbiol. 251: 8–14.

Allijn, I.E., W. Leong, J. Tang, A. Gianella, A.J. Mieszawska, F. Fay, et al. 2013. Gold nanocrystal labeling allows low-density lipoprotein imaging from the subcellular to macroscopic level. ACS Nano. 7: 9761–9770.

Anjugam Vandarkuzhali, S.A., N. Pugazhenthiran, R.V. Mangalaraja, P. Sathishkumar, B. Viswanathan and S. Anandan. 2018. Ultrasmall plasmonic nanoparticles decorated hierarchical mesoporous TiO. ACS Omega. 3: 9834–9845.

Arnedo, A., S. Espuelas and J.M. Irache. 2002. Albumin nanoparticles as carriers for a phosphodiester oligonucleotide. Int. J. Pharm. 244: 59–72.

Bai, H.P., X.X. Lu, G.M. Yang and Y.H. Yang. 2008. Hydrogen peroxide biosensor based on electrodeposition of zinc oxide nanoflowers onto carbon nanotubes film electrode. Chin. Chem. Lett. 19: 314–318.

Bayda, S., M. Hadla, S. Palazzolo, P. Riello, G. Corona, G. Toffoli, et al. 2018. Inorganic nanoparticles for cancer therapy: A transition from lab to clinic. Curr. Med. Chem. 25: 4269–4303.

Bergin, I.L., L.A. Wilding, M. Morishita, K. Walacavage, A.P. Ault, J.L. Axson, et al. 2016. Effects of particle size and coating on toxicologic parameters, fecal elimination kinetics and tissue distribution of acutely ingested silver nanoparticles in a mouse model. Nanotoxicology. 10: 352–360.

Biswas, R., J. Furtado and B. Bagchi. 2013. Layerwise decomposition of water dynamics in reverse micelles: A simulation study of two-dimensional infrared spectrum. J. Chem. Phys. 139: 144906.

Bowerman, S., R.J. Hickok and J. Wereszczynski. 2019. Unique dynamics in asymmetric macroH2A-H2A hybrid nucleosomes result in increased complex stability. J. Phys. Chem. B. 123: 419–427.

Breger, J.C., K. Susumu, G. Lasarte-Aragonés, S.A. Díaz, J. Brask and I.L. Medintz. 2020. Quantum Dot Lipase Biosensor Utilizing a Custom-Synthesized Peptidyl-Ester Substrate. ACS Sens.

Briggs, B.D. and M.R. Knecht. 2012. Nanotechnology meets biology: Peptide-based methods for the fabrication of functional materials. J. Phys. Chem. Lett. 3: 405–418.

Caballero-Quintana, I., D. Romero-Borja, J.L. Maldonado, J. Nicasio-Collazo, O. Amargós-Reyes and A. Jiménez-González. 2020. Interfacial energetic level mapping and nano-ordering of small molecule/fullerene organic solar cells by scanning tunneling microscopy and spectroscopy. Nanomaterials (Basel) 10.

Carazo, E., A. Borrego-Sanchez, C. Aguzzi, P. Cerezo and C. Viseras. 2017a. Use of Clays as nanocarriers of first-line tuberculostatic drugs. Curr. Drug Deliv. 14: 902–903.

Carazo, E., A. Borrego-Sánchez, F. García-Villén, R. Sánchez-Espejo, C. Aguzzi, C. Viseras, et al. 2017b. Assessment of halloysite nanotubes as vehicles of isoniazid. Colloids Surf. B Biointerfaces. 160: 337–344.

Carazo, E., A. Borrego-Sánchez, F. García-Villén, R. Sánchez-Espejo, P. Cerezo, C. Aguzzi, et al. 2018. Advanced inorganic nanosystems for skin drug delivery. Chem. Rec. 18: 891–899.

Cattani-Scholz, A. 2017. Functional organophosphonate interfaces for nanotechnology: A review. ACS Appl. Mater. Interfaces. 9: 25643–25655.

Duhachek, S.D., J.R. Kenseth, G.P. Casale, G.J. Small, M.D. Porter and R. Jankowiak. 2000. Monoclonal antibody–gold biosensor chips for detection of depurinating carcinogen–DNA adducts by fluorescence line-narrowing spectroscopy. Anal Chem. 72: 3709–3716.

El-Sayed, I.H., X. Huang and M.A. El-Sayed. 2005. Surface plasmon resonance scattering and absorption of anti-EGFR antibody conjugated gold nanoparticles in cancer diagnostics: Applications in oral cancer. Nano Lett. 5: 829–834.

Ezpeleta, I., M.A. Arangoa, J.M. Irache, S. Stainmesse, C. Chabenat, Y. Popineau, et al. 1999. Preparation of *Ulex europaeus* lectin-gliadin nanoparticle conjugates and their interaction with gastrointestinal mucus. Int. J. Pharm. 191: 25–32.

Fan, Z., X. Tao, X. Cui, X. Fan, X. Zhang and L. Dong. 2012. Metal-filled carbon nanotube based optical nanoantennas: bubbling, reshaping, and *in situ* characterization. Nanoscale. 4: 5673–5679.

Fire, A.Z. 2007. Gene silencing by double-stranded RNA (Nobel Lecture). Angew. Chem. Int. Ed. Engl. 46: 6966–6984.

Fukushima, D., U.H. Sk, Y. Sakamoto, I. Nakase and C. Kojima. 2015. Dual stimuli-sensitive dendrimers: Photothermogenic gold nanoparticle-loaded thermo-responsive elastin-mimetic dendrimers. Colloids Surf. B Biointerfaces. 132: 155–160.

Ge, Y., J.H. Priester, L.C. Van De Werfhorst, S.L. Walker, R.M. Nisbet, Y.-J. An, et al. 2014. Soybean plants modify metal oxide nanoparticle effects on soil bacterial communities. Environ. Sci. Technol. 48: 13489–13496.

Guchhait, B., R. Biswas and P.K. Ghorai. 2013. Solute and solvent dynamics in confined equal-sized aqueous environments of charged and neutral reverse micelles: A combined dynamic fluorescence and all-atom molecular dynamics simulation study. J. Phys. Chem. B. 117: 3345–3361.

Izadi, Z., A. Divsalar, A.A. Saboury and L. Sawyer. 2016. *β*-lactoglobulin-pectin Nano-particle-based oral drug delivery system for potential treatment of colon cancer. Chem. Biol. Drug Des. 88: 209–216.

Jackman, J.A., A. Rahim Ferhan and N.J. Cho. 2017. Nanoplasmonic sensors for biointerfacial science. Chem. Soc. Rev. 46: 3615–3660.

Kamaraj, C., P.R. Gandhi, G. Elango, S. Karthi, I.M. Chung and G. Rajakumar. 2018. Novel and environmental friendly approach; Impact of Neem (*Azadirachta indica*) gum nano formulation (NGNF) on *Helicoverpa armigera* (Hub.) and *Spodoptera litura* (Fab.). Int. J. Biol. Macromol. 107: 59–69.

Kashani, A.S., K. Kuruvinashetti, D. Beauet, S. Badilescu, A. Piekny and M. Packirisamy. 2018. Enhanced internalization of indian ayurvedic swarna bhasma (Gold Nanopowder) for effective interaction with human cells. J. Nanosci. Nanotechnol. 18: 6791–6798.

Khalid, K., X. Tan, H.F. Mohd Zaid, Y. Tao, C. Lye Chew, D.T. Chu, et al. 2020. Advanced in developmental organic and inorganic nanomaterial: A review. Bioengineered. 11: 328–355.

Lee, E.J., S.A. Khan and K.H. Lim. 2011. Gelatin nanoparticle preparation by nanoprecipitation. J. Biomater. Sci. Polym. Ed. 22: 753–771.

Li, C.Z., K.B. Male, S. Hrapovic and J.H. Luong. 2005. Fluorescence properties of gold nanorods and their application for DNA biosensing. Chem. Commun. (Camb). 31: 3924–3926.

Liu, X., P.Y. Lee, C.M. Ho, V.C. Lui, Y. Chen, C.M. Che, et al. 2010. Silver nanoparticles mediate differential responses in keratinocytes and fibroblasts during skin wound healing. Chem. Med. Chem. 5: 468–475.

Liu, F. and C.H. Tang. 2013. Soy protein nanoparticle aggregates as pickering stabilizers for oil-in-water emulsions. J. Agric. Food Chem. 61: 8888–8898.

Madhyastha. H., R. Madhyastha, Y. Nakajima, H. Daima, N.P. Navya and M. Masugi. 2019. An opinion on nano-medicine and toxicology-cellular crosstalk: Considerations and caveats. Materials Today: Proceedings. 10: 100–105.

Mahajan, K.D., G. Ruan, G. Vieira, T. Porter, J.J. Chalmers, R. Sooryakumar, et al. 2020. Biomolecular detection, tracking, and manipulation using a magnetic nanoparticle-quantum dot platform. J. Mater. Chem B. 8: 3534–3541.

Maki, W.C., N.N. Mishra, E.G. Cameron, B. Filanoski, S.K. Rastogi and G.K. Maki. 2008. Nanowire-transistor based ultra-sensitive DNA methylation detection. Biosens. Bioelectron. 23: 780–787.

Meles, S.K., D. Vadasz, R.J. Renken, E. Sittig-Wiegand, G. Mayer, C. Depboylu, et al. 2017. FDG PET, dopamine transporter SPECT, and olfaction: Combining biomarkers in REM sleep behavior disorder. Mov. Disord. 32: 1482–1486.

Metwally, A.A., S.H. El-Ahmady and R.M. Hathout. 2016. Selecting optimum protein nano-carriers for natural polyphenols using chemoinformatics tools. Phytomedicine. 23: 1764–1770.

Michael, F.H., K.L. Steven, M. Patrica, R.L. Pen, S. Nita, D.L. Sparks, et al. 2008. Nano minerals mineral nano Particles and earth systems. Science. 319: 1631–1635.

Mickoleit, F., C.B. Borkner, M. Toro-Nahuelpan, H.M. Herold, D.S. Maier, J.M. Plitzko, et al. 2018. *In vivo* coating of bacterial magnetic nanoparticles by magnetosome expression of spider silk-inspired peptides. Biomacromolecules. 19: 962–972.

Mieszawska, A.J., Y. Kim, A. Gianella, I. van Rooy, B. Priem, M.P. Labarre, et al. 2013. Synthesis of polymer-lipid nanoparticles for image-guided delivery of dual modality therapy. Bioconjug. Chem. 24: 1429–1434.

Mirab, F., Y. Wang, H. Farhadi and S. Majd. 2019. Preparation of gel-liposome nanoparticles for drug delivery applications. Conf. Proc. IEEE Eng. Med. Biol. Soc. 2019: 3935–3938.

Mirzaei, H. and M. Darroudi. 2017. Zinc oxide nanoparticles: Biological synthesis and biomedical applications. Ceram. Int. 43: 907–914.

Mishra, D., Iyyanki, T.S., Hubenak, J.R., Zhang, Q. and Mathur, A.B., 2017. Silk fibroin nanoparticles and cancer therapy. pp. 19–44. *In*: A.B. Mathur(ed.). Nanotechnology in Cancer, Elsevier.

Mohamed, M.A., S.A. Atty, H.A. Merey, T.A. Fattah, C.W. Foster and C.E. Banks. 2017. Titanium nanoparticles (TiO_2)/graphene oxide nanosheets (GO): An electrochemical sensing platform for the sensitive and simultaneous determination of benzocaine in the presence of antipyrine. Analyst. 142: 3674–3679.

Mohamed, M.B., V. Volkov, S. Link and M.A. El-Sayed. 2000. The 'lightning' gold nanorods: Fluorescence enhancement of over a million compared to the gold metal. Chem. Phys. Lett. 317: 517–523.

Moran, M., N. Rosell, G. Ruano, M. Busquets and M. Vinardell. 2015. Gelatin-based nanoparticles as DNA delivery systems: Synthesis, physicochemical and biocompatible characterization. Coll. Surf. B. 134: 156–168.

Mukherjee, S., R. Kotcherlakota, S. Haque, D. Bhattacharya, J.M. Kumar, S. Chakravarty, et al. 2020. Improved delivery of doxorubicin using rationally designed PEGylated platinum nanoparticles for the treatment of melanoma. Mater. Sci. Eng. C. 108: 110375.

Nagakura, T., H. Hirata, M. Tsujii, T. Sugimoto, K. Miyamoto, T. Horiuchi, et al. 2005. Effect of viscous injectable pure alginate sol on cultured fibroblasts. Plast. Reconstr. Surg. 116: 831–838.

Navya, P.N., H. Madhyastha, R. Madhyastha, Y. Nakajima, M. Maruyama, S.P. Srinivas, et al. 2019. Single step formation of biocompatible bimetallic alloy nanoparticles of gold and silver using isonicotinylhydrazide. Mater. Sci. Eng. C Mater. Biol. Appl. 96: 286–294.

Pandey, P.K., Preeti, K. Rawat, T. Prasad, and H.B. Bohidar. 2020. Multifunctional, fluorescent DNA-derived carbon dots for biomedical applications: Bioimaging, luminescent DNA hydrogels, and dopamine detection. J. Mater. Chem. B. 8: 1277–1289.

Park, E.J., S.W. Kim, C. Yoon, Y. Kim and J.S. Kim. 2016. Disturbance of ion environment and immune regulation following biodistribution of magnetic iron oxide nanoparticles injected intravenously. Toxicol. Lett. 243: 67–77.

Parker, R.N., W.A. Wu, T.B. McKay, Q. Xu and D.L. Kaplan. 2019. Design of silk-elastin-like protein nanoparticle systems with mucoadhesive properties. J. Funct. Biomater. 10: 23–28.

Patra, C.R., C.N. Rupasinghe, S.K. Dutta, S. Bhattacharya, E. Wang, M.R. Spaller, et al. 2019. Chemically-modified peptides targeting the PDZ domain of GIPC as a therapeutic approach for cancer. ACS Chem. Biol. 14: 2327.

Perez Herrera, M. and T. Vasanthan. 2018. Rheological characterization of gum and starch nanoparticle blends. Food Chem. 243: 43–49.

Raghav, R. and S. Srivastava. 2016. Immobilization strategy for enhancing sensitivity of immunosensors: L-Asparagine–AuNPs as a promising alternative of EDC–NHS activated citrate–AuNPs for antibody immobilization. Biosens Bioelectron. 78: 396-403.

Raphael, A.P., D. Sundh, J.E. Grice, M.S. Roberts, H.P. Soyer and T.W. Prow. 2013. Zinc oxide nanoparticle removal from wounded human skin. Nanomedicine (Lond). 8: 1751–1761.

Rezaei, L., M.S. Safavi and S.A. Shojaosadati. 2019. Protein nanocarriers for targeted drug delivery. pp. 199–218. *In*: S.S. Mohapatra, S. Ranjan, N. Dasgupta, R.K. Mishra and S. Thomas (eds). Micro and Nano Technologies: Characterization and Biology of Nanomaterials for Drug Delivery. Elsevier.

Rohiwal, S.S., N. Dvorakova, J. Klima, M. Vaskovicova, F. Senigl, M. Slouf, et al. 2020. Polyethylenimine based magnetic nanoparticles mediated non-viral CRISPR/Cas9 system for genome editing. Sci. Rep. 10: 4619.

Sato, K., M. Sawayanagi, K. Hosokawa and M. Maeda. 2004. Single-base mutation detection using neutravidin-modified polystyrene nanoparticle aggregation. Anal. Sci. 20: 893–894.

Schiek, M., F. Balzer, K. Al-Shamery, J.R. Brewer, A. Lützen and H.G. Rubahn. 2008. Organic molecular nanotechnology. Small. 4: 176–181.

Schirra, C.O., A. Senpan, E. Roessl, A. Thran, A.J. Stacy, L. Wu, et al. 2012. Second generation gold nanobeacons for robust K-Edge imaging with multi-energy CT. J. Mater. Chem. 22: 23071–23077.

Septiadi, D., W. Abdussalam, L. Rodriguez-Lorenzo, M. Spuch-Calvar, J. Bourquin, A. Petri-Fink, et al. 2018. Revealing the role of epithelial mechanics and macrophage clearance during pulmonary epithelial injury recovery in the presence of carbon nanotubes. Adv. Mater. 30: e1806181.

Sikora, K.N., J.M. Hardie, L.J. Castellanos-García, Y. Liu, B.M. Reinhardt, M.E. Farkas, et al. 2020. Dual mass spectrometric tissue imaging of nanocarrier distributions and their biochemical effects. Anal. Chem. 92: 2011–2018.

Son, S.J., X. Bai and S.B. Lee. 2007. Inorganic hollow nanoparticles and nanotubes in nanomedicine Part 1. Drug/gene delivery applications. Drug Discov. Today. 12: 650–656.

Sturm, R. 2017. Carbon nanotubes in the human respiratory tract-clearance modeling. Ann. Work Expo Health. 61: 226–236.

Sutariya, V., S.J. Kelly, R.G. Weigel, J. Tur, K. Halasz, N.S. Sharma, et al. 2019. Nanoparticle drug delivery characterization for fluticasone propionate and *in vitro* testing. Can. J. Physiol. Pharmacol. 97: 675–684.

Tanaka, Y., Y. Saijo, Y. Fujihara, H. Yamaoka, S. Nishizawa, S. Nagata, et al. 2012. Evaluation of the implant type tissue-engineered cartilage by scanning acoustic microscopy. J. Biosci. Bioeng. 113: 252–257.

Tanbour, R., A.M. Martins, W.G. Pitt and G.A. Hussein. 2016. Drug delivery systems based on polymeric micelles and ultrasound: A review. Curr. Pharm. Des. 22: 2796–2807.

Tanhaian, A., E. Mohammadi, R. Vakili-Ghartavol, M.R. Saberi, M. Mirzayi and M.R. Jaafari. 2020. *In silico* and *in vitro* investigation of a likely pathway for anti-cancerous effect of Thrombocidin-1 as a novel anticancer peptide. Protein Pept Lett. 27: 751–762.

Taniguchi, K., S. Urasawa and T. Urasawa. 1979. Virus-like particle, 35 to 40 nm, associated with an institutional outbreak of acute gastroenteritis in adults. J. Clin. Microbiol. 10: 730–736.

Thakkar, H.P., A.K. Baser, M.P. Parmar, K.H. Patel and R. Ramachandra Murthy. 2012. Vincristine-sulphate-loaded liposome-templated calcium phosphate nanoshell as potential tumor-targeting delivery system. J. Liposome. Res. 22: 139–147.

Umapathi, A., N.P. Nagaraju, H. Madhyastha, D. Jain, S.P. Srinivas, V.M. Rotello, et al. 2019. Highly efficient and selective antimicrobial isonicotinylhydrazide-coated polyoxometalate-functionalized silver nanoparticles. Colloids Surf. B Biointerfaces. 184: 11–22.

Vandivort, T.C., T.P. Birkland, T.P. Domiciano, S. Mitra, T.J. Kavanagh and WC. Parks. 2017. Stromelysin-2 (MMP-10) facilitates clearance and moderates inflammation and cell death following lung exposure to long multiwalled carbon nanotubes. Int. J. Nanomedicine. 12: 1019–1031.

Wang, Y., L. Zhang, S. Guo, A. Hatefi and L. Huang. 2013. Incorporation of histone derived recombinant protein for enhanced disassembly of core-membrane structured liposomal nanoparticles for efficient siRNA delivery. J. Controlled Release. 172: 179–189.

Wang, F., X. Zhang, Y. Lu, J. Yang, W. Jing, S. Zhang, et al. 2017. Continuously evolving 'chemical tongue' biosensor for detecting proteins. Talanta. 165: 182–187.

Xia, Z., G. Wang, K. Tao and J. Li. 2005. Preparation of magnetic dextran microspheres by ultrasonication. J. Mag. and Mag. Mat. 293: 182–186.

Yang, X.H., H.T. Fu, K. Wong, X.C. Jiang and A.B. Yu. 2013. Hybrid Ag@TiO_2 core-shell nanostructures with highly enhanced photocatalytic performance. Nanotechnology. 24: 415601.

Yang Q. and R.H. Guy. 2015. Characterization of skin barrier function using bioengineering and biophysical techniques. Pharm. Res. 32: 445–457.

You, C.C., O.R. Miranda, B. Gider, P.S. Ghosh, I.B. Kim, B. Erdogan, et al. 2007. Detection and identification of proteins using nanoparticle-fluorescent polymer 'chemical nose' sensors. Nat. Nanotechnol. 2: 318–323.

Zhang, X., Q. Guo and D. Cui. 2009. Recent advances in nanotechnology applied to biosensors. Sensors (Basel) 9: 1033–1053.

Zhao, Y., G. Chen, Z. Meng, G. Gong, W. Zhao, K. Wang, et al. 2019. A novel nanoparticle drug delivery system based on PEGylated hemoglobin for cancer therapy. Drug Deliv. 26: 717–723.

8

Microbial Synthesis of Nanoparticles and Their Applications

Meryam Sardar* **and Jahirul Ahmed Mazumder**

Department of Biosciences
Jamia Millia Islamia, New Delhi 110025, India
E-mail: msardar@jmi.ac.in; jahir.jmi57@gmail.com

INTRODUCTION

Nanoparticles find applications in diverse areas beneficial to human beings, and as a result this area of research is developing extensively (Silva 2008, Mandal and Ganguly 2011). Different types of nanoparticles have been described in literature according to their size, morphology, and chemical properties; the present chapter describes only metal nanoparticles in detail and their microbial synthesis.

Metal nanoparticles have been attracting considerable attention and are studied extensively because of their exceptional photochemical features, catalytic activity (Sreeju et al. 2017), anti-microbial activity (Rosarin and Mirunalini 2011), magnetic (Yamamoto et al. 2003), electronic and optical properties (McConnell et al. 2000). In 1831, Michael Faraday proposed that the red color of colloidal gold is due to the small size of metal particles (Brown et al. 2008).

Metal nanoparticle synthesis can be easily monitored, as some metal particles give a characteristic Surface Plasmon Resonance, which can be determined by UV-Visible spectroscopy (Cao et al. 2010, Gautam et al. 2013). These nanoparticles can be characterized by techniques such as X-ray diffraction, and the size and shape can be determined by electron microscopy. Due to their small size, fraction of

*For Correspondence: msardar@jmi.ac.in

atoms present on the surface are more as compared to inner core, thus making them highly reactive materials; this increases the chances of aggregation. To prevent aggregation, their surfaces are generally modified by various surfactants, polymers, and biomolecules (Cao et al. 2010, Gautam et al. 2013). Nanoparticles exist as connections between bulk materials and atomic structure (Rahman et al. 2009, Kulkarni et al. 2015), but have advanced properties as compared to element metal (Perni et al. 2014). Due to their high surface area to volume ratio, they act as excellent adsorbents, possess chemo-catalytic activity, and are ideal candidates for protein immobilization (Roduner 2006, Gautam et al. 2013).

SYNTHESIS OF NANOPARTICLES

Till date, many papers have been published on methods describing synthesis of metal nanoparticles, but broadly two approaches have been described: top-down and bottom-up (Yabu 2013, Wijesena et al. 2015). Examples of top-down approach include thermal decomposition, irradiation, diffusion, arc discharge, etc. (Yabu 2013, Wijesena et al. 2015), and bottom-up approach includes polyol synthesis method, seeded growth method, electrochemical synthesis, etc. (Singh et al. 2011). Mainly all physical methods are included in top-down process, and chemical and biological methods of synthesis come under bottom-up approach. So far, reduction of metals using chemicals is the most frequently used method, but the use of hazardous chemicals as reducing agents, such as hydroxylamine, sodium borohydride is harmful to humans, and also pollutes the environment (Binupriya et al. 2009). Thus, there is a demand to use eco-friendly approach for production of nanoparticles at a large scale (Kharissova et al. 2013, Mishra et al. 2013). Biological synthesis of nanoparticles with controlled morphology can be an alternative to chemical reduction methods, and needs more attention, as this process can be carried by using microbes, plants, and their biomolecules (Kharissova et al. 2013, Rauwel et al. 2015, Fariq et al. 2017).

Synthesis of nanoparticles by microbial cells is an exciting approach, and the microbial cells are viewed as bio-factories for the preparation of a number of different types of nanoparticles (Hulkoti and Taranath 2014, Fariq et al. 2017). Moreover, it is reported by several investigators that the nanoparticles synthesized by biological agents are less toxic compared to nanoparticles synthesized by chemical and physical means (Ayele et al. 2019, Fakhari et al. 2019). The first evidence of Ag nanoparticle synthesis was reported by Haefeli et al., using the bacteria *Pseudomonas stutzeri* AG259 (Haefeli et al. 1984). They suggested that the defence mechanism of the bacteria is involved in the process of synthesis (Haefeli et al. 1984). This result is further validated by many researchers, that to overcome the metal stress, the cellular machinery of the bacteria reduces the reactive metal ions into stable metal atoms (Slawson et al. 1994, Ali et al. 2019)

Microbes are classified into different categories based on their cellular composition and morphology, among them bacteria, and fungi, are extensively explored for the synthesis of nanoparticles. The various biomolecules (enzymes, proteins, peptides, amino acids, polysaccharides, and vitamins) present in microbes are involved in the synthesis, as well as the stabilization of the nanoparticles (Durán et al. 2011). Metal nanoparticles can be synthesized by microbes either by extracellular or intracellular means (Reverberi et al. 2017). Table 8.1 depicts the synthesis and applications of different types of nanoparticles by various kinds of microbes. The probable mechanism and the synthesis using bacteria, fungi, algae, and virus are mentioned below.

Table 8.1 Synthesis of metal nanoparticles using microbes and their applications

Microbes	***Types of Nanoparticles***	***Size* (nm)**	***Application***	***References***
		Bacteria		
Pseudomonas stutzeri AG259	Ag	200	Surface coating	(Klaus et al. 1999)
Stenotrophomonas sp. BHU-S7	Ag	12	Management of phytopathogens	(Mishra et al. 2017)
Bacillus species	Ag	140	Antimicrobial	(Ghiuță et al. 2018)
Shewanellaloihica PV-4	Cu	10–16	Antibacterial	(Lv et al. 2018)
Lactobacillus acidophilus	Se	2–15	Cytotoxic	(Alam et al. 2019)
Desulforibrio caledoiensis	CdS	40–50	Sensor	(Qi et al. 2016)
Shewanella sp. KR-12	Pb	3–8	Wastewater treatment	(Liu and Yen 2016)
Enterobacter cloacae Z0206	Se	100–300	NA	(Song et al. 2017)
Streptomyces sp. HBUM171191	$MnSO_4$, $ZnSO_4$,	10–20	NA	(Waghmare et al. 2011)
Escherichia coli	Quantum dots (QDs)	2.0–3.2	Cytotoxicity and cell imaging	(Bao et al. 2010)
Bacillus sp. (MTCC10650)	MnO_2	4–5	NA	(Sinha et al. 2011)
Lactobacillus sp. Beijerinck	Au crystals	20–50	Investigation of biomolecules	(Nair and Pradeep 2002)
Enterobacter sp.	Hg	2–5	NA	(Sinha and Khare 2011)
Rhodopseudomonas capsulata	Gold nanowires	10–20	NA	(He et al. 2008)
		Fungi		
Aspergillus oryzae	Ag	5–50	Bactericidal	(Binupriya et al. 2009)
Alternaria alternata	Ag	7–12	NA	(Kareem et al. 2019)
Aspergillus flavus	TiO_2	18	Nano fertilizer	(Raliya et al. 2015)
Fusarium oxysporum	Fe_3O_4	20–50	NA	(Bansal et al. 2005)
Arthroderma fulvum	Ag	15	Antifungal	(Xue et al. 2016)
Penicillium brevicompactum	Au	10–50	Anticancer activity	(Mishra et al. 2011)

Table 8.1 .... (*contd...*)

Table 8.1 (Contd.) Synthesis of metal nanoparticles using microbes and their applications

Microbes	*Types of Nanoparticles*	*Size* (nm)	*Application*	*References*
Verticillium sp.	Fe_3O_5	100–400	NA	(Bharde et al. 2006)
Aspergillus tubingensis	$Ca_3P_2O_8$	28.2	Cytotoxicity	(Abdullaeva et al. 2012)
Coriolus versicolor (L.) Quél.	CDs	100–200	NA	(Sanghi and Verma 2009)
Aspergillus terreus Thom	ZnO	54–82	Antifungal	(Baskar et al. 2013)
Fusarium solani	ZnO nanorods	60–95	NA	(Venkatesh et al. 2013)
		Algae		
Shewanella algae	Au	10–20	NA	(Konishi et al. 2007a)
Sargassum muticum	ZnO	30–57	NA	(Azizi et al. 2013)
Bifurcaria bifurcata	Cu	5–45	Antibacterial	(Abboud et al. 2014)
Shewanella algae	Pt	5	NA	(Konishi et al. 2007b)
Sargassum plagiophyllum	Ag	21–48	Antibacterial	(Dhas et al. 2014)
Cystophora moniliformis	Ag	75	NA	(Prasad et al. 2013)
Spirulina platensis	Au core-Ag shell	17–25	NA	(Govindaraju et al. 2008)
Chlorella vulgaris	Pd	5–20	NA	(Arsiya et al. 2017)
Sargassum muticum	Fe_3O_4	18	NA	(Mahdavi et al. 2013)
Kappaphycus alvarezii	Cu_2O	53	Antimicrobial	(Khanehzaei et al. 2014)
		Virus		
Tobacco mosaic virus	CDS/PbS	22	NA	(Shenton et al. 1999)
Cowpea Mosaic Virus (CPMV)	Au	2–5	NA	(Blum et al. 2004)
Engineered M13 bacteriophage	Au and CdSe nanocrystals/ nanowires	10	Electrical transport studies	(Huang et al. 2005)
Cloned virus (p8#9 and E4)	Metal nanowire	50	Electrochemical activity	(Lee et al. 2010)
M13 Virus	Perovskite nanomaterials	10–13	Solar Energy Conversion	(Nuraje et al. 2012)

NA: Not Available

Nanoparticle Biosynthesis by Bacteria

Different species of bacteria are present in the environment; some are pathogenic, while others are beneficial to humans. An important characteristic feature of bacteria is that they can adapt to extreme environmental conditions; this feature makes them favorable candidates for the synthesis of nanoparticles. Moreover, bacteria have a very high growth rate, therefore the synthesis of nanomaterial is fast and economical (Pantidos and Horsfall 2014). When subjected to metal ion stress, this metal is accumulated in nano-form either intracellularly or extracellularly (Malik et al. 2014, Gahlawat and Choudhury 2019).

Metal or metal ions can be synthesized either intracellularly or extracellularly by three different ways, as shown in Figure 8.1.

1. Metal ions (M) can be reduced by interacting with cell wall components.
2. Metal ions can interact with extracellular proteins, enzymes, and organic molecules.
3. Metal ions can also interact with intracellular proteins and other co-factors.

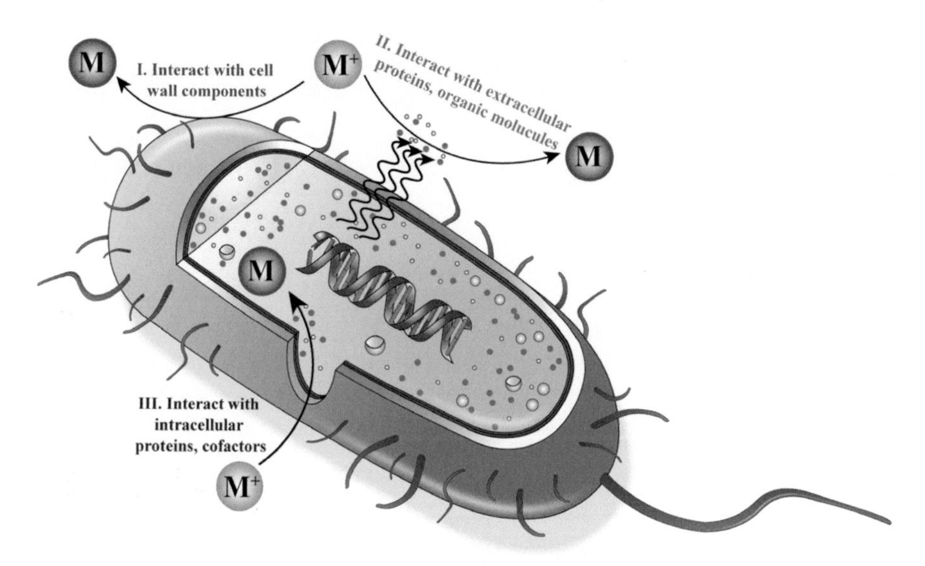

Figure 8.1 Schematic representation of the synthesis of nanoparticles by bacteria by both intracellular and extracellular process. (Reprinted from Fang et al. 2019).

Generally extracellular reduction is preferred over intracellular reduction, due to its low cost, simpler extraction, and higher efficiency. In the intracellular process, as shown in Figure 8.1, carboxyl groups present on the cell wall interact with the metal ions by electrostatic interactions. Then, the ions enter the cells and interact with cellular proteins and co-factors to produce nanoparticles (Kulkarni and Muddapur 2014). In addition, many studies have shown that not only the living bacteria, but also the dead entities of some bacteria are also capable of synthesizing metal nanoparticles (Sneha et al. 2010).

Beveridge and Murray (1980) studied the interaction of metal with the biomolecules/proteins present in the cell wall of *Bacillus subtilis*. They chemically modified the amine and carboxyl groups of the cell wall in order to neutralize the charges present on the cell wall. They found that the carboxyl group of proteins present in the cell wall of *Bacillus subtilis* play a major role in interaction with the metal. In another report, *Pseudomonas stutzeri* reduces an aqueous solution of $AgNO_3$ into silver nanoparticles of different sizes and shapes (Klaus et al. 1999). They further mentioned that temperature and pH play crucial roles in determining the morphology of the synthesized nanoparticle (Klaus et al. 1999). Gurunathan et al. (2009a, b) showed that by regulating the pH and temperature of the reaction mixture, silver nanoparticles of various sizes and morphologies can be synthesized. Further, they mentioned that silver nanoparticles having size of 50 nm were synthesized at room temperature, and at 60°C, the size reduced to 15 nm. Similarly, at acidic pH, large sized nanoparticles of 45 nm were formed, whereas when pH was increased to 10, the size of the nanoparticle decreased to approximately 15 nm. This size variation at different pH is due to the generation of different amounts of seed crystals. The rate of nucleation is slow at acidic pH and at low temperatures, due to fewer amounts of OH^- ions. Once the pH and temperature increases, the interaction of the ions increases, and thus rate of nucleation increases due to the presence of higher concentration of OH^- ions (Gurunathan et al. 2009a, b).

Actinomycetes are mainly gram-positive bacteria; they produce a number of biomolecules which have medical and industrial importance. Ahmad et al. (2003a) reported that an alkalotolerant actinomycete, *Rhodococcus* species, synthesized gold nanoparticles intracellularly. They reported that the secondary metabolites present in this species play an essential role. Recently, the biomass extract of endophytic *Streptomyces* sp. was used for the synthesis of copper nanoparticles, which exhibited good antibacterial, antioxidant, and larvicidal properties (Fouda et al. 2019).

Nanoparticle Biosynthesis by Fungi

Mycosynthesis is the field of study where a fungal system is used for the synthesis of metal nanomaterials. Synthesis by fungi takes place at room temperature and at neutral pH, hence considered a part of green chemistry (Chhipa 2019). A number of research papers describe the mode of synthesis and various applications of nanoparticles with different genera of fungi; a few are mentioned in Table 8.1. Fungi are reported to synthesize nanogold, nanosilver, and quantum dots, such as cadmium sulphide (CdS) (Sastry et al. 2003, Bhainsa and D'souza 2006), reason being they can be grown easily and produce different enzymes, which play a crucial role in the synthesis. Among fungi, the mycelia forms secrete a large amount of enzymes/proteins, and are generally considered better for the synthesis of nanomaterial (Ahmad et al. 2003b). There is one specific drawback associated with the fungal mode of synthesis, which is that genetic modification of a specific enzyme of fungal system is more difficult as compared to that in prokaryotes (Thakkar et al. 2010). When fungal mycelium is subjected to metal salt stress, it produces a number of biomolecules, such as enzymes, proteins, organic acids, and polysaccharides as a defence mechanism. These biomolecules in turn reduce the toxic metal ions into metallic solid nanoparticles by means of the catalytic effect of the enzymes and other biomolecules released during the process (Prathna et al. 2010, Mansoori 2013). Vetchinkina et al. (2018) studied the synthesis of different metal nanoparticles using various species of basidomycetes (*Pleurotus ostreatus,*

Lentinus edodes, *Ganoderma lucidum*, and *Grifola frondosa*). Transmission electron microscope (TEM) analysis revealed that depending on the type of fungal species and the metal salt used, the nanoparticles of various shapes and size can be obtained, as shown in Figures 8.2 to 8.5. Thus, by selecting the fungal strains, one can synthesize the nanoparticles of desired shape and size.

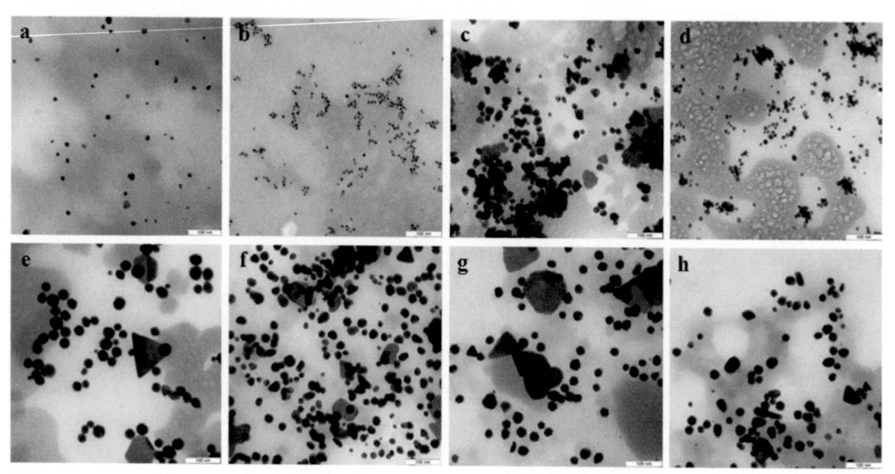

Figure 8.2 TEM of Au nanoparticles. Nanoparticles produced from $HAuCl_4$ with extracellular (a–d) and intracellular (e–h) extracts of *L. edodes* (a, e), *P. ostreatus* (b, f), *G. lucidum* (c, g), and *G. frondosa* (d, h). Bar marker –100 nm. (Reprinted from Vetchinkina et al. 2018).

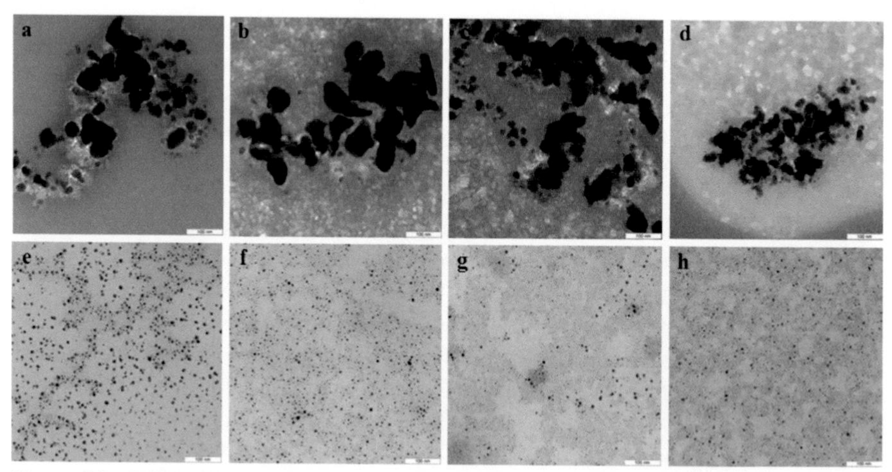

Figure 8.3 TEM of Ag nanoparticles. Nanoparticles produced from $AgNO_3$ with extracellular (a–d) and intracellular (e–h) extracts of *L. edodes* (a, e), *P. ostreatus* (b, f), *G. lucidum* (c, g), and *G. frondosa* (d, h). Bar marker –100 nm. (Reprinted from Vetchinkina et al. 2018).

Mukherjee et al. (2001) used *Verticillium* fungus, which intracellularly synthesized silver nanoparticles; they explained the role of enzymes of cell wall. These enzymes reduce $AuCl_4^-$ into Ag nanoparticles. Fungal isolate *Aspergillus fumigatus* also synthesized nanoparticles which were effective against bacteria

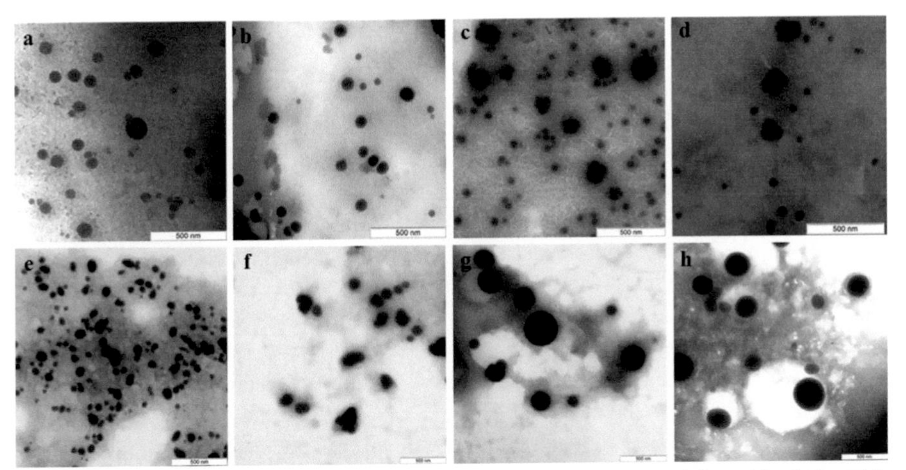

Figure 8.4 TEM of Se/SeO_2 nanoparticles. Nanoparticles produced from Na_2SeO_3 with extracellular (a–d) and intracellular (e–h) extracts of *L. edodes* (a, e), *P. ostreatus* (b, f), *G. lucidum* (c, g), and *G. frondosa* (d, h). Bar marker –500 nm. (Reprinted from Vetchinkina et al. 2018).

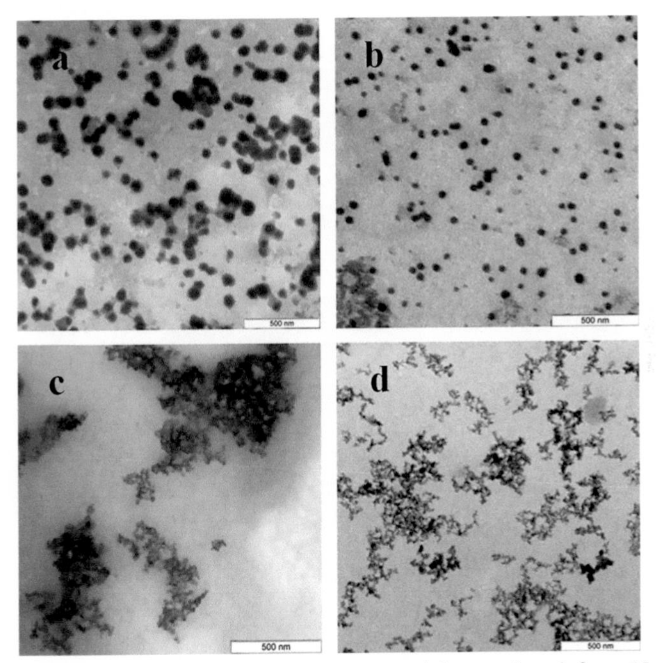

Figure 8.5 TEM of Si/SiO_2 nanoparticles. Nanoparticles produced from Na_2SiO_3 with extracellular extracts of *L. edodes* (a), *G. lucidum* (b), *P. ostreatus* (c), and *G. frondosa* (d). Bar marker –500 nm. (Reprinted from Vetchinkina et al. 2018).

(Sarsar et al. 2016). A non-pathogenic strain of *Alternaria* was used for the synthesis of Ag nanoparticles; it was suggested that proteins and secondary metabolites present on endophytic fungi facilitated the reduction of $AgNO_3$ and capping of Ag nanoparticles (Abdel-Hafez et al. 2016). Sastry et al. (2003) used *Verticillium* sp. for the synthesis of Ag nanoparticles, and they explained two mechanisms that led to the formation of Ag nanoparticles. The first mechanism is where there is an interaction between the Ag^+ and carboxylic groups of the fungal enzymes, followed by the reduction of Ag^+ ions into nuclei of the silver particles, which further grow to form nanoparticles. It has also been reported that extracellular cationic proteins of *Fusarium oxysporum* facilitate the formation of SiO_2 and TiO_2 nanoparticles (Bansal et al. 2005). Jain et al. (2011) suggested that protein molecules of *Aspergillus flavus* help in the synthesis and stabilization of Ag nanoparticles. It has also been reported that sulfur-containing amino acids (–SH group in cysteine) mediate the formation of nanoparticles; the electron from *β*-carbon of cysteine and a hydrogen radical helps in the reduction of Ag^+ into Ag nanoparticles by means of NADPH reductase enzyme (Mukherjee et al. 2008), as shown in Figure 8.6. It has also been speculated that temperature and pH of the reaction mixture play a crucial role in determining the morphology and dispersity of nanoparticle. The pH regulates protonation of amino acids, which helps in the stabilization and synthesis (Gericke and Pinches 2006).

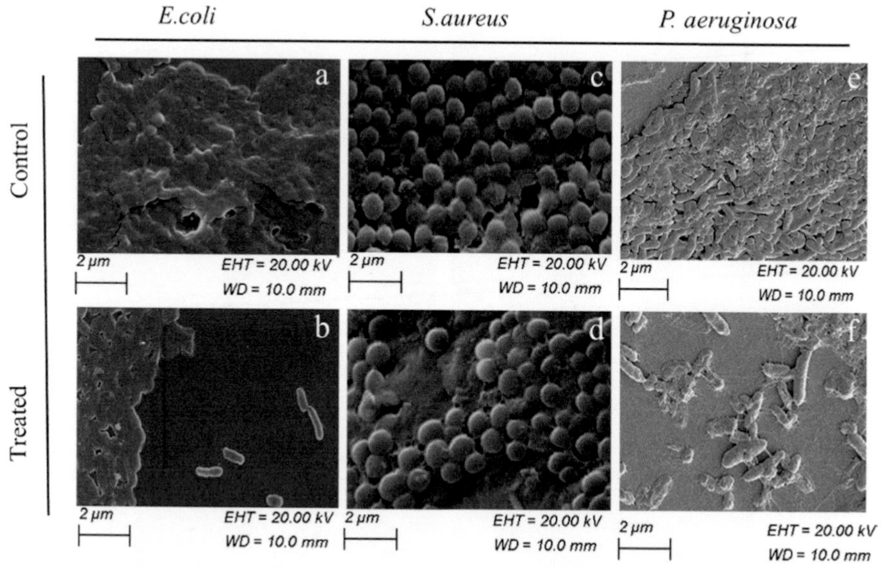

Figure 8.6 SEM images of *E. coli*, *S. aureus*, and *P. aeruginosa* with LA-Se nanoparticles at MIC concentration. All the values are means of triplicate ($n = 3$)±SD. ANOVA significant at $P \leq 0.05$. (Reprinted from Alam et al. 2019 with permission from SpringerLink).

Nanoparticle Biosynthesis by Algae

Not much research has been carried out in the field of synthesis of nanoparticles using algae as compared to plants and other microbes, as a result of which this area remains largely unexplored (Merin et al. 2010, Arya et al. 2018). Among algae, marine macroalgae are most commonly studied (Sangeetha et al. 2014,

Jin et al. 2016, Abdel-Raouf et al. 2017). Few examples of algae synthesizing different metal nanoparticles are highlighted in Table 8.1.

Azizi et al. (2013) studied that the sulphated polysaccharides in aqueous extract of *Sargassum muticum* were important in the synthesis of ZnO nanoparticle. Singaravelu and co-workers used marine algae, *Sargassum wightii* Greville for the synthesis of well-dispersed gold nanoparticles of size 8–12 nm (Singaravelu et al. 2007). Silver nanoparticles were synthesized using polysaccharides from different algae, and coated on cotton fabrics for antimicrobial activity (El-Rafie et al. 2013). Abboud et al. (2014) reported the synthesis of antibacterial CuO nanoparticles of 5–45 nm size using brown algae *Bifurcaria bifurcate*, and it was found that water-soluble di-terpenoids present in the algae extract were important for synthesis.

Virus in Nanoparticle Biosynthesis

Some viruses are naked viruses and some are enveloped, i.e., coated by lipid membranes. Those without external envelopes are more suitable for nanomaterial synthesis, because the amino acids present in capsid proteins can interact with metal (Krajina et al. 2018). Plant viruses are non-pathogenic to humans, and thus they are widely explored for the synthesis of nanomaterials (Bittner et al. 2013). Chemically cross-linked Cowpea chlorotic mottle virus (CCMV) cage was studied to synthesize TiO_2 nanoparticles (Klem et al. 2008), and it was found that the interior surface of the viral capsid regulates the morphology of the nanoparticle. Shenton et al. (1999) reported when metal sulphides, such as CdS and PbS were deposited on the outer surface of wild-type Tobacco Mosaic Virus (TMV), they formed metal nanotubes. Blum et al. (2004) incorporated cysteine into the capsid of Cowpea mosaic virus (CPMV), and treated it with gold salt, which led to the deposition of gold particles. Huang et al. (2005) used M13 bacteriophage to assemble various nanoparticles as arrays, hetero-nanoparticle, and nanowires. Such arrays have the ability to further act as templates that can nucleate nanowires into templates (Huang et al. 2005). Lee et al. (2010) used two cloned viruses (p8#9 and E4) to synthesize metal nanowires having diameter of below 50 nm. These nanowires were further used for electrochemical activities in lithium ion batteries.

APPLICATIONS OF NANOPARTICLES WITH SPECIFIC REFERENCE TO MICROBIAL NANOSYNTHESIS

Since the last ten years, the research on nanomaterials has increased, and many chemical companies are manufacturing nanoparticles, and these are commercially available for various applications (Bystrzejewska-Piotrowska et al. 2009, Dutton 2018, Ko and Huh 2019, Lim et al. 2019) Use of nanoparticles in medicine include tissue engineering, drug delivery, bioimaging, wound healing, and in antimicrobial therapy (Nitta and Numata 2013). Applications are not limited to just medicine; these nanomaterials have a positive impact on agriculture and food sector as well. This section aims to introduce several significant applications of nanotechnology in different inter-disciplinary fields.

Bio-medical Applications

Metal nanoparticles, carbon nanotubes, liposomes, etc. are extensively studied for bio-medical applications, like in drug delivery, dental implants, and diagnosis (Kubik et al. 2005, Saji et al. 2010, Mazumder et al. 2018). Early diagnosis of diseases is the need of the day; nanobodies (recombinantly produced antigen binding fragments) have the potential to be used as biomarkers as screening tool in diseases, such as cancer. The advantages of nanobodies are that they are highly stable, have high binding affinity, are highly specific, and have low toxicity (Revets et al. 2005). The composition, structure, and surface characteristics of nanoparticles make them ideal for drug delivery; their surface can be modified and drug can be conjugated for specific targets. Liposomes, micelles, dendrimers, nanospheres, and nanocapsules are mostly used in drug delivery (Liechty and Peppas 2012). Quantum dots (QDs) play a significant role in diagnosis; they are slowly replacing fluorescent dyes (Fluorescein isothiocyanate FITC), and have been used in fluorescence *in situ* hybridization (FISH) assays for diagnosis of genetic disorders (Davis 1997).

Nanoparticles acts as potent antimicrobial agents (Mazumder et al. 2019). Cationic metal nanoparticles generally bind negatively charged proteins and/or nucleic acids of the microorganisms, thus damaging the bacterial cell wall, membranes, and disrupting the DNA molecules (Galdiero et al. 2014). Silver, gold, iron, etc. ions of the metal nanoparticles interact with the electron donor groups, such as thiols, hydroxyls, and amines and thus inactivate them. It is reported that silver ions bind to DNA and block the transcription machinery; they also bind to ribosomes and inhibit translation and adenosine triphosphate (ATP) synthesis (Dallas et al. 2011). Alam et al. (2019) synthesized selenium nanoparticles using *Lactobacillus acidophilus*, a probiotic bacteria. TEM studies showed that the nanoparticles of 2–15 nm sizes were obtained using culture filtrate of this bacterium, and show potent antibacterial activity against different bacterial strains. Further, the inhibition of bacterial biofilm (shown in Figure 8.7) was studied, and the results indicate they have potent antibacterial activity and degrade the biofilm, and thus can be used in various applications.

Applications in Textile Industries

Nano-Tex, a US-based textile industry was the first to report the applications of nanomaterials in textile industry (Russell 2002). Nanoparticles can be conjugated to textiles to improve their properties. Coating of nanoparticles on fabrics can be achieved by different techniques, such as spraying, transfer printing, washing, rinsing, and padding (Kale and Meena 2012). Titanium dioxide and Zinc oxide nanoparticles when immobilized on cotton fabric provide UV-protection (Mao et al. 2009, Kale and Meena 2012). ZnO, TiO_2, zinc sulphide, hematite, etc. are photocatalytic materials, and their photocatalytic property as self-cleaning was exploited in the field of textile (Montazer and Pakdel 2011, Patra and Gouda 2013). It is reported that a nano-TiO_2 coated fabric is antibacterial and is a self-cleaning fabric; it also decolorizes the stains (Sundaresan et al. 2012). Liu et al. (2014) coated silver nanoparticles on cotton fabric, which exhibited excellent antibacterial properties against *Escherichia coli* and *Staphylococcus aureus* even after 50 laundering cycles of usage. Silver nanoparticles (synthesized using marine micro algae *Pterocladia capillacae* and *Jania rubens*) treated cotton fabrics showed excellent antibacterial activity (El-Rafie et al. 2013). Silver nanoparticles coated jute fiber showed antibacterial resistance against *Bacillus subtilis* and *Escherichia coli* and washing durability of these coated fibers were up to 15 home launderings (Lakshmanan and Chakraborty 2017). Gold nanoparticles when immobilized onto soybean knitted fabric revealed an ultraviolet protection factor (UPF) of +50 and an improved antimicrobial effect against *Staphylococcus aureus* and *Escherichia coli* (Silva et al. 2019).

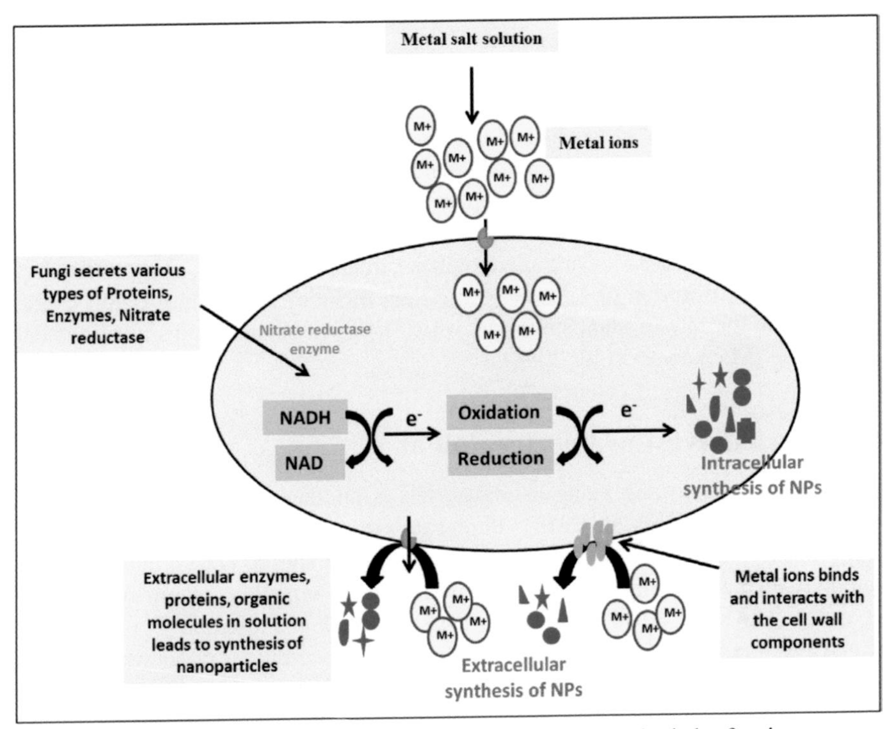

Figure 8.7 Possible mechanism of nanoparticles synthesis by fungi. (Reprinted from Khandel and Shahi 2018).

Agricultural Applications

Agriculture is the major source of food and nutrients for meeting the demands of the growing population of a country. Agriculture acts as an interface between humans and the environment, and any negative impacts of human on agriculture can affect the environment adversely (Tilman 1999). Pesticides generate free radicals in plants and increase the level of antioxidant enzymes (Singh et al. 2007). Continuous use of pesticides for protection of plants from pests can also lead to resistance of pests from pesticides (Zamani et al. 2011). Use of nanoparticles in agriculture can provide us the sustainable crop yield, improved variety, and increased productivity without affecting human health (Shojaei et al. 2019). Nanoparticles and nanocarriers can be used for enhancing plant growth, enhancing yield of crop, soil improvement, and water purification (Prasad et al. 2014). Nanoparticles, such as those of iron oxide, silver, and gold have been utilized as nanopesticides (Kah et al. 2013). Carbon nanotubes and alumina fibers can be used for nanofiltration, which desalinates irrigation water (Shanmuganathan et al. 2015) and increases yield of crops by requiring 25% less irrigation water than required when unfiltered water is used for irrigation. Liu and Lal (2014) studied the role of nanoparticles as fertilizers; they compared the effect of conventional phosphorus fertilizer and carboxy methyl cellulose-stabilized apatite nanoparticles on soybean (*Glycine max*) and found that the use of nanoparticles increased the plant growth rate (Liu and Lal 2014). Krishnaraj et al. (2012) also showed that when *Bacopa monnieri* was treated with nano silver, there was increase in plant growth. As nanomaterials are small in size, they can penetrate into cells and tissues and alter their mechanisms (Mailander and Landfester 2009). This increases plant growth and their metabolism, seed germination, shelf life, crop yield, crop length, and also helps in the development of microbial resistant plants (Pérez-de-Luque and Rubiales 2009, Ghormade et al. 2011, Sekhon 2014). Carbon nanotubes can enter into chlorophyll molecules, participate in absorption of UV rays more than those of normal absorption range, and increase the rate of photosynthesis, which helps to enhance nutritional content of the crop (Mackowski et al. 2008).

Nano-based Wastewater Treatment

Treatment of wastewater using nanomaterials is another important application of nanotechnology (Qu et al. 2013). Photocatalyst nanoparticles (titanium dioxide (TiO_2), zinc oxide (ZnO), ferric oxide (Fe_2O_3), zinc sulfide (ZnS), and cadmium sulfide (CdS)) are commonly used to degrade various organic contaminants present in water (Hu et al. 2013, Qu et al. 2013, Mondal and Sharma 2014). Eggins et al. (1997) reported 50% reduction of humic acid in drinking water when treated with TiO_2 nanoparticles. ZnO nanoparticles under visible light remove potassium cyanide and Cr(Vi) ions from water (Farrokhi et al. 2013). Chen et al. (2011) used CdS/Titanate nanotubes for oxidation of ammonia in water. A number of heavy metals can be removed using photocatalytic nanomaterials (Borgarello et al. 1986, Chen et al. 2011). Zinc oxide (ZnO) nanoparticles have an antimicrobial effect on *Escherichia coli* and *Staphylococcus aureus* present in wastewater (Dimapilis et al. 2018). Carbon nanomaterials are the most widely used materials for water filtration membranes because they have very unique mechanical, thermal, electrical, and chemical properties. Srivastava et al. (2004) show that carbon nanotubes can efficiently remove heavy hydrocarbons and bacterial contaminants from water. Khatoon et al. (2018) reported that silver nanoparticles can efficiently adsorb textile dyes and also remove microbial and other contaminants from water sample collected from Yamuna. It has also been reported that zinc-aluminum layered double hydroxide act as nanosorbent, and can be used to remove several yellow dyes from textile wastewater effluents (Abdolmohammad-Zadeh et al. 2013). Silver nanoparticles immobilized on silane modified glass were used to remove toxic textile dyes from aqueous solution and microbial contaminants from Yamuna river (Mazumder et al. 2019). It has been reported that the photocatalytic activity of gold nanoparticles was employed in the reduction of organic pollutants, such as methyl orange, bromocresol green, etc. from wastewater (Choudhary et al. 2017). Thus, nanotechnology can overcome the issues of wastewater treatment in the future.

Applications in Food Industries

Silver nanostructures are incorporated in food storage containers as antimicrobial agents. Nanoparticles have wide applications in the food industry—they are used in food packaging, food preservation, and as nano-sensors to detect food spoilage (Omanović-Mikličanin and Maksimović 2016).

CONCLUSIONS

The chapter highlights synthesis of metal nanoparticles using microbes and their applications. The microbial synthesis involves interaction of metal salts with the biomolecules. The biomolecules, such as carbohydrates, amino acids, and enzymes, which can donate the free electrons, reduce the metals, and the polymeric biomolecules serve as the stabilizing agents. Diversity of microorganisms is present in nature, and they can be genetically modified to express the desired biomolecules, and thus can be implemented for the cost-effective synthesis. These synthesized nanoparticles are used in medicine, wastewater treatment, and agriculture and food industry. One major concern using microorganisms is that one cannot control the size and morphology of the nanoparticles formed. Future research should be carried out to control the size and the mechanism of biosynthesis. This technology has the potential to be used for large scale synthesis.

ACKNOWLEDGMENTS

Author Jahirul Ahmed Mazumder acknowledges the Indian Council of Medical Research (ICMR), Government of India for financial support in the form of Senior Research Fellowship (SRF).

References

Abboud, Y., T. Saffaj, A. Chagraoui, A. El Bouari, K. Brouzi, O. Tanane, et al. 2014. Biosynthesis, characterization and antimicrobial activity of copper oxide nanoparticles (CONPs) produced using brown alga extract (*Bifurcaria bifurcata*). Appl. Nanosci. 4: 571–576.

Abdel-Hafez, S.I., N.A. Nafady, I.R. Abdel-Rahim, A.M. Shaltout, J.A. Daròs and M.A. Mohamed. 2016. Assessment of protein silver nanoparticles toxicity against pathogenic *Alternaria solani*. 3 Biotech. 6: 199.

Abdel-Raouf, N., N.M. Al-Enazi and I.B. Ibraheem. 2017. Green biosynthesis of gold nanoparticles using *Galaxaura elongata* and characterization of their antibacterial activity. Arab. J. Chem. 10: S3029–S3039.

Abdolmohammad-Zadeh, H., E. Ghorbani and Z. Talleb. 2013. Zinc–aluminum layered double hydroxide as a nano-sorbent for removal of reactive yellow 84 dye from textile wastewater effluents. J. Iran Chem. Soc. 10: 1103–1112.

Abdullaeva, Z., E. Omurzak, C. Iwamoto, H.S. Ganapathy, S. Sulaimankulova, C. Liliang, et al. 2012. Onion-like carbon-encapsulated Co, Ni, and Fe magnetic nanoparticles with low cytotoxicity synthesized by a pulsed plasma in a liquid. Carbon. 50: 1776–1785.

Ahmad, A., S. Senapati, M.I Khan, R. Kumar, R. Ramani, V. Srinivas, et al. 2003a. Intracellular synthesis of gold nanoparticles by a novel alkalotolerant actinomycete, *Rhodococcus* species. Nanotechnology. 14: 824.

Ahmad, A., P. Mukherjee, S. Senapati, D. Mandal, M.I. Khan, R. Kumar, et al. 2003b. Extracellular biosynthesis of silver nanoparticles using the fungus *Fusarium oxysporum*. Colloid Surf. B. 28: 313–318.

Alam, H., N. Khatoon, M.A. Khan, S.A. Husain, M. Saravanan and M. Sardar. 2019. Synthesis of selenium nanoparticles using probiotic bacteria *Lactobacillus acidophilus* and their enhanced antimicrobial activity against resistant bacteria. J. Clust. Sci. 31: 1003–1011.

Ali, J., N. Ali, L. Wang, H. Waseem and G. Pan. 2019. Revisiting the mechanistic pathways for bacterial mediated synthesis of noble metal nanoparticles. J. Microbiol. Methods. 159: 18–25.

Arsiya, F., M.H. Sayadi and S. Sobhani. 2017. Green synthesis of palladium nanoparticles using *Chlorella vulgaris*. Mater. 186: 113–115.

Arya, A., K. Gupta, K, T.S. Chundawat and D. Vaya. 2018. Biogenic synthesis of copper and silver nanoparticles using green alga *Botryococcus braunii* and its antimicrobial activity. Bioinorg. Chem. Appl. 1–9.

Ayele, A., R.S. Mujmdar, T. Addisu and Y. Woinue. 2019. Green synthesis of silver nanoparticles for various biomedical and agro industrial application. J. Nanosci. Nanotechnol. 5: 694–698.

Azizi, S., F. Namvar, M. Mahdavi, M. Ahmad and R. Mohamad. 2013. Biosynthesis of silver nanoparticles using brown marine macroalga, *Sargassum muticum* aqueous extract. Materials. 6: 5942–5950.

Bansal, V., D. Rautaray, A. Bharde, K. Ahire, A. Sanyal and M. Sastry. 2005. Fungus-mediated biosynthesis of silica and titania particles. J. Mater. Chem. 15: 2583–2589.

Bao, H., Z. Lu, X. Cui, Y. Qiao, J. Guo and C.M. Li. 2010. Extracellular microbial synthesis of biocompatible CdTe quantum dots. Acta Biomater. 6: 3534–3541.

Baskar, G., J. Chandhuru, K.S. Fahad and A.S. Praveen. 2013. Mycological synthesis, characterization and antifungal activity of zinc oxide nanoparticles. AJPTech. 3: 142–146.

Beveridge, T.J. and R.G. Murray. 1980. Sites of metal deposition in the cell wall of *Bacillus subtilis*. J. Bacteriol. Res. 141: 876–887.

Bhainsa, K.C. and S.F. D'souza. 2006. Extracellular biosynthesis of silver nanoparticles using the fungus *Aspergillus fumigatus*. Colloids Surf. B Biointerfaces. 47: 160–164.

Bharde, A., D. Rautaray, V. Bansal, A. Ahmad, I. Sarkar and M. Sastry. 2006. Extracellular biosynthesis of magnetite using fungi. Small. 2: 135–141.

Binupriya, A.R., M. Sathishkumar and S.I. Yun. 2009. Myco-crystallization of silver ions to nanosized particles by live and dead cell filtrates of *Aspergillus oryzae var. viridis* and its bactericidal activity toward *Staphylococcus aureus* KCCM 12256. Ind. Eng. Chem. Res. 49: 852–858.

Bittner, A.M., J.M. Alonso, M.L. Górzny and C. Wege. 2013. Nanoscale science and technology with plant viruses and bacteriophages. pp. 667–702. *In*: M. Mateu [ed.]. Structure and Physics of Viruses. Subcellular Biochemistry. Springer, Dordrecht.

Blum, A.S., C.M. Soto, C.D. Wilson, J.D. Cole, M. Kim, B. Gnade, et al. 2004. Cowpea mosaic virus as a scaffold for 3-D patterning of gold nanoparticles. Nano Lett. 4: 867–870.

Borgarello, E., N. Serpone, G. Emo, R. Harris, E. Pelizzetti and C. Minero. 1986. Light-induced reduction of rhodium (III) and palladium (II) on titanium dioxide dispersions and the selective photochemical separation and recovery of gold (III), platinum (IV), and rhodium (III) in chloride media. Inorg. Chem. 25: 4499–4503.

Brown, C.L., M.W. Whitehouse, E.R.T. Tiekink, and G.R. Bushell. 2008. Colloidal metallic gold is not bio-inert. Inflammopharmacology. 16: 133–137.

Bystrzejewska-Piotrowska, G., J. Golimowski and P.L. Urban. 2009. Nanoparticles: Their potential toxicity, waste and environmental management. Waste Manag. 29: 2587–2595.

Cao, A., R. Lu and G. Veser. 2010. Stabilizing metal nanoparticles for heterogeneous catalysis. Phys. Chem. Chem. Phys. 12: 13499–13510.

Chen, Y.C., S.L. Lo, H.H. Ou and C.H. Chen. 2011. Photocatalytic oxidation of ammonia by cadmium sulfide/titanate nanotubes synthesised by microwave hydrothermal method. Water Sci. Technol. 63: 550–557.

Chhipa, H. 2019. Mycosynthesis of nanoparticles for smart agricultural practice: A green and eco-friendly approach. pp. 87–109. *In*: A.K. Shukla and S. Iravani [eds]. Micro and Nano Technologies. Green Synthesis, Characterization and Applications of Nanoparticles. Elsevier.

Choudhary, B.C., D. Paul, T. Gupta, S.R. Tetgure, V.J. Garole, A.U. Borse, et al. 2017. Photocatalytic reduction of organic pollutant under visible light by green route synthesized gold nanoparticles. J. Environ. Sci. 55: 236–246.

Dallas, P., V.K. Sharma and R. Zboril. 2011. Silver polymeric nanocomposites as advanced antimicrobial agents: Classification, synthetic paths, applications, and perspectives. Adv. Colloid Interface Sci. 166: 119–135.

Davis, S.S. 1997. Biomedical applications of nanotechnology—implications for drug targeting and gene therapy. Trends Biotechnol. 15: 217–224.

Dhas, T.S., V.G. Kumar, V. Karthick, K.J. Angel and K. Govindaraju. 2014. Facile synthesis of silver chloride nanoparticles using marine alga and its antibacterial efficacy. Spectrochim. Acta A. 120: 416–420.

Dimapilis, E.A.S., C.S. Hsu, R.M.O. Mendoza and M.C. Lu. 2018. Zinc oxide nanoparticles for water disinfection. Sustain. Environ. Res. 28: 47–56.

Durán, N., P.D. Marcato, M. Durán, A. Yadav, A. Gade and M. Rai. 2011. Mechanistic aspects in the biogenic synthesis of extracellular metal nanoparticles by peptides, bacteria, fungi, and plants. Appl. Microbiol. Biotechnol. 90: 1609–1624.

Dutton, G. 2018. Nanoparticles for multiple applications: NanoComposix works at nanoscale to improve new drug development and diagnostics. Genet. Eng. Biotech. N. 38: 8–9.

Eggins, B.R., F.L. Palmer and J.A. Byrne, 1997. Photocatalytic treatment of humic substances in drinking water. Water Res. 31: 1223–1226.

El-Rafie, H.M., M.H. El-Rafie and M.K. Zahran. 2013. Green synthesis of silver nanoparticles using polysaccharides extracted from marine macro algae. Carbohydr. Polym. 96: 403–410.

Fakhari, S., M. Jamzad, and H.K. Fard. 2019. Green synthesis of zinc oxide nanoparticles: A comparison. Green Chem. Lett. Rev. 12: 19–24.

Fang, X., Y. Wang, Z. Wang, Z. Jiang and M. Dong. 2019. Microorganism assisted synthesized nanoparticles for catalytic applications. Energies. 12: 190.

Fariq, A., T. Khan and A. Yasmin. 2017. Microbial synthesis of nanoparticles and their potential applications in biomedicine. J. Appl. Biomed. 15: 241–248.

Farrokhi, M., J.K. Yang, S.M. Lee and M. Shirzad-Siboni. 2013. Effect of organic matter on cyanide removal by illuminated titanium dioxide or zinc oxide nanoparticles. J. Environ. Health Sci. 11: 23.

Fouda, A., S.E.D. Hassan, A.M. Abdo and M.S. El-Gamal. 2019. Antimicrobial, antioxidant and larvicidal activities of spherical silver nanoparticles synthesized by endophytic *Streptomyces* spp. Biol. Trace. Elem. Res. 1–18.

Gahlawat, G. and A.R. Choudhury. 2019. A review on the biosynthesis of metal and metal salt nanoparticles by microbes. RSC Adv. 9: 12944–12967.

Galdiero, S., A. Falanga, M. Cantisani, A. Ingle, M. Galdiero and M. Rai. 2014. Silver nanoparticles as novel antibacterial and antiviral agents. pp. 565–594. *In*: Torchilin, V. [ed.]. Handbook of Nanobiomedical Research: Fundamentals, Applications and Recent Developments, vol 1-4. World Scientific.

Gautam, S., P. Dubey and M.N. Gupta. 2013. A facile and green ultrasonic-assisted synthesis of BSA conjugated silver nanoparticles. Colloids Surf. B. 102: 879–883.

Gericke, M. and A. Pinches. 2006. Microbial production of gold nanoparticles. Gold Bull. 39: 22–28.

Ghiuță, I., D. Cristea, C. Croitoru, J. Kost, R. Wenkert and D. Munteanu. 2018. Characterization and antimicrobial activity of silver nanoparticles, biosynthesized using *Bacillus* species. Appl. Surf. Sci. 438: 66–73.

Ghormade, V., M.V. Deshpande and K.M. Paknikar. 2011. Perspectives for nano-biotechnology enabled protection and nutrition of plants. Biotechnol. Adv. 29: 792–803.

Govindaraju, K., S.K. Basha, V.G. Kumar and G. Singaravelu. 2008. Silver, gold and bimetallic nanoparticles production using single-cell protein (*Spirulina platensis*) Geitler. J. Mater. Sci. 43: 5115–5122.

Gurunathan, S., K. Kalishwaralal, R. Vaidyanathan, D. Venkataraman, S.R.K. Pandian and S.H. Eom. 2009a. Biosynthesis, purification and characterization of silver nanoparticles using *Escherichia coli.* Colloids Surf B Biointerfaces. 74: 328–335.

Gurunathan, S., K.J. Lee, K. Kalishwaralal, S. Sheikpranbabu, R. Vaidyanathan and S.H. Eom. 2009b. Antiangiogenic properties of silver nanoparticles. Biomaterials. 30: 6341–6350.

Haefeli, C., C. Franklin and K. Hardy. 1984. Plasmid-determined silver resistance in *Pseudomonas stutzeri* isolated from a silver mine. J. Bacteriol. 158: 389–392.

He, S., Y. Zhang, Z. Guo and N. Gu. 2008. Biological synthesis of gold nanowires using extract of *Rhodopseudomonas capsulata*. Biotechnol. Prog. 24: 476–480.

Hu, Y., X. Gao, L. Yu, Y. Wang, J. Ning and X.W. Lou. 2013. Carbon-coated CdS petalous nanostructures with enhanced photostability and photocatalytic activity. Angew. 52: 5636–5639.

Huang, Y., C.Y. Chiang, S.K. Lee, Y. Gao, E.L. Hu and A.M. Belcher. 2005. Programmable assembly of nanoarchitectures using genetically engineered viruses. Nano Lett. 5: 1429–1434.

Hulkoti, N.I. and T.C. Taranath. 2014. Biosynthesis of nanoparticles using microbes—A review. Colloids Surf B Biointerfaces. 121: 474–483.

Jain, N., A. Bhargava, S. Majumdar, J.C. Tarafdar and J. Panwar. 2011. Extracellular biosynthesis and characterization of silver nanoparticles using *Aspergillus flavus* NJP08: A mechanism perspective. Nanoscale. 3: 635–641.

Jin, J., C. Dupré, J. Legrand and D. Grizeau. 2016. Extracellular hydrocarbon and intracellular lipid accumulation are related to nutrient-sufficient conditions in pH-controlled chemostat cultures of the microalga *Botryococcus braunii* SAG 30.81. Algal Res. 17: 244–252.

Kah, M., S. Beulke, K. Tiede and T. Hofmann. 2013. Nanopesticides: State of knowledge, environmental fate, and exposure modeling. Crit. Rev. Env. Sci. Tec. 43: 1823–1867.

Kale, R. and C.R. Meena. 2012. Synthesis of titanium dioxide nanoparticles and application on nylon fabric using layer by layer technique for antimicrobial property. Adv. Appl. Sci. Res. 3: 3073–3080.

Kareem, S.O., O.T. Familola, A.R. Oloyede and E.O. Dare. 2019. Microbial synthesis and characterization of silver nanoparticles using *Alternaria alternata*. Appl. Env. Res. 4: 1–7.

Khandel, P. and S.K. Shahi. 2018. Mycogenic nanoparticles and their bio-prospective applications: Current status and future challenges. J. Nanostruc. Chem. 8: 369–391.

Khanehzaei, H., M.B. Ahmad, K. Shameli and Z. Ajdari. 2014. Synthesis and characterization of Cu@Cu2O core shell nanoparticles prepared in seaweed *Kappaphycus alvarezii* Media. Int. J. Electrochem. Sci. 9: 8189–8198.

Kharissova, O.V., H.R. Dias, B.I. Kharisov, O.B. Pérez and V.M.J. Pérez. 2013. The greener synthesis of nanoparticles. Trends Biotechnol. 31: 240–248.

Khatoon, N., H. Alam, N. Manzoor and M. Sardar. 2018. Removal of toxic contaminants from water by sustainable green synthesised non-toxic silver nanoparticles. IET Nanobiotechnol. 12: 1090–1096.

Klaus, T., R. Joerger, E. Olsson and C.G. Granqvist. 1999. Silver-based crystalline nano-particles, microbially fabricated. Proc. Natl. Acad. Sci. 96: 13611–13614.

Klem, M.T., M. Young and T. Douglas. 2008. Biomimetic synthesis of β-TiO_2 inside a viral capsid. J. Mater. Chem. 18: 3821–3823.

Ko, S. and C. Huh. 2019. Use of nanoparticles for oil production applications. J. Petrol Sci. Eng. 172: 97–114.

Konishi, Y., T. Tsukiyama, T. Tachimi, N. Saitoh, T. Nomura and S. Nagamine. 2007a. Microbial deposition of gold nanoparticles by the metal-reducing bacterium *Shewanella* algae. Electrochimica Acta. 53: 186–192.

Konishi, Y., K. Ohno, N. Saitoh, T. Nomura, S. Nagamine and T. Uruga. 2007b. Bioreductive deposition of platinum nanoparticles on the bacterium *Shewanella* algae. J. Biotechnol. 128: 648–653.

Krajina, B.A., A.C. Proctor, A.P. Schoen, A.J. Spakowitz and S.C. Heilshorn. 2018. Biotemplated synthesis of inorganic materials: An emerging paradigm for nanomaterial synthesis inspired by nature. Prog. Mater. Sci. 91: 1–23.

Krishnaraj, C., E.G. Jagan, R. Ramachandran, S.M. Abirami, N. Mohan and P.T. Kalaichelvan. 2012. Effect of biologically synthesized silver nanoparticles on *Bacopa monnieri* (Linn.) Wettst. plant growth metabolism. Process Biochem. 47: 651–658.

Kubik, T., K. Bogunia-Kubik and M. Sugisaka. 2005. Nanotechnology on duty in medical applications. Curr. Pharm. Biotechno. 6: 17–33.

Kulkarni, N. and U. Muddapur. 2014. Biosynthesis of metal nanoparticles: A review. J. Nanotechnol. 1–8.

Kulkarni, R.R., N.S. Shaiwale, D.N. Deobagkar and D.D. Deobagkar. 2015. Synthesis and extracellular accumulation of silver nanoparticles by employing radiation-resistant *Deinococcus radiodurans*, their characterization, and determination of bioactivity. Int. J. Nanomedicine. 10: 963.

Lakshmanan, A. and S. Chakraborty. 2017. Coating of silver nanoparticles on jute fibre by *in situ* synthesis. Cellulose. 24: 1563–1577.

Lee, Y.J., Y. Lee, D. Oh, T. Chen, G. Ceder and A.M. Belcher. 2010. Biologically activated noble metal alloys at the nanoscale: For lithium ion battery anodes. Nano Lett. 10: 2433–2440.

Liechty, W.B. and N.A. Peppas. 2012. Expert opinion: Responsive polymer nanoparticles in cancer therapy. Eur. J. Pharm. Biopharm. 80: 241–246.

Lim, J.M., T. Cai, S. Mandaric, S. Chopra, H. Han, S. Jang, et al. 2019. Drug loading augmentation in polymeric nanoparticles using a coaxial turbulent jet mixer: Yong investigator perspective. J. Colloid Interface Sci. 538: 45–50.

Liu, R. and R. Lal. 2014. Synthetic apatite nanoparticles as a phosphorus fertilizer for soybean (Glycine max). Sci. Rep. 4: 5686.

Liu, H., M. Lv. B. Deng, J. Li, M. Yu, Q. Huang, et al. 2014. Laundering durable antibacterial cotton fabrics grafted with pomegranate-shaped polymer wrapped in silver nanoparticle aggregations. Sci. Rep. 4: 5920.

Liu, C.L. and J.H. Yen. 2016. Characterization of lead nanoparticles formed by *Shewanella* sp. KR-12. J. Nanoparticle Res. 18: 30.

Lv, Q., B. Zhang, X. Xing, Y. Zhao, R. Cai, W. Wang. et al. 2018. Biosynthesis of copper nanoparticles using *Shewanella loihica* PV-4 with antibacterial activity: Novel approach and mechanisms investigation. J. Hazard. Mater. 347: 141–149.

Mackowski, S., S. Wörmke, A.J. Maier, T.H. Brotosudarmo, H. Harutyunyan, A. Hartschuh, et al. 2008. Metal-enhanced fluorescence of chlorophylls in single light-harvesting complexes. Nano Lett. 8: 558–564.

Mahdavi, M., F. Namvar, M. Ahmad and R. Mohamad. 2013. Green biosynthesis and characterization of magnetic iron oxide (Fe_3O_4) nanoparticles using seaweed (*Sargassum muticum*) aqueous extract. Molecules. 18: 5954–5964.

Mailander, V. and K. Landfester. 2009. Interaction of nanoparticles with cells. Biomacromolecules, 10: 2379–2400.

Malik, P., R. Shankar, V. Malik, N. Sharma and T.K. Mukherjee. 2014. Green chemistry based benign routes for nanoparticle synthesis. Journal of Nanoparticles. 1–14.

Mandal, G. and T. Ganguly. 2011. Applications of nanomaterials in the different fields of photosciences. Indian J. Phys. 85: 1229.

Mansoori, G.A. 2013. University of Illinois. Synthesis of nanoparticles by fungi. U.S. Patent 8, 394, 421.

Mao, Z., Q. Shi, L. Zhang and H. Cao. 2009. The formation and UV-blocking property of needle-shaped ZnO nanorod on cotton fabric. Thin Solid Films. 517: 2681–2686.

Mazumder, J.A., N. Khatoon, P. Batra and M. Sardar. 2018. Biosynthesized silver nanoparticles for orthodontic applications. Adv. Sci. Eng. Med. 10: 1169–1173.

Mazumder, J.A., M. Perwez, R. Noori and M. Sardar. 2019. Development of sustainable and reusable silver nanoparticle-coated glass for the treatment of contaminated water. Environ. Sci. Pollut. Res. Int. 26: 23070–23081.

McConnell, W.P., J.P. Novak, L.C. Brousseau, R.R. Fuierer, R.C. Tenent and D.L. Feldheim. 2000. Electronic and optical properties of chemically modified metal nanoparticles and molecularly bridged nanoparticle arrays. J. Phys. Chem. B. 104: 8930–8925.

Merin, D.D., S. Prakash and B.V. Bhimba. 2010. Antibacterial screening of silver nanoparticles synthesized by marine micro algae. Asian Pac. J. Trop. Med. 3: 797–799.

Mishra, A., S.K. Tripathy, R. Wahab, S.H. Jeong, I. Hwang, Y.B. Yang, et al. 2011. Microbial synthesis of gold nanoparticles using the fungus *Penicillium brevicompactum* and their cytotoxic effects against mouse mayo blast cancer C_2 C_{12} cells. Appl. Microbiol. Biotechnol. 92: 617–630.

Mishra, A., R. Ahmad, V. Singh, M.N. Gupta and M. Sardar. 2013. Preparation, characterization and biocatalytic activity of a nanoconjugate of alpha amylase and silver nanoparticles. J. Nanosci. Nanotechnol. 13: 5028–5033.

Mishra, S., B.R. Singh, A.H. Naqvi and H.B. Singh. 2017. Potential of biosynthesized silver nanoparticles using *Stenotrophomonas* sp. BHU-S7 (MTCC 5978) for management of soil-borne and foliar phytopathogens. Sci. Rep. 7: 45154.

Mondal, K. and A. Sharma. 2014. Photocatalytic oxidation of pollutant dyes in wastewater by TiO_2 and ZnO nano-materials—A mini-review. pp. 36–72. *In:* A. Misra and J.R. Bellare [eds.]. Nanoscience and Technology for Mankind. The National Academy of Sciences India (NASI), Allahabad, India.

Montazer, M. and E. Pakdel. 2011. Functionality of nano titanium dioxide on textiles with future aspects: Focus on wool. J. Photoch. Photobio. C. 12: 293–303.

Mukherjee, P., A. Ahmad, D. Mandal, S. Senapati, S.R. Sainkar, M.I. Khan, et al. 2001. Bioreduction of $AuCl_4^-$ ions by the fungus, *Verticillium* sp. and surface trapping of the gold nanoparticles formed. Angew. 40: 3585–3588.

Mukherjee, P., M. Roy, B.P. Mandal, G.K. Dey, P.K. Mukherjee, J. Ghatak, et al. 2008. Green synthesis of highly stabilized nanocrystalline silver particles by a non-pathogenic and agriculturally important fungus *T. asperellum*. Nanotechnology. 19: 075103.

Nair, B. and T. Pradeep. 2002. Coalescence of nanoclusters and formation of submicron crystallites assisted by *Lactobacillus* strains. Cryst. Growth Des. 2: 293–298.

Nitta, S.K. and K. Numata. 2013. Biopolymer-based nanoparticles for drug/gene delivery and tissue engineering. Int. J. Mol. 14: 1629–1654.

Nuraje, N., X. Dang, J. Qi, M.A. Allen, Y. Lei and A.M. Belcher. 2012. Biotemplated synthesis of perovskite nanomaterials for solar energy conversion. Adv. Mater. 24: 2885–2889.

Omanović-Mikličanin, E. and M. Maksimović. 2016. Nanosensors applications in agriculture and food industry. Bull. Chem. Technol. Bosnia Herzegovina. 47: 59–70.

Pantidos, N. and L.E. Horsfall. 2014. Biological synthesis of metallic nanoparticles by bacteria, fungi and plants. J. Nanomed. Nanotechnol. 5: 233.

Patra, J.K. and S. Gouda. 2013. Application of nanotechnology in textile engineering: An overview. J. Eng. Technol. Res. 5: 104–111.

Pérez-de-Luque, A. and D. Rubiales. 2009. Nanotechnology for parasitic plant control. Pest Manag. Sci. 6: 540–545.

Perni, S., V. Hakala and P. Prokopovich. 2014. Biogenic synthesis of antimicrobial silver nanoparticles capped with L-cysteine. Colloid Surf. A. 460: 219–224.

Prasad, T.N., V.S.R. Kambala and R. Naidu. 2013. Phyconanotechnology: synthesis of silver nanoparticles using brown marine algae *Cystophora moniliformis* and their characterisation. J. Appl. Phycol. 25: 177–182.

Prasad, R., V. Kumar and K.S. Prasad. 2014. Nanotechnology in sustainable agriculture: Present concerns and future aspects. Afr. J. Biotechnol. 13: 705–713.

Prathna, T.C., L. Mathew, N. Chandrasekaran, A.M. Raichur and A. Mukherjee. 2010. Biomimetic synthesis of nanoparticles: Science, technology & applicability. *In:* A. Mukherjee [ed]. Biomimetics Learning from Nature. Intechopen.

Qi, P., D. Zhang, Y. Zeng and Y. Wan. 2016. Biosynthesis of CdS nanoparticles: A fluorescent sensor for sulfate-reducing bacteria detection. Talanta. 147: 142–146.

Qu, X., P.J. Alvarez and Q. Li. 2013. Applications of nanotechnology in water and wastewater treatment. Water Res. 47: 3931–3946.

Rahman, I.A., P. Vejayakumaran, C.S. Sipaut, J. Ismail and C.K. Chee. 2009. Size-dependent physicochemical and optical properties of silica nanoparticles. Mater. Chem. Phys. 114: 328–332.

Raliya, R., P. Biswas and J.C. Tarafdar. 2015. TiO_2 nanoparticle biosynthesis and its physiological effect on mung bean (*Vigna radiata* L.). Biotechnol. Rep. 5: 22–26.

Rauwel, P., S. Küünal, S. Ferdov and E. Rauwel. 2015. A review on the green synthesis of silver nanoparticles and their morphologies studied via TEM. Adv. Mater. Sci. Eng. 2015.

Reverberi, A., M. Vocciante, E. Lunghi, L. Pietrelli and B. Fabiano. 2017. New trends in the synthesis of nanoparticles by green methods. Chem. Eng. Trans. 61: 667–672.

Revets, H., P. De. Baetselier and S. Muyldermans. 2005. Nanobodies as novel agents for cancer therapy. Expert Opin. Biol. Th. 5: 111–124.

Roduner, E. 2006. Size matters: Why nanomaterials are different. Chem. Soc. Rev. 35: 583–592.

Rosarin, F.S. and S. Mirunalini. 2011. Nobel metallic nanoparticles with novel biomedical properties. J. Bioanal. Biomed. 3: 85–91.

Russell, E., 2002. Nanotechnologies and the shrinking world of textiles. Textile Horizons. 9: 7–9.

Saji, V.S., H.C. Choe and K.W. Yeung. 2010. Nanotechnology in biomedical applications: A review. Int. J. Nan. Biomat. 3: 119–139.

Sangeetha, S., N.B. Dhayanithi and N. Sivakumar. 2014. Antibacterial activity of *Sargassum longifolium* and *Gracilaria corticata* from Gulf of Mannar against selected common shrimp pathogens. Int. J. Pharma. Bio. Sci. 5: 76–82.

Sanghi, R. and P. Verma. 2009. A facile green extracellular biosynthesis of CdS nanoparticles by immobilized fungus. Chem. Eng. 155: 886–891.

Sarsar, V., M.K. Selwal and K.K. Selwal. 2016. Biogenic synthesis, optimisation and antibacterial efficacy of extracellular silver nanoparticles using novel fungal isolate *Aspergillus fumigatus* MA. IET Nanobiotechnol. 10: 215–221.

Sastry, M., A. Ahmad, M.I. Khan and R. Kumar. 2003. Biosynthesis of metal nanoparticles using fungi and actinomycete. Curr. Sci. 85: 162–170.

Sekhon, B.S. 2014. Nanotechnology in agri-food production: An overview. Nanotechnol. Sci. Appl. 7: 31–53.

Shanmuganathan, S., S. Vigneswaran, T.V. Nguyen, P. Loganathan and J. Kandasamy. 2015. Use of nanofiltration and reverse osmosis in reclaiming micro-filtered biologically treated sewage effluent for irrigation. Desalination. 364: 119–125.

Shenton, W., T. Douglas, M. Young, G. Stubbs and S. Mann. 1999. Inorganic–organic nanotube composites from template mineralization of tobacco mosaic virus. Adv. Mat. 11: 253–256.

Shojaei, T.R., M.A.M. Salleh, M. Tabatabaei, H. Mobli, M. Aghbashlo, S.A. Rashid, et al. 2019. Applications of nanotechnology and carbon nanoparticles in agriculture. pp. 247–277. *In:* S.A. Rashid, R.N.I.R. Othman and M.Z. Hussein [eds]. Synthesis, Technology and Applications of Carbon Nanomaterials. Elsevier.

Silva, G.A. 2008. Nanotechnology approaches to crossing the blood-brain barrier and drug delivery to the CNS. BMC Neurosci. 9(Suppl 3): S4.

Silva, I.O., R. Ladchumananandasivam, J.H.O. Nascimento, K.K.O. Silva, F.R. Oliveira, A.P. Souto, et al. 2019. Multifunctional chitosan/gold nanoparticles coatings for biomedical textiles. Nanomaterials. 9: 1064.

Singaravelu, G., J.S. Arockiamary, V.G. Kumar and K. Govindaraju. 2007. A novel extracellular synthesis of monodisperse gold nanoparticles using marine alga, *Sargassum wightii Greville*. Colloids Surf. B. 57: 97–101.

Singh, C., I. Ahmad and A. Kumar. 2007. Pesticides and metals induced Parkinson's disease: Involvement of free radicals and oxidative stress. Cell. Mol. Biol (Noisy-le-grand). 53: 19–28.

Singh, M., S. Manikandan and A.K. Kumaraguru. 2011. Nanoparticles: A new technology with wide applications. Research Journal of Nanoscience and Nanotechnology. 1: 1–11.

Sinha, A. and S.K Khare. 2011. Mercury bioaccumulation and simultaneous nanoparticle synthesis by *Enterobacter* sp. cells. Bioresour. Technol. 102: 4281–4284.

Sinha, A., V.N. Singh, B.R. Mehta and S.K. Khare. 2011. Synthesis and characterization of monodispersed orthorhombic manganese oxide nanoparticles produced by *Bacillus* sp. cells simultaneous to its bioremediation. J. Hazard. Mater. 192: 620–627.

Slawson, R.M., E.M. Lohmeier-Vogel, H. Lee and J.T. Trevors, 1994. Silver resistance in *Pseudomonas stutzeri*. Biometals. 7: 30–40.

Sneha, K., M. Sathishkumar, J. Mao, I.S. Kwak and Y.S. Yun. 2010. *Corynebacterium glutamicum*-mediated crystallization of silver ions through sorption and reduction processes. Chem. Eng. 162: 989–996.

Song, D., X. Li, Y. Cheng, X. Xiao, Z. Lu, Y. Wang, et al. 2017. Aerobic biogenesis of selenium nanoparticles by *Enterobacter cloacae* Z0206 as a consequence of fumarate reductase mediated selenite reduction. Sci. Rep. 7:3239.

Sreeju, N., A. Rufus and D. Philip. 2017. Studies on catalytic degradation of organic pollutants and anti-bacterial property using biosynthesized CuO nanostructures. J. Mol. Liq. 242: 690–700.

Srivastava, A., O.N. Srivastava, S. Talapatra, R. Vajtai and P.M. Ajayan. 2004. Carbon nanotube filters. Nat. Mater. 3: 610.

Sundaresan, K., A. Sivakumar, C. Vigneswaran and T. Ramachandran. 2012. Influence of nano titanium dioxide finish, prepared by sol-gel technique, on the ultraviolet protection, antimicrobial, and self-cleaning characteristics of cotton fabrics. J. Ind. Textil. 41: 259–277.

Thakkar, K.N., S.S. Mhatre and R.Y. Parikh. 2010. Biological synthesis of metallic nanoparticles. Nanomed-Nanotechnol. 6: 257–262.

Tilman, D. 1999. Global environmental impacts of agricultural expansion: The need for sustainable and efficient practices. Proc. Natl. Acad. Sci. 96: 5995–6000.

Venkatesh, K.S., N.S. Palani, S.R. Krishnamoorthi, V. Thirumal and R. Ilangovan. 2013. Fungus mediated biosynthesis and characterization of zinc oxide nanorods. In AIP Conference Proceedings. 1536: 93–94.

Vetchinkina, E., E. Loshchinina, M. Kupryashina, A. Burov, T. Pylaev and V. Nikitina. 2018. Green synthesis of nanoparticles with extracellular and intracellular extracts of basidiomycetes. Peer J. 6: e5237.

Waghmare, S.S., A.M. Deshmukh, S.W. Kulkarni and L.A. Oswaldo. 2011. Biosynthesis and characterization of manganese and zinc nanoparticles. Univers. J. Environ. Res. Technol. 1: 64–69.

Wijesena, R.N., N. Tissera, Y.Y. Kannangara, Y. Lin, G.A. Amaratunga and K.N. de Silva. 2015. A method for top down preparation of chitosan nanoparticles and nanofibers. Carbohydr. Polym. 117: 731–738.

Xue B, D. He, S. Gao, D. Wang, K. Yokoyama and L. Wang. 2016. Biosynthesis of silver nanoparticles by the fungus *Arthroderma fulvum* and its antifungal activity against genera of *Candida*, *Aspergillus* and *Fusarium*. Int. J. Nanomedicine. 11: 1899–1906.

Yabu, H. 2013. Bottom-up approach to creating three-dimensional nanoring arrays composed of au nanoparticles. Langmuir. 29: 1005–1009.

Yamamoto, Y., T. Miura, Y. Nakae, T. Teranishi, M. Miyake and H. Hori 2003. Magnetic properties of the noble metal nanoparticles protected by polymer. Physica B. 329: 1183–1184.

Zamani, S., J.J. Sendi, and M. Ghadamyari. 2011. Effect of *Artemisia Annua* L. (Asterales: Asteraceae) Essential oil on mortality, development, reproduction and energy reserves of *Plodia Interpunctella* (HÃ¼bner). (Lepidoptera: Pyralidae). J. Biofertil. Biopestici. 2: 105.

9

Microbial Synthesis of Nanomaterials and Their Biotechnological Applications

Indramani Kumar, Moumita Mondal, Kranti Tanguturu and Natarajan Sakthivel*

Department of Biotechnology
School of Life Sciences, Pondicherry University
Kalapet, Puducherry 605014, India
Tel: +91 413 2654430; Fax: +91 413 2654300
Email: kumar44mani@gmail.com; moumita91mondal@gmail.com
kind_krt@yahoo.com; puns2005@gmail.com

INTRODUCTION

Nanomaterials have gained great attraction due to their unique properties of large surface area-volume ratio, size, and shape. These have immense applications in biomedical, industrial, and agricultural fields (Menon et al. 2017, Zikalala et al. 2018). Nanomaterials are synthesized by physical, chemical, and biological methods. Unlike biological synthesis, synthesis of nanomaterials by physical and chemical methods involves high temperature, pressure, and use of toxic chemicals, such as hydrazine or potassium bitartrate, which cause cellular toxicity (Kim et al. 2015, Hussain and Frazier 2002) and environmental toxicity (Iravani et al. 2014, Kumar et al. 2019). Microbial synthesis of nanomaterials is cost effective and environment-friendly (Kulkarni and Muddapur 2014). Microbes such as bacteria, fungi, yeast, virus, and actinomycetes act as reducing and stabilizing agents. Synthesis by the microbial route is categorized as a bottom-up process.

*For Correspondence: puns2005@gmail.com

The characteristic features of nanoparticles synthesized by microorganisms are regulated by varying pH, temperature, metal and culture concentrations, reaction time, and genetic make-up (Patil and Kim 2018, Ng et al. 2013). In this chapter, we discuss the microbial synthesis of silver, gold, copper, iron, zinc, platinum, and palladium nanoparticles, the mechanisms based on enzymatic and non-enzymatic reduction and their biotechnological applications.

MICROBIAL SYNTHESIS OF NANOMATERIALS

Microbes are considered to be potent green nanofactories for the synthesis of metallic and bi-metallic nanoparticles owing to their abundance and diversification. Both intracellular and extracellular routes of microbial synthesis are largely mediated by enzymes, sugars, and proteins that participate in the bioreduction of nanomaterials (Prabhu and Poulose 2012).

Diversity of Microbes and Different Types of Nanomaterials

Different microbes, such bacteria, fungi, algae, yeast, and actinomycetes have been extensively used as biological agents for the synthesis of nanomaterials (Kumar and Yadav 2009, Sathyavathi et al. 2010). These diverse groups of microbes have several advantages over physiochemical methods, such as simple scale-up, easy biomass handling, biomass recovery, and downstream processing, and therefore, offer economic viability (Thakkar et al. 2010). Microbial cultivation methods with controlled rate of synthesis facilitate the achievement of desired properties of nanomaterials (Narayanan and Sakthivel 2010). The details of microbes, characteristics, and biological properties of different types of nanoparticles of silver (Table 9.1), gold (Table 9.2), copper (Table 9.3), iron (Table 9.4), zinc (Table 9.5), platinum, and palladium (Table 9.6) are described.

BIOCHEMICAL MECHANISMS THAT MEDIATE MICROBIAL SYNTHESIS OF NANOMATERIALS

An array of enzymes and other agents are involved in the microbes-mediated synthesis of nanomaterials.

Enzymatic Reduction

Enzymes play a key role in the microbial reduction of metal ions and synthesis of nanomaterials. Identification of enzymes and understanding the mechanisms of enzymatic microbial reduction is a major challenge. Metal nanoparticles can be synthesized on the cell wall, outside of cells, and in the periplasmic space of the microbes, where, many enzymes are involved in the reduction process by transporting the electrons from electron donors to electron accepting of metal ions in several microorganisms (Zhang et al. 2005, Lloyd 2003).

Table 9.1 Microbial synthesis of silver nanoparticles and their biological properties

Microorganism	***Species***	***Shape***	***Range/Average Size* (nm)**	***Biological Property***	***References***
Bacteria	*Bacillus brevis* NCIM 2533	Spherical	42–92	Antibacterial	Saravanan et al. 2018a
	Bacillus persicus, B. pumilis and *B. licheniformis*	Triangle, hexagonal and spherical	77–92	Antibacterial; antiviral	Elbeshehy et al. 2015
	Pseudomonas deceptionensis DC5	Spherical	–	Antibacterial; antifungal; biolfilm inhibition	Jo et al. 2016
	Sporasarcina koreensis DC4	Spherical	102	Antibacterial; biolfilm inhibition	Singh et al. 2016
Actinomycetes	*Streptomyces* sp. (DPUA 1549, 1747 and 1748)	Spherical	1–40	Antibacterial; anticancer	Silva-Vinhote et al. 2017
Fungi	*Phenerochaete chrysosporium* MTCC-787	Spherical and oval	34–90	Antibacterial	Saravanan et al. 2018b
	Trichoderma viridae	Spherical, rectangular, penta and hexagonal	2–5, 40–65, 50–100	Antibacterial	Kumari et al. 2017
	Apergillus flavus	Spherical	10–35	–	Jain et al. 2011
	Aspergillus terreus Bios PTK 6	Spherical	8–20	Antibacterial; antifungal	Balakumaran et al. 2016
	Basidomycete sp.	Spherical, round	15–25	Antibacterial	Gudikandula et al. 2017
	Rhizopus oryzae	Spherical	7.1	Antibacterial	Ramalingam et al. 2016
Yeast	*Saccharomyces cerevisiae*	Spherical	2–20	–	Korbekandi et al. 2016

Table 9.2 Microbial synthesis of gold nanoparticles and their biological properties.

Microorganism	*Species*	*Shape*	*Range/Average Size* (nm)	*Biological property*	*References*
Bacteria	*Bacillus subtilis* ANR88	Hexagonal	40–60	–	Rane et al. 2017
	Bacillus marisflavi YCIS MN	Spherical	14	Catalaytic degradation	Nadaf and Kanase 2019
	Geobacillus sp.	Quasi-hexagonal	5–50	–	Correa-Llantén et al. 2013
	Sporasarcina koreensis DC4	Spherical	92.4	Catalaytic degradation	Singh et al. 2016
Actinomycete	*Streptomyces viridogens*	Spherical, rod	18–20	Antibacterial	Balagurunathan et al. 2011
	Gordonia amarae	Spherical	15–40	Detection of copper	Bennur et al. 2016
Fungi	*Penicillium rugulosum*	Spherical	20–40	Stabilization with bacterial genomic DNA	Mishra et al. 2012
	Endophytic Gx2, Gx3 and ARA	Spherical	15–30	Antibacterial; anticancer	Nachiyar et al. 2015
	Magnusiomyces ingens LH-F1	Spherical, traingle and hexagonal	80.1	Catalytic degradation	Zhang et al. 2016
Yeast	*Saccharomyces cerevisiae*	Spherical	13	Anticancer	Attia et al. 2016

Table 9.3 Microbial synthesis of copper nanoparticles and their biological properties

Microorganism	***Species***	***Shape***	***Range/Average Size* (nm)**	***Biological Property***	***References***
Bacteria	*Pseudomonas fluorescens*	Spherical	49	–	Shantkriti and Rani 2014
	Shewanella oneidensis MR-1	–	20–50	Catalysis	Kimber et al. 2018
	Bacillus cereusNCIM 2458, SWSD1	Spherical	26–97, 11–33	Antibacterial; anticancer	Tiwari et al. 2016
	Stereum hirsutum	Spherical	5–20	–	Cuevas et al. 2015
	Salmonella typhimurium	–	40–60	–	Ghorbani et al. 2015
Actinomycete	*Streptomyces capillispiralis* Ca-1	Spherical	3.6–59	Antibacterial; antifungal; insecticidal; larvicidal	Saad et al. 2018
	Streptomyces sp.	–	100–150	Antibacterial; antifungal	Usha et al. 2010
Fungi	*Fusarium oxysporum*	–	93–115	Bioremediation	Majumder 2012
	Streptomyces griseus	Spherical	30–50	Increased soil macronutrients; antifungal	Ponmurugan et al. 2016
	Trichoderma koningiopsis	Spherical	87.5	Bioremediation	Salvadori et al. 2014a
Yeast	*Rhodotorula mucilaginosa*	Spherical	10.5	Bioremediation	Salvadori et al. 2014b
	Yarrowia lipolytica AUMC 9256	Spherical	15.62	Bioremediation	El-Sayed 2018

Table 9.4 Microbial synthesis of iron nanoparticles and their characteristics and biological properties

Microorganism	*Species*	*Shape*	*Range/Average Size* (nm)	*Biological Property*	*References*
Bacteria	*Shewanella oneidensis*	Rectangular, rhombic, hexagonal	40–50	–	Perez-Gonzalez et al. 2010
	Bacillus subtilis, *B. pasteurii* and *B. lichniformis*	Blade	37.4, 53.5 and 98.7	–	Daneshvar and Hosseini 2018
	Bacillus cereus strain HMH1	Spherical	29.3	Anticancer	Fatemi et al. 2018
Fungi	*Aspergillus oryzae*	Spherical	10–24.6	–	Raliya 2013
	Aspergillus niger YESM1	Spherical	18	Superparamagnetism	Abdeen et al. 2016
Yeast	Saccharomyces cerevisiae and *Cryptococcus humicola*	Spherical	8–9	–	Vainshtein et al. 2014

Table 9.5 Microbial synthesis of zinc nanoparticles and their biological properties

Microorganism	***Species***	***Shape***	***Range/Average Size* (nm)**	***Biological Properties***	***References***
Bacteria	*Aeromonas hydrophila*	Spherical, oval	42–64	Antibacterial, antifungal	Jayaseelan et al. 2012
	Pseudomonas aeruginosa	Spherical	35–80	Antioxidant	Singh et al. 2014
	Rhodococcus pyridinivorans NT2	Hexagonal, spherical	100–200	Antibacterial, anticancer, catalytic	Kundu et al. 2014
	Bacillus lichniformis	Nanoflower	200 with nanopetals, 40 in width and 400 in length	Catalytic	Tripathi et al. 2014
Actinomycetes	*Streptomyces* sp.	Spherical	20–50	Anticancer, antibacterial	Balraj et al. 2017
	Streptomyces sp. (MA30)	–	15–30	Antibacterial, antioxidant and cytotoxic	Shanmugasundaram and Balagurunathan 2017
Fungi	*Aspergillus fumigatus*	Spherical	60–80	Antibacterial	Rajan et al. 2016
	Aspergillus niger	Spherical	53–69	Antibacterial, dye degradation	Kalpana et al. 2018
Yeast	*Candida albicans*	Quasi-spherical	~20	Catalytic	Mashrai et al. 2017
	Pichia kudriavzevii	Hexagonal	~10–61	Antibacterial, antioxidant	Moghaddam et al. 2017

Table 9.6 Microbial synthesis of platinum and palladium nanoparticles and their biological properties

Nanoparticle	*Microorganism*	*Species*	*Shape*	*Range/Average Size* (nm)	*Biological Property*	*References*
Platinum	Bacteria	*Pseudomonas aeruginosa* SM1	Circular disk	450	–	Srivastava and Constanti 2012
		Shewanella loihica PV-4	–	2–10	Catalytic degradation	Ahmed et al. 2018
	Actinomycete	*Streptomyces* sp.	Spherical	20–50	Anticancer	Baskaran et al. 2016
	Fungi	*Fusarium oxysporum*	Spherical	5–30	–	Syed and Ahmad 2012
	Yeast	*Saccharomyces boulardii*	Spherical	120	Anticancer	Borse et al. 2015
Palladium	Bacteria	*Pseudomonas aeruginosa* SM1	Polygonal	22.1	–	Srivastava and Constanti 2012
		Shewanella loihica PV-4	–	2–10	Catalytic degradation	Ahmed et al. 2018
	Yeast	*Saccharomyces cerevisiae*	Hexagonal	32	Photocatalytic degradation	Sriramulu and Sumathi 2018

Intracellular Reduction

The intracellular mechanism for the synthesis of nanoparticles involves specific ion transportation into the cell wall. Intracellular reduction involves electrostatic attraction of the negative charges of cell wall with positive charges of metal ions. Enzymes present within the cell reduce the metal ions to metal nanoparticles and the enzymes diffuse to the exterior through the cell wall (Khandel and Shahi 2016). Mukherjee et al. (2001) demonstrated the intracellular mechanism using *Verticillium* sp. The synthesis of nanoparticles involves trapping, reduction, and capping. The mechanism is mediated by electrostatic attraction of fungal cell with metal ions, and thereby, trapping the ions. Enzymes present within the cell wall reduce the metal ions to metal nanoparticles. Nair and Pradeep (2002) reported the synthesis of gold nanoparticles using *Lactobacillus* strains, where the nucleation of metal ions takes place. Because of the interactions between the bacterial cell surface and metal clusters, which lead to the synthesis of nanoclusters, they get diffused through the cell wall of bacteria. Gold ions are probably reduced by enzymes and sugars. Mabbett et al. (2004) reported the synthesis of palladium nanoparticles using *Desulfovibrio desulfuriacans* ATCC 29577. The reduction of palladium ions to palladium nanoparticles was in the presence of sodium pyruvate, H_2 or formate as an electron donor. In addition to that, cytochrome C_3 and hydrogenase were also involved in the reduction of palladium ions.

Extracellular Reduction

The extracellular enzymes for the microbial synthesis of metal nanoparticles act as reducing agents as well as shuttle of electrons which is involved in the reduction of metal ions (Subbaiya et al. 2017). The cofactors Nicotinamide Adenine Dinucleotide Hydrogen (NADH) and NADH-dependent enzymes play an important role for the synthesis of metal nanoparticles, because of electron transfer from NADH by NADH-dependent reductase enzyme, which acts as electron carrier, resulting in the bioreduction of metal ions (Mukherjee et al. 2008). The extracellular bimetallic Au-Ag alloy nanoparticles were synthesized from *F. oxysporum*. The cofactor NADH secreted by *F. oxysporum* enabled the enzymes present to reduce both gold and silver ions, and the quantity of the cofactor determined the composition of bimetallic nanoparticles (Senapati et al. 2005). Similarly, He et al. (2007) reported that the nitrate reductase enzyme and cofactor NADH were involved in the extracellular synthesis of gold nanoparticles from *Rhodopseudomonas capsulata*. In this process, the reduction of gold ions to gold nanoparticles was initiated by the electrons transferred from NADH by NADH-dependent reductase enzyme as an electron carrier. Interestingly, Narayanan and Sakthivel (2011a) reported the synthesis of gold nanostructure using *Sclerotium rolfsii* extract. They found that the presence of NADPH dependent enzyme in the cell-free filtrate was involved in the reduction of gold ions to gold nanoparticles within 10–15 minutes at ambient temperature. Similarly, Velmurugan et al. (2014) reported the synthesis of silver nanoparticles using cell-free extract of *Bacillus subtilis* EWP-46. The enzyme nitrate reductase was probably involved in the reduction of silver ions to silver nanoparticles. Khan and Ahmad (2014a) reported the extracellular synthesis of gold nanoparticles from *Thermomonospora* sp., where sulfite reductase enzyme was involved in the reduction of gold ions to gold nanoparticles. El-Batal et al. (2015) optimized the laccase enzyme production from fungus *Pleurotus ostreatus* using solid state fermentation and reported the synthesis of the gold nanoparticles using laccase enzyme. Manivasagan et al. (2015a) reported the α-amylase production in the synthesis of gold nanoparticles using *Streptomyces* sp. MBRC-82. Gold nanoparticles were also synthesized using purified α-NADPH-dependent sulfite reductase from *E. coli* (Gholami-Shabani et al. 2015). Microbial enzymes involved in the synthesis of nanoparticles are presented in Table 9.7.

Non-enzymatic Reduction

In non-enzymatic reduction, organic functional groups of cell wall of microbes are involved for the synthesis of nanomaterials (Melaiye et al. 2005). Non-enzymatic mechanism was reported for the synthesis of silver nanoparticles in *Corynebacterium* sp. SH09 (Zhang et al. 2007), *Aeromonas* SH10 (Wang et al. 2012), *Gordonia amicalis* (Sowani et al. 2016), *Leuconostoc lactis* (Saravanan et al. 2017), *Yarrowia lipolytica* (Apte et al. 2013), and *Lactobacillus* sp. (Milanowski et al. 2017), and gold nanoparticles in *G. amicalis* (Sowani et al. 2016). Non-enzymatic reduction of gold ions for the synthesis of gold nanoparticles using killed cells of *Shewanella oneidensis* in the presence of an electron donor was also reported (Rajeshkumar et al. 2016).

BIOTECHNOLOGICAL APPLICATIONS OF MICROBIAL NANOMATERIALS

Microbial nanomaterials have been applied in many applications in medicine (Verma et al. 2017), industry (Moustafa 2017, Nadaf and Kanase 2019), and agriculture (Kamaraj et al. 2018, Spagnoletti et al. 2019).

Biomedical Applications

Microbial nanomaterials have been employed for a broad range of medical applications in drug delivery, antimicrobial treatment, and cancer therapy.

Antimicrobial Agents

Emergence of multidrug resistance in various microbial systems is a serious concern, for which the development of new drugs is quintessential. Metallic nanoparticles are considered to be the most suited nanomaterials as antimicrobial agents (Rai et al. 2009). Different nanoparticles, such as silver, gold, copper, zinc, platinum, and palladium synthesized from microbes exhibited broad-spectrum antimicrobial activities. Among these, silver nanoparticles have been explored well

Table 9.7 Enzymes involved in the microbial synthesis of different nanoparticles

Microorganism	*Enzyme*	*Nano-particles*	*Location*	*Size* (nm)	*Shape*	*References*
			Bacteria			
Bacillus clausii	Nitrate reductase	Ag	Intracellular	30–80	Spherical	Mukherjee et al. 2018
Acinetobacter sp. SW30	Protein, lignin peroxidase	Se	Intracellular	10±2	Spherical	Wadhwani et al. 2017
Morganella psychrotolerans	Ag reductase	Ag	Extracellular	100–150	Nanoplates	Ramanathan et al. 2010
Stenotrophomonas maltophilia	Chromium reductase	Ag	Extracellular	93	Cuboidal	Oves et al. 2013
Shewanella oneidensis	c-Type cytochromes	Ag	Extracellular	24.4±0.8	Spherical	Ng et al. 2013
Escherichia coli	Nitrate reductase	Ag	Intracellular	5–70	Spherical	Wing-ShanáLin 2014
Escherichia coli	Sulfite reductase	Au	Extracellular	10	Spherical	Gholami-Shabani et al. 2015
			Fungi			
Phanerochaete chrysosporium	Ligninase	Au	Intracellular	10–100	Spherical	Sanghi et al. 2011
Rhizopous oryzae	Cytoplasmic proteins	Au	Intracellular/ Extracellular	~15	–	Das et al. 2012
Lentinus edodes	Laccase, tyrosinase, Mn-peroxidase	Au	Intracellular/ Extracellular	5–50	Spherical	Vetchinkina et al. 2014
Pleurotus ostreatus	Laccase	Ag	Extracellular	22–39	–	El-Batal et al. 2015
Fusarium oxysporum	Nitrate reductase	Ag	Extracellular	10–20	Spherical	Talekar et al. 2014
			Actinomycetes			
Thermomonospora sp.	Sulfite reductase	Au	Extracellular	2–6	Spherical	Khan and Ahmad 2014a
Streptomyces sp. MBRC-82	α-amylase	Au	Extracellular	20–80	Spherical	Manivasagan et al. 2015a

as antimicrobial agents (Ramalingam et al. 2014, Sudha et al. 2013). Verma et al. (2017) reported the synthesis of silver nanoparticles from the culture supernatant of *E. coli*, *S. typhimurium*, *B. thuringiensis*, and *S. aureus* under UV light. They found that synthesized silver nanoparticles showed significant antibacterial activities against bacterial pathogens, such as *E. coli*, *S. typhimurium*, *B. thuringiensis*, and *S. aureus*. Venkatesan et al. (2016) synthesized the silver nanoparticles using algal extract of *Ecklonia cava*, which showed good antimicrobial activity towards *E. coli* and *S. aureus*. Rajeshkumar et al. (2014) reported the cadmium sulfide nanoparticles synthesized from microorganisms and showed their antibacterial activity towards *Klebsiella planticola, Serratia nematodiphila, Planomicrobium* sp., *E. coli*, and *Vibrio* sp. Thomas et al. (2014) synthesized the silver nanoparticles by using soil *Bacillus* sp. and explained the combined effect of silver nanoparticles with standard antibiotics against multidrug-resistant, biofilm-forming, coagulase negative *Staphylococci* isolated from clinical samples. The more synergistic activity of silver nanoparticles was observed with chloramphenicol against *Salmonella typhi*. Gajbhiye et al. (2009) demonstrated the synergistic effect of synthesized silver nanoparticles with fluconazol against fungi such as *P. herbarum, Phoma* sp., *Fusarium semitectum*, *Candida albicans*, and *Trichoderma* sp. Similarly, Musarrat et al. (2010) synthesized the silver nanoparticles using fungal strain of *Amylomyces rouxii*. The synthesized silver nanoparticles showed the antifungal activity against *C. albicans* and *Fusarium oxysporum*, and antibacterial activity against *S. aureus, E. coli, Pseudomonas aeruginosa, B. subtilis, Shigella dysenteriae* type I, and *Citrobacter* sp. Rajeshkumar et al. (2014) reported microbial synthesis of cadmium sulfide nanoparticles that exhibited the fungicidal activity against *Aspergillus niger* and *A. flavus*.

Anticancer Agents

Cancer is one of the most common diseases, which leads to death all over the world (Deepa et al. 2013). Conventional methods, such as chemotherapy, surgery, and radiation have been employed for the treatment of cancer. However, these methods have some side effects. Early diagnosis and efficient drug delivery to the affected organs are important for the treatment of cancer (Jabir et al. 2012). Nanoparticles, being smaller in size, can effectively cross biological barriers and target only the unhealthy cells (Gaikwad et al. 2013). Maharani et al. (2016) synthesized the silver nanoparticles using *E. coli* VM1 and evaluated their cytotoxicity against cancerous cells (HeLa and A549) and normal (Vero) cells in a dose-dependent manner. The cell viability was decreased in case of HeLa and A549 cells, but the nanoparticles were less toxic towards normal cells. Similarly, Manivasagan et al. (2015b) reported the synthesis of gold nanoparticles using *Nocardiopsis* sp., and reported their anticancer activity against HeLa cells in a dose-dependent manner, and also concluded that the HeLa cells died due to apoptosis. Some of the gold nanomaterials-based drugs are under clinical trials for human cancer diagnostics and therapy (Singh et al. 2018). Borse et al. (2015) reported the synthesis of platinum nanoparticles from *Saccharomyces boulardii* and their evaluation for anticancer activity against MCF-7 and A431 cell lines.

Drug Delivery

Drug delivery to targeted sites through nanomaterials is accurate and safe, and therefore leads to controlled release and highest therapeutic effect of nanoparticles. To reach target cells, the targeted nanocarriers must pass through blood-tissue barrier. Moreover, targeted nanocarriers once in contact with cytoplasmic targets, various antibodies, and their parts and ligands, transfer the nanoparticles across the cellular barrier via specific endocytotic and transcytotic transport mechanisms (Maharani et al. 2016, Häfeli et al. 2009). Kumar et al. (2008) reported the synthesis of gold nanoparticles using fungus, *Helminthosporum solani.* The synthesized gold nanoparticles were conjugated with doxorubicin, which was taken up by HEK293 cells showing cytotoxicity comparable to doxorubicin. Khan et al. (2014b) reported the synthesis of gadolinium oxide nanoparticles from fungus, *Humicola* sp. and conjugated with taxol to increase its potency towards antitumor cells. Nanoparticles have many advantages, such as specific targeted site, biodistribution, biocompability, and safety when compared to conventional drug delivery systems (Saravanakumar and Wang 2018).

Industrial Applications

Various dyes extensively used in textile, rubber, paper, and plastic industries and the wastewater discharges from textile, paper, leather, food, plastic, and cosmetic industries are the causes for serious environmental pollution (Kang et al. 2000, Saratale et al. 2011, Hsueh and Chen 2008). The microbial nanomaterials have enormous ability to treat such kind of pollutants.

Degradation of Dyes

The microbial nanomaterials have been used for the degradation of different dyes. Saravanan et al. (2017) synthesized silver nanoparticles using exopolysaccharides (EPS) of *Leuconostoc lactis* and used them for the degradation of azo dyes, such as methyl orange and congo red. Narayanan and Sakthivel (2011c) reported the synthesis of silver-bionanocomposites using the fungus, *Cylindrocladium floridanum.* The synthesized silver-bionanocomposites were used for the degradation of 4-nitrophenol in the presence of sodium borohydride. Similarly, Narayanan and Sakthivel (2011b) synthesized the gold nanocomposites using *C. floridanum* for the catalytic reduction of 4-nitrophenol. Srivastava and Mukhopadhyay (2014) synthesized the tin oxide (SnO_2) nanoparticles from bacteria *Erwinia herbicola*, where bacterial protein and biomolecules were involved in the reduction and stabilization of nanoparticles. They observed that synthesized SnO_2 nanoparticles exhibited excellent photocatalytic activity for the degradation of organic dyes, such as erichrome black T, methylene blue, and methyl orange, with percentages of degradation as 94.0, 93.3, and 97.8 respectively. Gold nanoparticles synthesized from cell-free extract of *Bacillus marisflavi* were used for the catalytic dye degradation of congo red and methylene blue (Nadaf and Kanase 2019).

Wastewater Treatment

Moustafa (2017) reported the extracellular synthesis of silver nanoparticles from fungi *Penciillium citreonigum* Dierck and *Scopulaniopsos brumptii* Salvanet-Duval. TEM analysis confirmed that the particles synthesized by *P. citreonigum* Dierck and *S. brumptii* Salvanet-Duval were spherical in shape, with sizes in the range of 6–26 and 4.24–23.2 nm, respectively. The synthesized nanoparticles showed antibacterial activity at concentrations of 550.7 and 676.9 mg/L, with contact times of 15, 60, and 120 minutes. Polyurethane foam was used as silver carrier and nano-silver solution for the removal of pathogenic bacteria from wastewater. Das et al. (2009) reported the synthesis of gold nanoparticles on the surface of *Rhizopus oryzae* with an average size of 10 nm. These nanoparticles showed high adsorption capacity against organophosphorous pesticides, such as malathion, parathion, chlorpyrifos, dimethoate, and gamma-benzene hexachloride. Synthesized gold nanoparticles also showed broad-spectrum antimicrobial activity towards *P. aeruginosa, E. coli, B. subtilis, S. aureus, Salmonella* sp., *S. cerevisiae, and C. albicans*. Nanoparticles synthesized with various bacterial species were used for wastewater treatment to remove heavy metals (Zhou et al. 2016, Xiao et al. 2017), toxic textile dyes (Qu et al. 2017), pesticides (Gurunathan et al. 2015), pathogenic microorganisms (Moustafa 2017), pharmaceutical and personal care products (Orłowski et al. 2018, Martins et al. 2017).

Agricultural Applications

Biological entities have been entitled 'bio-laboratories' due to the efficient synthesis of various nanomaterials (Sharma et al. 2019). Employment of nanomaterials produced with the help of microbes, such as Cu, Au, TiO_2, MnO_2, Ag-$AgCl_2$, and ZnO in agriculture has a beneficial effect on plants. Microbial nanomaterials that suppress the various plant pathogens and promote plant growth have been reported (Ponmurugan et al. 2016).

Microbial Nanomaterials as Biocontrol Agents

Copper nanoparticles were synthesized from *Streptomyces griseus*. These nanoparticles were evaluated for their fungicidal activity against *Poria hypolateritia*, which causes the red-rot disease in tea plant. These nanoparticles also increased the soil macronutrients and leaf yield in tea plant (Ponmurugan et al. 2016). The gold nanoparticles synthesized from the microbes, such as *Pseudomonas aeruginosa, Trichoderma atroviride*, and *Streptomyces* sp. from tea leaves showed antifungal properties and plant growth promotion activities, and therefore, were used as biocontrol agents and biofertilizers. Such conjugated nanoparticles were assumed to confer increased stability of biocontrol agents (Balasubramanian and Punnusamy 2012). Using tobacco mild green mosaic virus as a nanocarrier, Chariou and Steinmetz (2017) reported the delivery of pesticides to plant parasitic nematodes. Kamaraj et al. (2018) reported the bio-pesticidal effect of titanium dioxide nanoparticles synthesized from *Trichoderma viride*. These nanoparticles showed larvicidal, antifeedant, and pupicidal activities towards *Helicoverpa armigera.* Moreover, they reported the 100% mortality rate on first and second instar larvae, but 92.34% on third instar larvae of *H. armigera* after treatment of 100 ppm of synthesized titanium oxide nanoparticles. However, these nanoparticles did not show toxicity towards *Eudrilus eugeniae* at 100 ppm nanoparticles in filter paper and artificial soil assays.

Microbial Nanomaterials for Plant Growth Promotion

Raliya et al. (2014) reported the synthesis of magnesium dioxide nanoparticles using *Aspergillus flavus*, which promoted the growth of root-shoot and chlorophyll content in *Cyamopsis tetragonoloba* at 15 mg/L of concentration. Similarly, Raliya et al. (2015) synthesized titanium dioxide nanoparticles using *A. flavus* and reported the plant growth conditions of *Vigna radiata* by spraying the nanoparticles (10 mg/L) on leaves (14 days old). They found that there was a significant increase in shoot length by 17.02%, root length by 49.6%, root area by 43%, root nodule by 67.5%, chlorophyll content by 46.4%, and total soluble leaf protein by 94%. Also, they observed that microbial population increased in the rhizosphere by 21.4–48.1% with increase in enzyme activities for acid phosphatase, alkaline phosphatase, phytase, and dehydrogenase in the rhizosphere by 67.3%, 72%, 64%, and 108.7%, respectively. Spagnoletti et al. (2019) reported the synthesis of silver-silver chloride nanoparticles (Ag/AgCl nanoparticles) using the soil fungus *Macrophomina phaseolina*. The synthesized silver nanoparticles exhibited the antimicrobial activity towards both gram-positive and gram-negative bacteria. Further, these nanoparticles also promoted seed germination of *Glycine max* in a dose-dependent manner. Oxidative damage was not observed in seeds of *Glycine max* because of Ag/AgCl nanoparticles. Zinc oxide nanoparticles synthesized from *Aspergillus niger* were used to degrade the bismarck brown dye and test the seed germination potential of *Pisum sativum* in the presence of untreated and degraded dye. Almost 90% degradation of Bismarck brown dye was observed using zinc oxide nanoparticles. The degraded dye was then used to evaluate the effect of seed germination and plant growth of *P. sativum*. The positive control (normal water), negative control (untreated dye without nanoparticles), and degraded dye (treated with nanoparticles) were used. They found 100% germination in positive control and the degraded dye, but 40% germination was observed in the untreated dye. The length of plumule and radicle was observed as 20 ± 0.64 and 6 ± 0.68 in positive control, 6.5 ± 0.59 and 3 ± 0.02 in untreated dye, and 14 ± 0.79 and 5.2 ± 0.14 in treated dye, respectively (Kalpana et al. 2018).

CONCLUSION

In recent years, there has been enormous interest, progress, and success in synthesizing nanomaterials from diverse microbial sources, including both prokaryotes and eukaryotes. Release of toxic chemicals into the environment has caused severe ecological damage. As the biological process possesses challenges, optimizing the growth and reaction parameters during synthesis may resolve the issues pertinent to dispersity, particle size, shape, and stability. More research at genetic and proteomic level is required to elucidate the mechanisms that mediate the biological synthesis of nanomaterials. With the advent of developments in nanobiotechnology, synthesis of novel microbial nanomaterials at commercial scale can be achieved to exploit their potential in the fields of biomedicine, agriculture, and industry.

ACKNOWLEDGMENTS

We thank the University Grants Commission, New Delhi, for financial support through University Research Fellowship to Indramani Kumar and Kranti Tanguturu, and Rajiv Gandhi National Fellowship to Moumita Mondal. We also thank UGC-SAP and DST-FIST programs coordinated by Prof. N. Sakthivel for providing infrastructure facilities.

References

Abdeen, M., S. Sabry, H. Ghozlan, A.A. El-Gendy and E.E. Carpenter. 2016. Microbial synthesis of Fe and Fe_3O_4 magnetic nanoparticles using *Aspergillus niger* YESM1 and supercritical condition of ethanol. J. Nanomaters. 2016: 7.

Ahmed, E., S. Kalathil, L. Shi, O. Alharbi and P. Wang. 2018. Synthesis of ultra-small platinum, palladium and gold nanoparticles by *Shewanella loihica* PV-4 electro-chemically active biofilms and their enhanced catalytic activities. J. Saudi. Chem. Soc. 22: 919–929.

Apte, M., D. Sambre, S. Gaikawad, S. Joshi, A. Bankar, A.R. Kumar, et al. 2013. Psychrotrophic yeast *Yarrowia lipolytica* NCYC 789 mediates the synthesis of antimicrobial silver nanoparticles via cell-associated melanin. AMB Express. 3: 32.

Attia, Y.A., Y.E. Farag, Y.M.A. Mohamed, A.T. Hussien and T. Youssef. 2016. Photo-extracellular synthesis of gold nanoparticles using Baker's yeast and their anticancer evaluation against *Ehrlich ascites* carcinoma cells. New. J. Chem. 40: 9395–9402.

Balagurunathan, R., M. Radhakrishnan, R.B. Rajendran and D. Velmurugan. 2011. Biosynthesis of gold nanoparticles by actinomycete *Streptomyces viridogens* strain HM10. Indian J. Biochem. Biophys. 48: 331–335.

Balakumaran, M.D., R. Ramachandran, P. Balashanmugam, D.J. Mukeshkumar and P.T. Kalaichelvan. 2016. Mycosynthesis of silver and gold nanoparticles: Optimization, characterization and antimicrobial activity against human pathogens. Microbiol. Res. 182: 8–20.

Balasubramanian, M.G. and P. Punnusamy. 2012. US 2012/0108425.

Balraj, B., N. Senthilkumar, C. Siva, R. Krithikadevi, A. Julie, I. Vetha Potheher, et al. 2017. Synthesis and characterization of zinc oxide nanoparticles using marine *Streptomyces* sp. with its investigations on anticancer and antibacterial activity. Res. Chem. Intermediat. 43: 2367–2376.

Baskaran, B., A. Muthukumarasamy, S. Chidambaram, A. Sugumaran, K. Ramachandran and T.R. Manimuthu. 2016. Cytotoxic potentials of biologically fabricated platinum nanoparticles from *Streptomyces* sp. on MCF-7 breast cancer cells. IET Nanobiotechnol. 11: 241–246.

Bennur, T., Z. Khan, R. Kshirsagar, V. Javdekar and S. Zinjarde. 2016. Biogenic gold nanoparticles from the Actinomycete *Gordonia amarae*: application in rapid sensing of copper ions. Sens. Actuators B. Chem. 233: 684–690.

Borse, V., A. Kaler and U.C. Banerjee. 2015. Microbial synthesis of platinum nanoparticles and evaluation of their anticancer activity. Int. J. Emerging Trends Electr. Electron. 11: 26–31.

Correa-Llantén, D.N., S.A. Muñoz-Ibacache, M.E. Castro, P.A. Muñoz, and J.M. Blamey. 2013. Gold nanoparticles synthesized by *Geobacillus* sp. strain ID17 a thermophilic bacterium isolated from Deception Island, Antarctica. Microb. Cell Fact. 12: 75.

Chariou, P.L. and N.F. Steinmetz. 2017. Delivery of pesticides to plant parasitic nematodes using tobacco mild green mosaic virus as a nanocarrier. ACS nano. 11: 4719–4730.

Cuevas, R., N. Durán, M.C. Diez, G.R. Tortella, and O. Rubilar. 2015. Extracellular biosynthesis of copper and copper oxide nanoparticles by *Stereum hirsutum*, a native white-rot fungus from chilean forests. J. Nanomaters. 16: 57.

Daneshvar, M. and R.M. Hosseini. 2018. From the iron boring scraps to superparamagnetic nanoparticles through an aerobic biological route. J. Hazard Mater. 357: 393–400.

Das, S.K., A.R. Das and A.K. Guha. 2009. Gold nanoparticles: Microbial synthesis and application in water hygiene management. Langmuir. 25: 8192–8199.

Das, S.K., J. Liang, M. Schmidt, F. Laffir and E. Marsili. 2012. Biomineralization mechanism of gold by zygomycete fungi *Rhizopous oryzae*. ACS Nano. 6: 6165–6173.

Deepa, S., K. Kanimozhi and A. Panneerselvam. 2013. Antimicrobial activity of extracellularly synthesized silver nanoparticles from marine derived actinomycetes. Int. J. Curr. Microbiol. Appl. Sci. 2: 223–230.

El-Batal, A.I., N.M. ElKenawy, A.S. Yassin and M.A. Amin. 2015. Laccase production by *Pleurotus ostreatus* and its application in synthesis of gold nanoparticles. Biotechnol. Rep. 5: 31–39.

El-Sayed, M.T. 2018. Bioremediation and extracellular synthesis of copper nanoparticles from wastewater using *Yarrowia lipolytica* AUMC 9256. Egyptian J. Botany. 58: 563–579.

Elbeshehy, E.K.F., A.M. Elazzazy and G. Aggelis. 2015. Silver nanoparticles synthesis mediated by new isolates of *Bacillus* spp., nanoparticle characterization and their activity against bean yellow mosaic virus and human pathogens. Front. Microbiol. 6: 453.

Fatemi, M., N. Mollania, M. Momeni-Moghaddam and F. Sadeghifar. 2018. Extracellular biosynthesis of magnetic iron oxide nanoparticles by *Bacillus cereus* strain HMH1: Characterization and *in vitro* cytotoxicity analysis on MCF-7 and 3T3 cell lines. J. Biotechnol. 270: 1–11.

Gajbhiye, M., J. Kesharwani, A. Ingle, A. Gade and M. Rai. 2009. Fungus-mediated synthesis of silver nanoparticles and their activity against pathogenic fungi in combination with fluconazole. Nanomed. Nanotechnol. Biol. Med. 5: 382–386.

Gaikwad, S., A. Ingle, A. Gade, M. Rai, A. Falanga, N. Incoronato, et al. 2013. Antiviral activity of mycosynthesized silver nanoparticles against herpes simplex virus and human parainfluenza virus type 3. Int. J. Nanomedicine. 8: 4303.

Gholami-Shabani, M., M. Shams-Ghahfarokhi, Z. Gholami-Shabani, A. Akbarzadeh, G. Riazi, S. Ajdari, et al. 2015. Enzymatic synthesis of gold nanoparticles using sulfite reductase purified from *Escherichia coli*: A green eco-friendly approach. Process Biochem. 50: 1076–1085.

Ghorbani, H.R., F.P. Mehr and A.K. Poor. 2015. Extracellular synthesis of copper nanoparticles using culture supernatants of *Salmonella typhimurium*. Orient. J. Chem. 31: 527–529.

Gudikandula, K., P. Vadapally and M.A.S. Charya. 2017. Biogenic synthesis of silver nanoparticles from white rot fungi: Their characterization and antibacterial studies. OpenNano. 2: 64–78.

Gurunathan, S., H.J. Park, W.J. Han and H.J. Kim. 2015. Comparative assessment of the apoptotic potential of silver nanoparticles synthesized by *Bacillus tequilensis* and *Calocybe indica* in MDA-MB-231 human breast cancer cells: targeting p53 for anticancer therapy. Int. J. Nanomedicine. 10: 4203.

Häfeli, U.O., J.S. Riffle, L. Harris-Shekhawat, A. Carmichael-Baranauskas, F. Mark, J.P. Dailey, et al. 2009. Cell uptake and *in vitro* toxicity of magnetic nanoparticles suitable for drug delivery. Mol. Pharm. 6: 1417–1428.

He, S., Z. Guo, Y. Zhang, S. Zhang, J. Wang and N. Gu. 2007. Biosynthesis of gold nanoparticles using the bacteria *Rhodopseudomonas capsulata*. Mater. Lett. 61: 3984–3987.

Hsueh, C.C. and B.Y. Chen. 2008. Exploring effects of chemical structure on azo dye decolorization characteristics by *Pseudomonas luteola*. J. Hazard. Mater. 154: 703–710.

Hussain, S.M. and J.M. Frazier. 2002. Cellular toxicity of hydrazine in primary rat hepatocytes. Toxicol. Sci. 69: 424–432.

Iravani, S., H. Korbekandi, S.V. Mirmohammadi and B. Zolfaghari. 2014. Synthesis of silver nanoparticles: Chemical, physical and biological methods. Res. Pharm. Sci. 9: 385.

Jabir, N.R., S. Tabrez, G.M. Ashraf, S. Shakil, G.A. Damanhouri and M.A. Kamal. 2012. Nanotechnology-based approaches in anticancer research. Int. J. Nanomed. 7: 4391.

Jain, N., A. Bhargava, S. Majumdar, J. Tarafdar and J. Panwar. 2011. Extracellular biosynthesis and characterization of silver nanoparticles using *Aspergillus flavus* NJP08: A mechanism perspective. Nanoscale. 3: 635–641.

Jayaseelan, C., A.A. Rahuman, A.V. Kirthi, S. Marimuthu, T. Santhoshkumar, A. Bagavan, et al. 2012. Novel microbial route to synthesize ZnO nanoparticles using *Aeromonas hydrophila* and their activity against pathogenic bacteria and fungi. Spectrochim. Acta A Mol. Biomol. Spectrosc. 90: 78–84.

Jo, J.H., P. Singh, Y.J. Kim, C. Wang, R. Mathiyalagan, C.G. Jin, et al. 2016. *Pseudomonas deceptionensis* DC5-mediated synthesis of extracellular silver nanoparticles. Artif. Cells Nanomed. Biotechnol. 44: 1576–1581.

Kalpana, V., B.A.S. Kataru, N. Sravani, T. Vigneshwari, A. Panneerselvam and V.D. Rajeswari. 2018. Biosynthesis of Zinc oxide nanoparticles using culture filtrates of *Aspergillus niger*: Antimicrobial textiles and dye degradation studies. OpenNano. 3: 48–55.

Kamaraj, C., B. Govindasamy, D. Paramasivam, A. Dilipkumar, A. Dhayalan, A. Vadivel, et al. 2018. Bio-pesticidal effects of *Trichoderma viride* formulated titanium dioxide nanoparticle and their physiological and biochemical changes on *Helicoverpa armigera* (Hub.). Pestic. Biochem. Physiol. 149: 26–36.

Kang, S.F., C.H. Liao and S.T. Po. 2000. Decolorization of textile wastewater by photo-Fenton oxidation technology. Chemosphere. 41: 1287–1294.

Khan, S.A. and A. Ahmad. 2014a. Enzyme mediated synthesis of water-dispersible, naturally protein capped, monodispersed gold nanoparticles; their characterization and mechanistic aspects. RSC Adv. 4: 7729–7734.

Khan, S.A., S. Gambhir and A. Ahmad. 2014b. Extracellular biosynthesis of gadolinium oxide (Gd_2O_3) nanoparticles, their biodistribution and bioconjugation with the chemically modified anticancer drug taxol. Beilstein J. Nanotechnol. 5: 249–257.

Khandel, P. and S.K. Shahi. 2016. Microbes mediated synthesis of metal nanoparticles: current status and future prospects. Int. J. Nanomater. Biostruct. 6: 1–24.

Kimber, R.L., E.A. Lewis, F. Parmeggiani, K. Smith, H. Bagshaw, T. Starborg, et al. 2018. Biosynthesis and characterization of copper nanoparticles using *Shewanella oneidensis*: Application for click chemistry. Small. 14: 1703145.

Kim, S.P., S.G. Lee, M.Y. Choi and H.C. Choi. 2015. Highly sensitive hydrazine chemical sensor based on CNT-PdPt nanocomposites. J. Nanomater. 16: 298.

Korbekandi, H., S. Mohseni, R. Mardani Jouneghani, M. Pourhossein and S. Iravani. 2016. Biosynthesis of silver nanoparticles using *Saccharomyces cerevisiae.* Artif. Cells Nanomed. Biotechnol. 44: 235–239.

Kulkarni, N. and U. Muddapur. 2014. Biosynthesis of metal nanoparticles: A review. J. Nanotechnol. 2014: 1–8.

Kumar, S.A., Y.A, Peter and J.L. Nadeau. 2008. Facile biosynthesis, separation and conjugation of gold nanoparticles to doxorubicin. Nanotechnol. 19: 495101.

Kumar, V. and S.K. Yadav. 2009. Plant-mediated synthesis of silver and gold nanoparticles and their applications. J. Chem. Technol. Biotechnol. 84: 151–157.

Kumar, I.M. Mondal and N. Sakthivel. 2019. Green synthesis of phytogenic nanoparticles. pp. 37–73. *In*: A.K. Shukla and S. Iravani [eds]. Green Synthesis, Characterization and Applications of Nanoparticles. Elsevier.

Kumari, M., S. Pandey, V.P. Giri, A. Bhattacharya, R. Shukla, A. Mishra, et al. 2017. Tailoring shape and size of biogenic silver nanoparticles to enhance antimicrobial efficacy against MDR bacteria. Microb. Pathog. 105: 346–355.

Kundu, D., C. Hazra, A. Chatterjee, A. Chaudhari, S. Mishra. 2014. Extracellular biosynthesis of zinc oxide nanoparticles using *Rhodococcus pyridinivorans* NT2: multifunctional textile finishing, biosafety evaluation and in vitro drug delivery in colon carcinoma. J. Photochem. Photobiol. B. 140: 194–204.

Lloyd, J.R. 2003. Microbial reduction of metals and radionuclides. FEMS Microbiol. Rev. 27: 411–425.

Mabbett, A.N., P. Yong, J.P.G. Farr and L.E. Macaskie. 2004. Reduction of Cr(VI) by "palladized" biomass of *Desulfovibrio desulfuricans* ATCC 29577. Biotechnol. Bioeng. 87: 104–109.

Maharani, V., A. Sundaramanickam and T. Balasubramanian. 2016. *In vitro* anticancer activity of silver nanoparticle synthesized by *Escherichia coli* VM1 isolated from marine sediments of Ennore southeast coast of India. Enzyme. Microb. Technol. 95: 146–154.

Majumder, D.R. 2012. Bioremediation: copper nanoparticles from electronic-waste. Int. J. Eng. Sci. and Tech. 4: 4380–4389.

Manivasagan, P., J. Venkatesan, K.H, Kang, K. Sivakumar, S.J. Park and S.K. Kim. 2015a. Production of α-amylase for the biosynthesis of gold nanoparticles using *Streptomyces* sp. MBRC-82. Int. J. Biol. Macromol. 72: 71–78.

Manivasagan, P., M.S. Alam, K.H. Kang, M. Kwak and S.K. Kim. 2015b. Extracellular synthesis of gold bionanoparticles by *Nocardiopsis* sp. and evaluation of its antimicrobial, antioxidant and cytotoxic activities. Bioprocess. Biosyst. Eng. 38: 1167–1177.

Martins, M., C. Mourato, S. Sanches, J.P. Noronha, M.T.B. Crespo and I.A.C. Pereira. 2017. Biogenic platinum and palladium nanoparticles as new catalysts for the removal of pharmaceutical compounds. Water. Res. 108: 160–168.

Mashrai, A., H. Khanam and R.N. Aljawfi. 2017. Biological synthesis of ZnO nanoparticles using *C. albicans* and studying their catalytic performance in the synthesis of steroidal pyrazolines. Arab. J. Chem. 10: S1530–S1536.

Melaiye, A., Z. Sun, K. Hindi, A. Milsted, D. Ely, D.H. Renekar, et al. 2005. Silver (I)–imidazole cyclophane gem-diol complexes encapsulated by electrospun tecophilic nanofibers: Formation of nanosilver particles and antimicrobial activity. J. Am. Chem. Soc. 127: 2285–2291.

Menon, S., S. Rajeshkumar and V. Kumar, V. 2017. A review on biogenic synthesis of gold nanoparticles, characterization, and its applications. Res. Effi. Tech. 3: 516–527.

Mishra, A., S.K. Tripathy and S.I. Yun. 2012. Fungus mediated synthesis of gold nanoparticles and their conjugation with genomic DNA isolated from *Escherichia coli* and *Staphylococcus aureus*. Process. Biochem. 47: 701–711.

Milanowski, M., P. Pomastowski, V. Railean-Plugaru, K. Rafińska, T. Ligor and B. Buszewski. 2017. Biosorption of silver cations onto *Lactococcus lactis* and *Lactobacillus casei* isolated from dairy products. PloS One. 12: e0174521.

Moghaddam, A.B., M. Moniri, S. Azizi, R.A. Rahim, A.B. Ariff, W.Z. Saad, et al. 2017. Biosynthesis of ZnO nanoparticles by a new *Pichia kudriavzevii* yeast strain and evaluation of their antimicrobial and antioxidant activities. Molecules. 22: 872.

Moustafa, M.T. 2017. Removal of pathogenic bacteria from wastewater using silver nanoparticles synthesized by two fungal species. Water. Sci. 31: 164–176.

Mukherjee, P., A. Ahmad, D. Mandal, S. Senapati, S.R. Sainkar, M.I. Khan, et al. 2001. Fungus-mediated synthesis of silver nanoparticles and their immobilization in the mycelial matrix: A novel biological approach to nanoparticle synthesis. Nano. Lett. 1: 515–519.

Mukherjee, P., M. Roy, B.P. Mandal, G.K. Dey, P.K. Mukherjee, J. Ghatak, et al. 2008. Green synthesis of highly stabilized nanocrystalline silver particles by a non-pathogenic and agriculturally important fungus *T. asperellum*. Nanotechnol. 19: 075103.

Mukherjee, K., R. Gupta, G. Kumar, S. Kumari, S. Biswas and P. Padmanabhan. 2018. Synthesis of silver nanoparticles by *Bacillus clausii* and computational profiling of nitrate reductase enzyme involved in production. J. Genetic. Eng. Biotechnol. 16: 527–536.

Musarrat, J., S. Dwivedi, B.R. Singh, A.A. Al-Khedhairy, A. Azam and A. Naqvi. 2010. Production of antimicrobial silver nanoparticles in water extracts of the fungus *Amylomyces rouxii* strain KSU-09. Bioresour. Technol. 101: 8772–8776.

Nachiyar, V., S. Sunkar and P. Prakash. 2015. Biological synthesis of gold nanoparticles using endophytic fungi. Der Pharma Chem. 7: 31–38.

Nadaf, N.Y. and S.S. Kanase. 2019. Biosynthesis of gold nanoparticles by *Bacillus marisflavi* and its potential in catalytic dye degradation. Arab. J. Chem. 12: 4806–4814.

Nair, B. and T. Pradeep. 2002. Coalescence of nanoclusters and formation of submicron crystallites assisted by Lactobacillus strains. Cryst. Growth. Des. 2: 293–298.

Narayanan, K.B. and N. Sakthivel. 2010. Biological synthesis of metal nanoparticles by microbes. Adv. Colloid Interface Sci. 156: 1–13.

Narayanan, K.B. and N. Sakthivel. 2011a. Facile green synthesis of gold nanostructures by NADPH-dependent enzyme from the extract of *Sclerotium rolfsii.* Colloids Surf. A Physicochem. Eng. Asp. 380: 156–161.

Narayanan, K.B. and N. Sakthivel. 2011b. Synthesis and characterization of nano-gold composite using *Cylindrocladium floridanum* and its heterogeneous catalysis in the degradation of 4-nitrophenol. J. Hazard. Mater. 189: 519–525.

Narayanan, K.B. and N. Sakthivel. 2011c. Heterogeneous catalytic reduction of anthropogenic pollutant, 4-nitrophenol by silver-bionanocomposite using *Cylindrocladium floridanum*. Bioresour. Technol. 102: 10737–10740.

Ng, C.K., K. Sivakumar, X. Liu, M. Madhaiyan, L. Ji, L. Yang, et al. 2013. Influence of outer membrane c-type cytochromes on particle size and activity of extracellular nanoparticles produced by *Shewanella oneidensis*. Biotechnol. Bioeng. 110: 1831–1837.

Orłowski, P., A. Kowalczyk, E. Tomaszewska, K. Ranoszek-Soliwoda, and A. Węgrzyn. 2018. Antiviral activity of tannic acid modified silver nanoparticles: Potential to activate immune response in herpes genitalis. Viruses. 10: 524.

Oves, M., M.S. Khan, A. Zaidi, A.S. Ahmed, F. Ahmed, E. Ahmad, et al. 2013. Antibacterial and cytotoxic efficacy of extracellular silver nanoparticles biofabricated from chromium reducing novel OS_4 strain of *Stenotrophomonas maltophilia*. PloS one. 8: e59140.

Patil, M.P. and G.D. Kim. 2018. Marine microorganisms for synthesis of metallic nanoparticles and their biomedical applications. Colloids Surf. B Biointerfaces. 172: 487–495.

Perez-Gonzalez, T., C. Jimenez-Lopez, A.L. Neal, F. Rull-Perez, A. Rodriguez-Navarro, A. Fernandez-Vivas, et al. 2010. Magnetite biomineralization induced by *Shewanella oneidensis*. Geochim. Cosmochim. Ac. 74: 967–979.

Ponmurugan, P., K. Manjukarunambika, V. Elango and B.M. Gnanamangai. 2016. Antifungal activity of biosynthesized copper nanoparticles evaluated against red root-rot disease in tea plants. J. Exp. Nanosci. 11: 1019–1031.

Prabhu, S. and E.K. Poulose. 2012. Silver nanoparticles: Mechanism of antimicrobial action, synthesis, medical applications, and toxicity effects. Int. Nano Lett. 2: 32.

Qu, Y., W. Shen, X. Pei, F. Ma, S. You, S. Li, et al. 2017. Biosynthesis of gold nanoparticles by *Trichoderma* sp. WL-Go for azo dyes decolorization. J. Environ. Sci. 56: 79–86.

Rai, M., A. Yadav and A. Gade. 2009. Silver nanoparticles as a new generation of antimicrobials. Biotechnol. Adv. 27: 76–83.

Rajan, A., E. Cherian and G. Baskar. 2016. Biosynthesis of zinc oxide nanoparticles using *Aspergillus fumigatus* JCF and its antibacterial activity. Int. J. Mod. Sci. Technol. 1: 52–57.

Rajeshkumar, S., M. Ponnanikajamideen, C. Malarkodi, M. Malini and G. Annadurai. 2014. Microbe-mediated synthesis of antimicrobial semiconductor nanoparticles by marine bacteria. J. Nanostructure. Chem. 4: 96.

Rajeshkumar, S., C. Malarkodi, M. Vanaja and G. Annadurai. 2016. Anticancer and enhanced antimicrobial activity of biosynthesizd silver nanoparticles against clinical pathogens. J. Mol. Struct. 1116: 165–173.

Raliya, R. 2013. Rapid, low-cost, and ecofriendly approach for iron nanoparticle synthesis using *Aspergillus oryzae* TFR9. J. Nanoparticles. 2013: 1–4.

Raliya, R., J. Tarafdar, S. Singh, R. Gautam, K. Choudhary, V.G. Maurino, et al. 2014. MgO nanoparticles biosynthesis and its effect on chlorophyll contents in the leaves of clusterbean (*Cyamopsis tetragonoloba* L.). Adv. Sci. Eng. Med. 6: 538–545.

Raliya, R., P. Biswas and J. Tarafdar. 2015. TiO_2 nanoparticle biosynthesis and its physiological effect on mung bean (*Vigna radiata* L.). Biotechnol. Rep. 5: 22–26.

Ramanathan, R., A.P. O'Mullane, R.Y. Parikh, P.M. Smooker, S.K. Bhargava and V. Bansal. 2010. Bacterial kinetics-controlled shape-directed biosynthesis of silver nanoplates using *Morganella psychrotolerans*. Langmuir. 27: 714–719.

Ramalingam, V., R. Rajaram, C. Premkumar, P. Santhanam, P. Dhinesh, S. Vinothkumar, et al. 2014. Biosynthesis of silver nanoparticles from deep sea bacterium *Pseudomonas aeruginosa* JQ989348 for antimicrobial, antibiofilm, and cytotoxic activity. J. Basic. Microbiol. 54: 928–936.

Ramalingam, B., T. Parandhaman and S.K. Das. 2016. Antibacterial effects of biosynthesized silver nanoparticles on surface ultrastructure and nanomechanical properties of gram-negative bacteria viz. *Escherichia coli* and *Pseudomonas aeruginosa*. ACS Appl. Mater. Interfaces. 8: 4963–4976.

Rane, A.N., V.V. Baikar, V. Ravi Kumar and R.L. Deopurkar. 2017. Agro-industrial wastes for production of biosurfactant by *Bacillus subtilis* ANR 88 and its application in synthesis of silver and gold nanoparticles. Front. Microbiol. 8: 492.

Saad, E., S.S. Salem, A. Fouda, M.A. Awad, M.S. El-Gamal and A.M. Abdo. 2018. New approach for antimicrobial activity and bio-control of various pathogens by biosynthesized copper nanoparticles using endophytic actinomycetes. J. Radiat. Res. Appl. Sci. 11: 262–270.

Salvadori, M.R., R.A. Ando, C.A. Oller Do Nascimento and B. Corrêa. 2014a. Bioremediation from wastewater and extracellular synthesis of copper nanoparticles by the fungus *Trichoderma koningiopsis*. J. Environ. Sci. Health A. 49: 1286–1295.

Salvadori, M.R., R.A. Ando, C.A.O. do Nascimento and B. Corrêa. 2014b. Intracellular biosynthesis and removal of copper nanoparticles by dead biomass of yeast isolated from the wastewater of a mine in the Brazilian Amazonia. PLoS One. 9: e87968.

Sanghi, R., P. Verma and S. Puri. 2011. Enzymatic formation of gold nanoparticles using *Phanerochaete chrysosporium*. Adv. Chem. Eng. Sci. 1: 154.

Saratale, R.G., G.D. Saratale, J.S. Chang and S.P. Govindwar. 2011. Bacterial decolorization and degradation of azo dyes: A review. J. Taiwan. Inst. Chem. Eng. 42: 138–157.

Saravanakumar, K. and M.H. Wang. 2018. Trichoderma based synthesis of anti-pathogenic silver nanoparticles and their characterization, antioxidant and cytotoxicity properties. Microb. Pathog. 114: 269–273.

Saravanan, C., R. Rajesh, T. Kaviarasan, K. Muthukumar, D. Kavitake and P.H. Shetty. 2017. Synthesis of silver nanoparticles using bacterial exopolysaccharide and its application for degradation of azo-dyes. Biotechnol. Rep. 15: 33–40.

Saravanan, M., S. Arokiyaraj, T. Lakshmi and A. Pugazhendhi. 2018a. Synthesis of silver nanoparticles from *Phenerochaete chrysosporium* (MTCC-787) and their antibacterial activity against human pathogenic bacteria. Microb. Pathog. 117: 68–72.

Saravanan, M., S.K. Barik, D. MubarakAli, P. Prakash and A. Pugazhendhi. 2018b. Synthesis of silver nanoparticles from *Bacillus brevis* (NCIM 2533) and their antibacterial activity against pathogenic bacteria. Microb. Pathog. 116: 221–226.

Sathyavathi, R., M.B. Krishna, S.V. Rao, R. Saritha and D.N. Rao. 2010. Biosynthesis of silver nanoparticles using *Coriandrum sativum* leaf extract and their application in nonlinear optics. Adv. Sci. Lett. 3: 138–143.

Senapati, S., A. Ahmad, M.I. Khan, M. Sastry and R. Kumar. 2005. Extracellular biosynthesis of bimetallic Au-Ag alloy nanoparticles. Small. 1: 517–520.

Shanmugasundaram, T. and R. Balagurunathan. 2017. Bio-medically active zinc oxide nanoparticles synthesized by using extremophilic actinobacterium, *Streptomyces* sp. (MA30) and its characterization Artif. Cell. Nanomed. Biotechnol. 45: 1521–1529.

Shantkriti, S. and P. Rani. 2014. Biological synthesis of copper nanoparticles using *Pseudomonas fluorescens*. Int. J. Curr. Microbiol. App. Sci. 3: 374–383.

Sharma, D., S. Kanchi and K. Bisetty. 2019. Biogenic synthesis of nanoparticles: A review Arab. J. Chem. 12: 3576–3600.

Silva-Vinhote, N.M., N.E.D. Caballero, T. de Amorim Silva, P.V. Quelemes, A.R. de Araujo, A.C.M. de Moraes, et al. 2017. Extracellular biogenic synthesis of silver nanoparticles by Actinomycetes from amazonic biome and its antimicrobial efficiency. African. J. Biotechnol. 16: 2072–2082.

Singh, B.N., A.K.S. Rawat, W. Khan, A.H. Naqvi and B.R. Singh. 2014. Biosynthesis of stable antioxidant ZnO nanoparticles by *Pseudomonas aeruginosa* rhamnolipids. PLoS One. 9: e106937.

Singh, P., H. Singh, Y.J. Kim, R. Mathiyalagan, C. Wang and D.C. Yang. 2016. Extracellular synthesis of silver and gold nanoparticles by *Sporosarcina koreensis* DC4 and their biological applications. Enzyme Microb. Technol. 86: 75–83.

Singh, P., S. Pandit, V. Mokkapati, A. Garg, V. Ravikumar and I. Mijakovic. 2018. Gold nanoparticles in diagnostics and therapeutics for human cancer. Int. J. Mol. Sci. 19: 1979.

Sowani, H., P. Mohite, H. Munot, Y. Shouche, T. Bapat, A.R. Kumar, et al. 2016. Green synthesis of gold and silver nanoparticles by an actinomycete *Gordonia amicalis* HS-11: mechanistic aspects and biological application. Process Biochem. 51: 374–383.

Spagnoletti, F.N., C. Spedalieri, F. Kronberg and R. Giacometti. 2019. Extracellular biosynthesis of bactericidal Ag/AgCl nanoparticles for crop protection using the fungus *Macrophomina phaseolina*. J. Environ. Manage. 231: 457–466.

Sriramulu, M. and S. Sumathi. 2018. Biosynthesis of palladium nanoparticles using *Saccharomyces cerevisiae* extract and its photocatalytic degradation behaviour. Adv. Nat. Sci. Nanosci. Nanotech. 9: 025018.

Srivastava, S.K. and M. Constanti. 2012. Room temperature biogenic synthesis of multiple nanoparticles (Ag, Pd, Fe, Rh, Ni, Ru, Pt, Co, and Li) by *Pseudomonas aeruginosa* SM1. J. Nanopart. Res. 14: 831.

Srivastava, N. and M. Mukhopadhyay. 2014. Biosynthesis of SnO_2 nanoparticles using bacterium *Erwinia herbicola* and their photocatalytic activity for degradation of dyes. Ind. Eng. Chem. Res. 53: 13971–13979.

Subbaiya, R., M. Saravanan, A.R. Priya, K.R. Shankar, M. Selvam, M. Ovais, et al. 2017. Biomimetic synthesis of silver nanoparticles from *Streptomyces atrovirens* and their potential anticancer activity against human breast cancer cells. IET Nanobiotechnol. 11: 965–972.

Sudha, S.S., K. Rajamanickam and J. Rengaramanujam. 2013. Microalgae mediated synthesis of silver nanoparticles and their antibacterial activity against pathogenic bacteria. Indian. J. Exp. Biol. 52: 393–399.

Syed, A. and A. Ahmad. 2012. Extracellular biosynthesis of platinum nanoparticles using the fungus *Fusarium oxysporum*. Colloids Surf. B Biointerfaces. 97: 27–31.

Talekar, S., G. Joshi, R. Chougle, B. Nainegali, S. Desai, A. Joshi, et al. 2014. Preparation of stable cross-linked enzyme aggregates (CLEAs) of NADH-dependent nitrate reductase and its use for silver nanoparticle synthesis from silver nitrate. Catal. Commun. 53: 62–66.

Thakkar, K.N., S.S. Mhatre and R.Y. Parikh. 2010. Biological synthesis of metallic nanoparticles. Nanomed. Nanotechnol. Biol. Med. 6: 257–262.

Thomas, R., A.P. Nair, K.R. Soumya, J. Mathew and E.K. Radhakrishnan. 2014. Antibacterial activity and synergistic effect of biosynthesized Ag NPs with antibiotics against multidrug-resistant biofilm-forming coagulase-negative staphylococci isolated from clinical samples. Appl. Biochem. Biotechnol. 173: 449–460.

Tiwari, M., P. Jain, R.C. Hariharapura, K. Narayanan, U. Bhat, N. Udupa, et al. 2016. Biosynthesis of copper nanoparticles using copper-resistant *Bacillus cereus*, a soil isolate. Process Biochem. 51: 1348–1356.

Tripathi, R., A.S. Bhadwal, R.K. Gupta, P. Singh, A. Shrivastav and B. Shrivastav. 2014. ZnO nanoflowers: Novel biogenic synthesis and enhanced photocatalytic activity. J. Photochem. Photobiol. B. 141: 288–295.

Usha, R., E. Prabu, M. Palaniswamy, C.K. Venil and R. Rajendran. 2010. Synthesis of metal oxide nano particles by *Streptomyces* sp. for development of antimicrobial textiles. Global J. Biotechnol. Biochem. 5: 153–160.

Vainshtein, M., N. Belova, T. Kulakovskaya, N. Suzina and V. Sorokin. 2014. Synthesis of magneto-sensitive iron-containing nanoparticles by yeasts. J. Ind. Microbiol. Biotechnol. 41: 657–663.

Velmurugan, P., M. Iydroose, M.H.A.K. Mohideen, T.S. Mohan, M. Cho and B.T. Oh. 2014. Biosynthesis of silver nanoparticles using *Bacillus subtilis* EWP-46 cell-free extract and evaluation of its antibacterial activity. Bioprocess. Biosyst. Eng. 37: 1527–1534.

Venkatesan, J., S.K. Kim and M.S. Shim. 2016. Antimicrobial, antioxidant, and anticancer activities of biosynthesized silver nanoparticles using marine algae *Ecklonia cava*. Nanomaterials. 6: 235.

Verma, S.K., E. Jha, B. Sahoo, P.K. Panda, A. Thirumurugan, S.K.S. Parashar, et al. 2017. Mechanistic insight into the rapid one-step facile biofabrication of antibacterial silver nanoparticles from bacterial release and their biogenicity and concentration-dependent *in vitro* cytotoxicity to colon cells. RSC Adv. 7: 40034–40045.

Vetchinkina, E.P., E.A. Loshchinina, A.M. Burov, L.A. Dykman and V.E. Nikitina. 2014. Enzymatic formation of gold nanoparticles by submerged culture of the basidiomycete *Lentinus edodes*. J. Biotechnol. 182: 37–45.

Wadhwani, S.A., M. Gorain, P. Banerjee, U.U. Shedbalkar, R. Singh, G.C. Kundu, et al. 2017. Green synthesis of selenium nanoparticles using *Acinetobacter* sp. SW30: Optimization, characterization and its anticancer activity in breast cancer cells. Int. J. Nanomed. 12: 6841.

Wang, H., H. Chen, Y. Wang, J. Huang, T. Kong, W. Lin, et al. 2012. Stable silver nanoparticles with narrow size distribution non-enzymatically synthesized by *Aeromonas* sp. SH10 cells in the presence of hydroxyl ions. Curr. Nanosci. 8: 838–846.

Wing-Shaná Lin, I. 2014. Biosynthesis of silver nanoparticles from silver (i) reduction by the periplasmic nitrate reductase c-type cytochrome subunit NapC in a silver-resistant *E. coli*. Chem. Sci. 5: 3144–3150.

Xiao, X., Q.Y. Liu, X.R. Lu, T.T. Li, X.L. Feng, Q. Li, et al. 2017. Self-assembly of complex hollow CuS nano/micro shell by an electrochemically active bacterium *Shewanella oneidensis* MR-1. Int. Biodeterior. Biodegradation. 116: 10–16.

Zhang, H., Q. Li, Y. Lu, D. Sun, X. Lin, X. Deng, et al. 2005. Biosorption and bioreduction of diamine silver complex by *Corynebacterium*. J. Chem. Technol. Biotechnol. 80: 285–290.

Zhang, H., Q. Li, H. Wang, D. Sun, Y. Lu and N. He. 2007. Accumulation of silver (I) ion and diamine silver complex by *Aeromonas* SH10 biomass. Appl. Biochem. Biotechnol. 143: 54–62.

Zhang, X., Y. Qu, W. Shen, J. Wang, H. Li, Z. Zhang, et al. 2016. Biogenic synthesis of gold nanoparticles by yeast *Magnusiomyces ingens* LH-F1 for catalytic reduction of nitrophenols. Colloids Surf. A Physicochem. Eng. Asp. 497: 280–285.

Zhou, H., H. Pan, J. Xu, W. Xu and L. Liu. 2016. Acclimation of a marine microbial consortium for efficient Mn (II) oxidation and manganese containing particle production. J. Hazard. Mater. 304: 434–440.

Zikalala, N., K. Matshetshe, S. Parani and O.S. Oluwafemi. 2018. Biosynthesis protocols for colloidal metal oxide nanoparticles. Nano-Structures & Nano-Objects. 16: 288–299.

10

Biosynthesis of Metallic Nanoparticles by Extremophiles and Their Applications

Sachin Paudel and Debora F. Rodrigues*

Civil and Environmental Engineering Department
University of Houston, TX 77041
Email: spaudel4@uh.edu; dfrigirodrigues@uh.edu

INTRODUCTION

The term "Nanotechnology" has been overweening and become synonymous with materials that are innovative and highly promising in the nanoscale size. Nanomaterial technology and its applications are emerging as an underlying technology in various disciplines, such as physics, chemistry, biology, medicine, material science, and environmental technology. Nanomaterials are materials in the size of 10^{-9} m in at least one of their dimensions (Hatchett and Josowicz 2008). Nanomaterials or nanoparticles are comparable to the size of cellular organelles, including nano-sized proteins, DNA molecules, and viruses. These particles have gained research interest in different technological applications due to their unusual physicochemical properties attributed to their size. The catalytic, electronic, magnetic, chemical, photoelectrochemical, and optical properties of these nanomaterials make them an attractive subject for investigations (Pető et al. 2002, Hulkoti and Taranath 2014). The dramatic surge in research studies regarding nanoparticles indicates its transcending future applications (Figure 10.1). Various

*For Correspondence: dfrigirodrigues@uh.edu

methods of top-down and bottom-up nanoparticle syntheses have been widely applied. Methodologies involved in top-down approaches of synthesis are typically tedious, energy-intensive, and expensive. Conversely, bottom-up approaches have been identified as promising alternatives to nanoparticles synthesis (Wang et al. 2016). Top-down methods require gradual removal of pieces of the bulk material to obtain nanoscale materials; whereas, bottom-up approaches take advantage of naturally occurring forces, such as Van der Waals force, electrostatic force, and a variety of interatomic and/or intermolecular forces between atoms or molecules to generate nanoparticles. These naturally occurring forces are triggered using chemical or electrochemical reactions involved in precipitation of nanostructures, sol-gel processing, laser pyrolysis, chemical vapor deposition (CVD), plasma or flame spraying synthesis, and atomic or molecular condensation (Daraio and Jin 2012). Biosynthesis of nanoparticles, more recently, has also been considered a bottom-up approach, since it involves the exploitation of interatomic forces triggered by biochemical reactions and metabolic activities (Sarikaya et al. 2003).

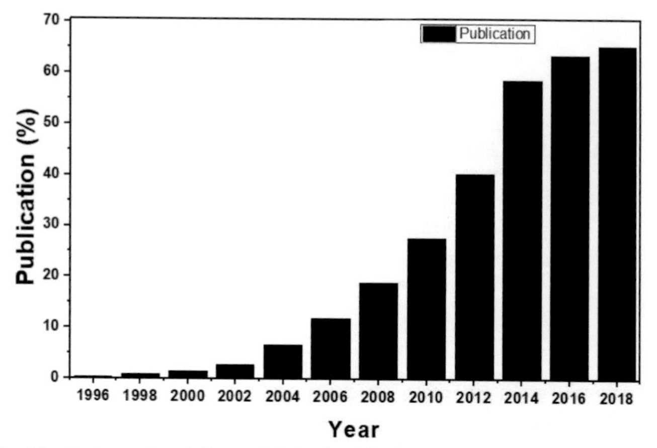

Figure 10.1 Evolution of articles published with the keyword "nanomaterial" in title in the last two decades. This data is collected from Google Scholar, which covers the literature published globally in the field of science and technology.

In the case of bottom-up approaches of chemical synthesis of nanoparticles, the nanoparticles are grown in liquid medium containing various reducing and stabilizing chemicals. In this case, the toxicity of chemicals in the final nanoparticles product is of paramount concern (Thakkar et al. 2010). Biosynthesis of nanoparticles, on the other hand, is viewed as green synthesis of nanoparticles, which occurs in natural environmental conditions and, therefore, has less possibility of having chemical toxicity. The microorganisms are used as possible "nano-factories" for the development of clean, non-toxic, and environmentally friendly methods for producing nanoparticles. Biosynthesis of metal nanoparticles by microorganisms involves the reduction of toxic metals as a defense mechanism of organisms. In the microbial nanoparticles synthesis, the nanoparticles are a by-product of resistance mechanisms developed by microorganisms towards a specific metal (Prasad et al. 2016). In the case of biosynthetic nanoparticles, microorganisms interact with the metal ions in their environment, and then turn the metal ions into nanoparticles or elemental atoms through metabolic activity (Shankar et al. 2016). Figure 10.2 illustrates the generic process for the biosynthesis of nanoparticles by microorganisms. The location of nanoparticles formation is an important aspect to be considered. Intracellular synthesis comprises transporting metal ions into the microbial cells to form the nanoparticles; whereas, extracellular nanoparticles synthesis involves reducing the metal ions with the presence of enzymes secreted by the cells (Mann 1993, Zhang et al. 2011a, Prasad et al. 2016). Biosynthesis of metal and metal-based nanoparticles is frequently studied using bacteria, fungi, and algae. Figure 10.3 represents the percentage of different types of nanoparticles produced by different groups of microorganisms, as presented in the current literature.

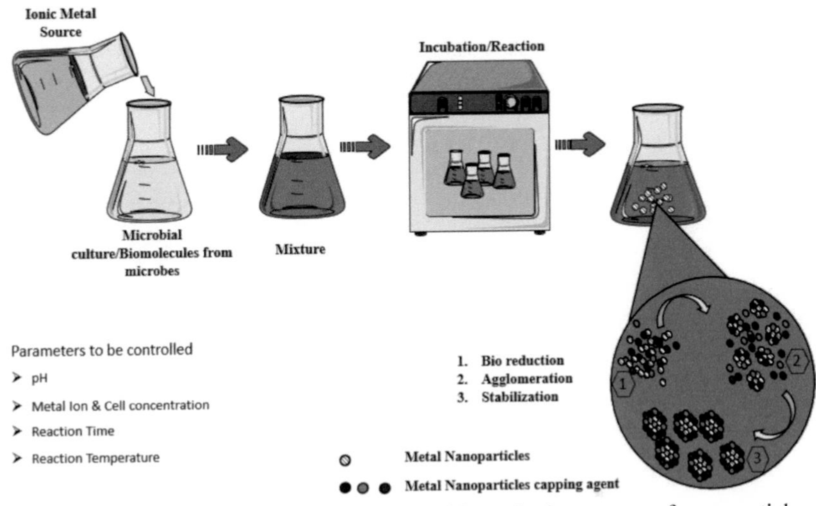

Figure 10.2 A generic representation of the biosynthesis process of nanoparticles by microorganisms. The production of nanoparticles typically requires controlled environmental conditions, such as pH, temperature, specific metal, and cell concentrations, as well as different reaction times for the formation of the nanoparticles. The first step of the reaction involves mixing the cells or cell extracts with a metal salt or ion source under specific conditions, followed by allowing the reaction to happen. During the formation of the nanoparticles intracellularly or extracellularly, the expected steps are initial bio-reduction of the metal ions, followed by their agglomeration, and stabilization by capping agents.

Recently, close attention has been given to the biosynthesis of nanoparticles by extremophiles (Tian and Hua 2010, Gabani and Singh 2013, Oren 2013, Zhang et al. 2015). Extremophiles are microorganisms able to grow and survive under extreme environmental conditions. The extreme conditions include, but are not limited to, high or low temperatures, salinities, pH, and pressures. Extremophiles may be defined as acidophilic (optimal growth between pH 1–5); alkaliphiles (optimal growth above pH 9); halophiles (optimal growth between 0.3–5 M salt concentrations); psychrophiles (optimal growth between 0–15°C), thermophiles (optimal growth between 60–80°C), and hyperthermophiles (optimal growth above 80°C) (Rampelotto 2013). Due to the ability of extremophiles to survive in harsh

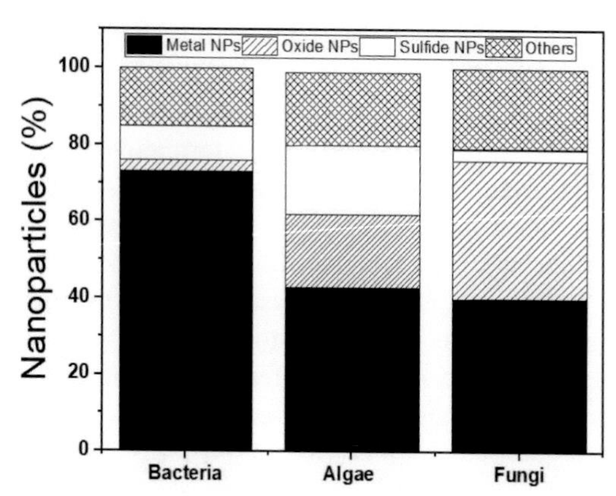

Figure 10.3 Percentage of different types of nanoparticles produced by different groups of microorganisms as described in the literature. Literature data collected between 1996 and 2018. This data is collected from Google Scholar, which covers the literature published globally in the field of science and technology.

conditions, these organisms are of significant biotechnological interest, since they produce extremozymes (enzymes produced under extreme conditions), which are able to transform substrates under extreme conditions (Colombo et al. 1995, Zhang et al. 2015). An example of the synthesis mechanism is given in Figure 10.4. This chapter aims to present the biosynthesis of nanoparticles using various halophiles, acidophiles, alkaliphiles psychrophiles, and thermophiles.

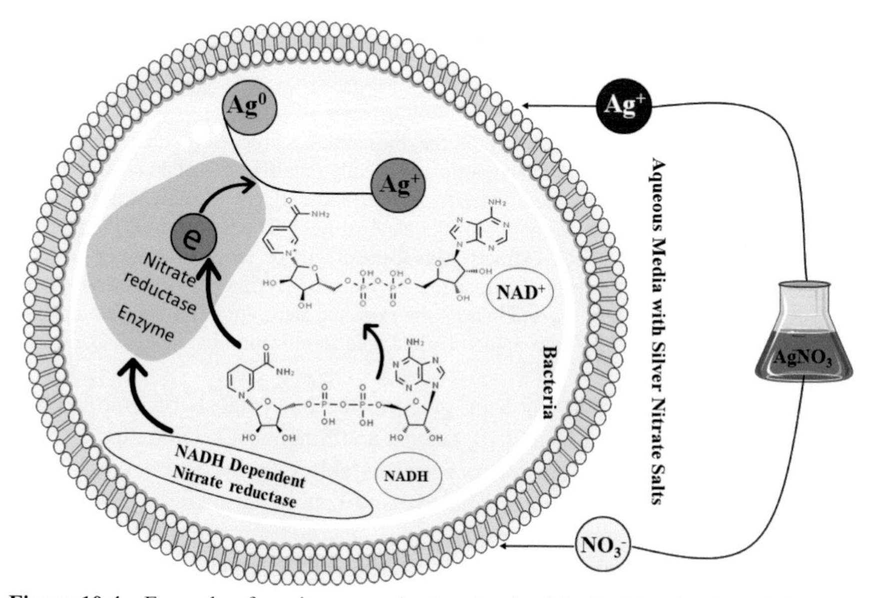

Figure 10.4 Example of a primary mechanism involved in the bioreduction of silver salt to elemental silver. Nitrate reductase reduces nitrate to nitrite using NADH as an electron donor. In the presence of silver, the electron is transferred from NADH to the nitrate reductase enzyme to reduce the silver ions.

NANOPARTICLE BIOSYNTHESIS BY HALOPHILES

Halophilic microorganisms have a higher affinity towards saline environments, and flourish where higher salt concentrations are maintained. They are typically found at places with salt concentration regimes between 0.3–5 M (1.4–30% of salt concentrations) (Ollivier et al. 1994). Halophiles are excellent natural sources of metal tolerant microbes, as many of them are naturally found in locations containing metallic pollutants from volcanic eruptions, natural weathering of rocks, anthropogenic activities, such as mining, combustion of fuel, and urban and industrial sewage discharges. Due to their ability to tolerate metals from different sources, these types of microorganisms have been abundantly exploited to biosynthesize elemental metallic nanoparticles of gold (Au), silver (Ag), and selenium (Se). There is great interest in biosynthesis of metal nanoparticles by extremophiles. For example, silver and gold nanoparticles, when biosynthesized, have great application in biomedicine and pharmaceutical industries for drug delivery purpose. These nanoparticles, when biosynthesized, are stable, have high dispersibility, and are biocompatible. Several unique biomolecules produced by halophiles, such as enzymes, bio-surfactants, and exopolysaccharides show biological activities in harsh environments and have been exploited for metal nanoparticles biosynthesis (Poli et al. 2010, Siddiqi et al. 2018). Typically, polysaccharides and bio-surfactants act as capping and stabilizing agents of synthesized nanoparticles (Mata et al. 2009, Siddiqi et al. 2018). In the next section, we will discuss the biosynthesis of nanoparticles by halophilic bacteria, fungi, and algae, and potential mechanisms associated with the biosynthesis process.

Biosynthesis of Nanoparticles by Halophilic Bacteria

The most frequently reported nanoparticle syntheses by halophilic bacteria are related to metallic nanoparticles. For instance, a silver tolerant halophilic bacterium *Idiomarina* sp. PR-58-8 was reported to synthesize silver (Ag) nanoparticles intracellularly (Seshadri et al. 2012). Intracellular biosynthesis is possible when the ions get inside the cells via active transport, and enzymes inside the cell reduce the metal ion into elemental metal or nanoparticles (Liang et al. 2011, Ovais et al. 2018). The average particle size of the nanoparticles produced by *Idiomarina* sp. PR-58-8, as per TEM analysis, was found to be 26 nm (Seshadri et al. 2012). This microorganism was able to synthesize the Ag nanoparticles within 48 hours of incubation with 5 mM $AgNO_3$. During the biosynthesis, an elevated level of thiols was observed and was thought to be produced during Ag^+ transport into the cell. The authors claimed Ag^+ tolerance of *Idiomarina* sp. is attributed to the presence of these thiol groups, which was suggested to assist in the tolerance of oxidative stress induced by Ag^+ ions. Cysteine, a thiol-containing amino acid, has been described to be able to neutralize the effects of silver ions by reducing and complexing Ag^+ ions into Ag nanoparticles (Holt and Bard 2005, Liu et al. 2012, Behra et al. 2013). However, the precise mechanism of how thiol groups or cysteine are involved in the reduction of Ag^+ to Ag nanoparticles is not elucidated yet. Similarly,

Ag nanoparticles were also synthesized intracellularly by a marine *Pseudomonas* sp. within 48 hours of incubation with 1 mM $AgNO_3$. The Ag nanoparticles produced by this microorganism were polydisperse, and with sizes ranging from 20–100 nm. These nanoparticles were stable in solution for several months (Muthukannan and Karuppiah 2011).

Another important metallic nanoparticle produced by halophilic bacteria is gold. This type of nanoparticle has been described to be biosynthesized by *Halomomas salina* extracellularly in both acidic and alkaline conditions, yielding Au nanoparticles of anisotropic and spherical shapes, respectively, with particle sizes ranging from 30–100 nm (Shah et al. 2012). Studies have indicated the involvement of NADH and NADH-dependent nitrate reductase enzymes, which are known to be secreted extracellularly by *Halomomas salina*. These reductases can act as scaffoldings or nucleating agents, responsible for the reduction of Au^{+3} to Au^0 and formation of Au nanoparticles (He et al. 2007). In addition to gold and silver nanoparticles, halophilic microorganisms, such as *Bacillus megaterium* strains, were also described to produce selenium (Se) nanoparticles intracellularly and extracellularly at 37°C in a media supplemented with 0.25 mM of Se(IV) (Mishra et al. 2011). The biosynthesis demonstrated the allotropic form of Se with an average particle size of 200 nm and spherical shape. Biosynthesis of elemental Se nanoparticles might have been possible due to the involvement of biomolecules in the metal reduction process. However, the mechanism involved in the reduction of selenite to selenium remains unknown. Intracellular biosynthesis of Se nanoparticles is also understood to be driven by enzymatic reaction with the involvement of thiol reduction (Kessi and Hanselmann 2004, Kessi 2006). Selenite reaction with glutathione forms selenoid-glutathione, which is further reduced by NADPH to seleno-persulfide. The dismutation of seleno-persulfide ultimately produces glutathione and elemental Se particles (Kessi and Hanselmann 2004).

Biosynthesis of Nanoparticles by Halophilic Fungi

Biosynthesis of metallic nanoparticles is not limited to halophilic bacteria. Fungi and algae found in the saline environment have also been exploited for metal nanoparticle synthesis. Common fungi *Aspergillus niger* and *Penicillium fellutanum* isolated from a halophilic environment were also used to biosynthesize Ag nanoparticles extracellularly (Kathiresan et al. 2009, Kathiresan et al. 2010). *Penicillium fellutanum* demonstrated the biosynthesis of Ag nanoparticles within 24 hours of incubation, with media supplemented by 1 mM $AgNO_3$ with 0.3% NaCl, yielding spherical particles of 5–25 nm. The exact mechanism for Ag nanoparticles biosynthesis is unclear; however, a single prominent band of protein (MW 70 kDa), attributed by the authors to a nitrate reductase enzyme, was observed in sodium dodecyl sulfate-polyacrylamide gel electrophoresis (SDS–PAGE) analysis. Furthermore, *Aspergillus niger* biosynthesized Ag nanoparticles also presented a single prominent band of protein of MW 70 kDa in the SDS–PAGE analysis when producing spherical nanoparticles with a size range of 5–35 nm (Kathiresan et al. 2010). These observations indicate that the nitrate reductase enzyme could be involved in the biosynthesis of Ag nanoparticles by halophilic fungi; however, claiming its direct relationship to the synthetic pathway of the nanoparticle needs to be further investigated. In addition to Ag nanoparticles synthesis, a dimorphic tropical marine fungus isolate of *Yarrowia lipolytica* (NCIM 3589) collected from an oil-polluted field was exploited for the biosynthesis of Au nanoparticles (Agnihotri et al. 2009). Since NADH and NADH-dependent nitrate reductase are known to be responsible for the reduction of metals, *Yarrowia lipolytica* was also suggested to reduce gold following the same pathway and produce triangular and hexagonal shaped Au nanoparticles. The production of nanoparticles by *Yarrowia lipolytica* was reported to have an inverse relationship of cell number and Au salt concentration on the scale of nanoparticles yielded (Pimprikar et al. 2009). When the cell number remained the same (10^{11} cells/mL) and the Au salt concentration (1, 2 and 3 mM) increased, an increment in nanoparticles sizes (15, 18, and 20 nm, respectively) was observed. On the other hand, when the cell number increased (10^9, 10^{10}, 10^{11} cells/mL) and the salt concentration remained constant (1 mM), it resulted in decreased sizes of nanoparticles produced (22, 18, 15 nm, respectively). *Yarrowia lipolytica* has also been exploited for the biosynthesis of sulfide and oxide nanostructures of cadmium. Spherical nanostructures of CdS and CdO with particle sizes of 45 nm and 20 nm, respectively, were synthesized (Pawar et al. 2012). *Yarrowia lipolytica* has been described to have different reductases; hence, cadmium salts are suggested to be reduced by reductase enzymes to form the nanoparticles. In addition to the role of reductases found in *Y. lipolytica*, biomolecules, such as glutathione and phytochelatins produced by the enzyme phytochelatin synthase, are also considered to be involved in the reduction and stabilization of the nanoparticles, but direct evidence of their involvement is lacking (Dameron et al. 1989, Kowshik et al. 2002, Pawar et al. 2012). Unlike halophilic bacteria, fungi are still very understudied, and there are no reports on selenium nanoparticles produced by this group of microorganisms.

Biosynthesis of Nanoparticles by Halophilic Algae

Algae are also described to be involved in the biosynthesis of metallic nanoparticles. Biosynthesis of nanoparticles by halophilic algae was recently reported, and mostly described to happen extracellularly. Among all algae investigated, brown algae are known to have better biosorption and metal ion uptake compared to some fungi and other types of algae (Davis et al. 2003). The higher metal uptake by brown algae is due to the presence of mucilaginous polysaccharides (alginate and sulfated fucoidans) rich cell walls (Romera et al. 2007, Mata et al. 2008). Mucilaginous polysaccharides contain major functional groups responsible for metal uptake, in particular, carboxylic groups, which account for 60–70% of the dry algal biomass (Shankar et al. 2016). Different types of nanoparticles have been described to be produced by brown algae, such as those of Ag, Au, CuO, and ZnO (Abboud et al. 2014, Azizi et al. 2014, Dahoumane et al. 2017). The most studied nanoparticle biosynthesis by brown algae pertains to Ag and Au. The extracts from marine brown algae *Sargassum longifolium* were used for the synthesis of spherical

Ag nanoparticles, which exhibited excellent antifungal activity (Rajeshkumar et al. 2014). Biosynthesis was carried out by reacting 10 mL of algal extracts with 90 mL of 1 mM $AgNO_3$ at room temperature for 32 hours while shaking at 120 rpm. This biosynthesis yielded spherical-shaped Ag nanoparticles.

Other *Sargassum* species, i.e., *Sargassum wightii,* have been reported to produce Au nanoparticles extracellularly (Singaravelu et al. 2007, Shah et al. 2012, Shankar et al. 2016). *Sargassum wightii* has been described to biosynthesize gold nanoparticles of spherical shape with an average particle size of 8–100 nm (Ollivier et al. 1994, Oza et al. 2012). This brown algae is known to be rich in reductases and nitrate reductases, which were suggested to serve as a scaffold or nucleating agents for NADH-dependent reduction of Au^{3+} to Au^0 (He et al. 2007, Oza et al. 2012). The same enzyme was also suggested to serve as a capping agent, ensuring the complete formation of thermodynamically stable nanostructures (He et al. 2007, Oza et al. 2012). Since the biosynthesis of nanoparticles by algae is mostly extracellular, studies with algae have also shown that biosynthesis can be conducted using cell-free extracts of algae containing biomolecules, which is promising for large scale industrial applications.

Overall, halophilic microorganisms have a lot to offer in the development of nanoscience. Various halophilic prokaryotic and eukaryotic organisms have been investigated concerning their nanoparticle synthesis ability. Most of these studies are on biosynthesis and characterization of silver and gold nanoparticles, which are listed in Table 10.1, but there are a few studies on other metal oxides and metal sulfides nanoparticles as well. Their application in biomedicine has been explored; however, there is a need for studies that focus more on mechanisms involved in the biosynthesis process by these groups of organisms. Understanding the mechanisms involved in biosynthesis would open new avenues for scaling up green synthesis of nanoparticles.

NANOPARTICLE BIOSYNTHESIS BY THERMOPHILES AND PSYCHROPHILES

Thermophiles refer to organisms that not only tolerate high temperatures, but also require it for their growth and survival. At these high temperatures, most other organisms are dead, while thermophiles thrive. Thermophiles survive at temperatures ranging from 50°C to as high as 120°C. Typically, thermophiles are found in naturally hot environments, such as volcanic areas, deep-sea hydrothermal vents, and deep oil and gas reservoirs (Mehta and Satyanarayana 2013). Also, human-made hot habitats of self-heated compost piles and biological waste treatment facilities have been reported to be favorable environments for the growth of thermophilic microorganisms (Fujio and Kume 1991, Mehta and Satyanarayana 2013). Since they survive in hot conditions, they contain enzymes that function at high temperatures, which are of high industrial significance (Mehta and Satyanarayana 2013, Tiquia-Arashiro 2014). As opposed to the thermophiles, there are also psychrophiles, which grow and survive in cold environments. Psychrophiles are found in cold

Table 10.1 Nanoparticle biosynthesis by halophiles

Halophiles	*NP*	*Site of Synthesis*	*Temp* (°C)	*pH*	*Size* (nm)	*Shape*	*References*
Bacteria							
Idiomarina sp. PR-58-8	Ag	Intracellular	30		26		Seshadri et al. 2012
Pseudomonas sp. 591786	Ag	Intracellular	30		20–100	Spherical	Muthukannan and Karuppiah 2011
Halomomas salina	Au	Extracellular	30	9	30–100	Spherical	Shah et al. 2012
Bacillus megaterium BSB6	Se	Intra/Extracellular	37	7.5	200		Mishra et al. 2011
Bacillus megaterium BSB12	Se	Intra/Extracellular	37	7.5	200		Mishra et al. 2011
Fungi							
Yarrowia lipolytica	Au	Intracellular	30	7 and 9	15	Triangular/ Hexagonal	Agnihotri et al. 2009
Aspergillus niger	Ag	Extracellular	25		5–35	Spherical	Kathiresan et al. 2010
Penicilium fellutanum	Ag	Extracellular	5	6	5–25	Spherical	Kathiresan et al. 2009
Algae							
Sargassum wightii	Au	Extracellular			8–12		Singaravelu et al. 2007
Sargassum wightii	Au	Extracellular	30	10	30–100	Spherical	Oza et al. 2012
Sargassum longifolium	Ag	Extracellular		6.2-8.4		Spherical	Rajeshkumar et al. 2014
Pterocladia capillacae	Ag	Extracellular	70	10	7	Spherical	El-Rafie et al. 2013
Jania rubins	Ag	Extracellular	70	10	12	Spherical	El-Rafie et al. 2013
Ulva faciata	Ag	Extracellular	70	10	7	Spherical	El-Rafie et al. 2013
Colpmenia sinuosa	Ag	Extracellular	70	10	20	Spherical	El-Rafie et al. 2013

NP: nanoparticle

temperatures ranging from –20 to 15°C, and have optimum growth conditions up to 15°C, and cannot grow beyond 20°C (D'Amico et al. 2006). These organisms are typically found in everlasting cold environments of polar regions, deep-sea, glaciers, mountains, and arctic and alpine soils. Psychrophiles manage to survive in icy environments by optimizing cell processes, such as cell membrane structures, enzyme functions, and nutrient transport (Wang et al. 2017). Both thermophiles and psychrophiles produce unique types of enzymes and metabolic products and possess immense industrial importance. Thermophiles and psychrophiles are also important microorganisms in diverse industries, where they are used for biomedicine, food processing, detergents, and environmental applications, such as biofuel production and bioremediation (Sarmiento et al. 2015). Thermophiles are in the industry due to their ability to produce thermostable enzymes, which have operational stability and denaturant tolerance at extremely hot temperatures (Turner et al. 2007). Psychrophiles are critical for industrial applications due to their cell membranes and proteins, which have structural flexibility (global flexibility) that augments catalytic function. Some thermophiles and psychrophiles have also developed the ability to reduce metal ions as a self-defense mechanism to quell stress provided by metals and heavy metal toxicity. Researchers have exploited this ability of thermo- and psychrophiles for biosynthesis of metallic nanoparticles. In this section, we will discuss various thermophiles and psychrophiles in action for the biosynthesis of metallic nanoparticles. It is essential to point out that psychrophile fungi or thermophile and psychrophile algae able to produce nanoparticles have not been described in the literature. The lack of data for these groups of organisms probably suggests that more studies with these groups of microorganisms need to be done.

Biosynthesis of Nanoparticles by Thermophilic Bacteria

Gram-positive thermophilic bacteria, including aerobes, denitrifiers, and facultative anaerobes that grow at temperatures ranging from 45 to 75°C, which are from the genus *Geobacillus,* are considered biorefineries for biofuel and chemical production because of their catabolic versatility (Hussein et al. 2015). *Geobacillus stearothermophilus*, which has optimum growth at pH 7 and 60°C, has been reported to biosynthesize Ag nanoparticles (Mohammed Fayaz et al. 2011). The formation of polydisperse spherical Ag nanoparticles of 25 nm average particle size was confirmed by incubating *G. stearothermophilus* with 1 mM Ag^+ ion ($AgNO_3$) for 48 hours at 27°C under dark conditions. The position of –N–H and CO-stretching vibrations from the FTIR spectra of sliver nitrate solution interaction with *G. stearothermophilus* indicated the presence of amide linkages between amino acid residues. To further confirm the involvement of enzyme in synthesis, SDS–PAGE analysis depicted the protein bands of molecular mass ranging from 12–98 kDa. Additionally, NADH-dependent reductase has been observed in the outer membrane of *Geobacter sulfurreducens* (Neal et al. 2004). Another isolate from a geothermal hot spring located in Iran, *Ureibacillus thermosphaericus,* demonstrated great Ag nanoparticles biosynthesizing potential at temperatures around 60–80°C with silver ion concentrations of 0.001–0.1 M (Juibari et al. 2011).

Maximum silver nanoparticle synthesis occurred at 80°C, when the cell supernatant was reacted with 0.01 M $AgNO_3$ within 24 hours. The particles produced were spherical, and with an average particle size between 10–100 nm.

In the case of Au nanoparticles, *Geobacillus stearothermophilus* has also been reported to be able to biosynthesize these nanoparticles with an average size of 25 nm (Mohammed Fayaz et al. 2011). Another *Geobacillus* sp. strain ID17 collected in Antarctica in the Deception Island (active volcano) has been reported to produce gold nanoparticles intracellularly (Correa-Llantén et al. 2013). Quasi hexagonal-shaped Au nanoparticles of particle size ranging from 5–50 nm were synthesized with the assistance of an NADH-dependent Au^{3+} reductase. This synthesis yielded Au nanoparticles with proteins coating the surface of the nanoparticle, which is significant to the biomedicine field. A biological molecule on the surface of Au nanoparticles can act as a substrate for catalysis or further functionalization. A study has reported the toxicity of cationic Au nanoparticles at specific concentrations since negatively charged cellular membranes interacted with cationic nanoparticles (Goodman et al. 2004). Few other thermophilic bacteria strains, such as *Thermomonospora* sp. and *Thermus scotoductus* SA-01 have also been successful in reducing the Au^{3+} ions to produce Au nanoparticles (Ahmad et al. 2003, Erasmus et al. 2014). In addition to Ag and Au, nanoparticles of Se, Pd, and Cu were also described to be produced by thermophilic microorganisms. For instance, another thermophilic strain of *Geobacillus weigelii* GWE intracellularly synthesized Se nanoparticles with spherical shape and size ranging from 40–160 nm. *G. weigelii* were able to produce Se nanoparticles at temperatures ranging from 60–100°C and pH ranging from 4–8. The synthesis mechanism was suggested to be via intracellular NADPH/NADH-dependent reductase, which triggered the reduction of Se^{4+} ions to elemental Se nanoparticles in aerobic environment. *Caldicellulosiruptor saccharolyticus*, a species of thermophilic-anaerobic cellulolytic bacteria, was used for palladium nanoparticle synthesis and H_2 production from wastewater treatment under extreme thermophilic conditions (Shen et al. 2015). Methyl orange and diatrizoate degradation in wastewater by *C. saccharolyticus* was accelerated with the addition of palladium. Copper nanoparticles can be biosynthesized by *Thermoanaerobacter* sp. X513 from oxidized copper salts by an extracellular metal-reduction process (Jang et al. 2015). The chelating and capping agents coat the biosynthesized copper nanoparticles. The production of heterogeneous bacterial organic matter provides the nanoparticle ability for oxidation resistance under aqueous and dry film conditions. Biosynthesis of elemental copper nanoparticles is important, since it is an efficient method to synthesize nanoparticles with the ability to resist oxidation, which otherwise is thermodynamically stable in oxide phase (Jeong et al. 2008).

Biosynthesis of Nanoparticles by Thermophilic Fungi

Nanoparticle biosynthesis is not limited to thermophilic bacteria and can be extended to thermophilic fungi. Thermophilic fungus, *Humicola* sp., is typically found in mushroom compost in the soil and can synthesize Ag nanoparticles extracellularly.

When the fungus reacts with Ag^+ ions, it reduces silver in the precursor solution and forms extracellular nanoparticles, as monitored by ultraviolet-visible spectroscopy (Syed et al. 2013). Thermophilic fungi are also reported to produce Au nanoparticles either intracellularly, extracellularly, or via their extracts (Molnár et al. 2018). *Humicola* sp. is also said to produce cerium oxide nanoparticles when exposed to an aqueous solution of cerium (III) nitrate hexahydrate ($CeN_3O_9.6H_2O$), resulting in the extracellular formation of CeO_2 nanoparticles, containing Ce (III) and Ce (IV) as mixed oxidation states (Khan and Ahmad 2013). The cerium oxide nanoparticles produced by this fungus are naturally capped by proteins secreted by the fungus, which prevent the agglomeration and improve the dispersibility of the particles (Khan and Ahmad 2013). *Humicola* sp. has also been known to produce Gadolinium oxide (Gd_2O_3) nanoparticles. This thermophilic fungus synthesizes Gd_2O_3 nanoparticles at 50°C when $GdCl_3$ is dissolved in water along with fungal biomass. The $GdCl_3$ gets ionized into Gd^{+3} and $3Cl^-$. The Gd^{+3} ions are attracted to anionic proteins secreted by *Humicola* sp., and reductase enzymes secreted by the fungus act on Gd^{+3} and convert to Gd^{+2} ions, resulting in the formation of Gd_2O_3 (Khan et al. 2014). There are also reports of other thermophilic fungi exploited for the biosynthesis of metal nanoparticles of Au, Ag, Cd, and others. They are listed in Table 10.2.

Table 10.2 Nanoparticle biosynthesis by thermophiles

Thermophiles	*NP*	*Site of Synthesis*	*Temp (°C)*	*pH*	*Size (nm)*	*Shape*	*References*
			Bacteria				
Geobacillus sp. strain ID17	Au	Intra-cellular	65	7	5–50	Quasi hexagonal shape	Correa-Llantén et al. 2013
Geobacillus weigelii GWE1	Se	Intra-cellular	60, 80,100	4–8	40–160	spherical	Correa-Llantén et al. 2014
Ureibacillus thermosphaericus	Ag	Extra-cellular	68–80		10–100	Spherical	Juibari et al. 2011
Thermoanaerobacter BKH1	SiO_2			7	15	Spherical	Show et al. 2015
Thermoanaerobacter X513	CdS	Extra-cellular	65	7	<10	Cubic/ Hexagon	Moon et al. 2013
Thermoanaerobacter sp. X513	Cu	Extra-cellular	65		3–70		Jang et al. 2015
			Fungi				
Humicola sp.	Ag	Extra-cellular	50	9	5–25	Spherical	Syed et al. 2013
Humicola sp.	CeO_2	Extra-cellular	50	9	12–20	Spherical	Khan and Ahmad 2013
Humicola sp.	Gd_2O_3	Extra-cellular	50	9	3–8	Quasai-Spherical	Khan et al. 2014

NP: Nanoparticle

Biosynthesis of Nanoparticles by Psychrophilic Bacteria

Psychrophilic bacteria are important biofactories to synthesize nanoparticles. Most of the literature on biosynthesis by psychrophiles is about silver nanoparticles. A silver resistant psychrophilic bacterium *Morganella psychroolerans,* which is typically found in seafood and grows optimally at 20°C, demonstrated the ability to biosynthesize Ag nanoparticles when incubated in media supplemented with 5 mM $AgNO_3$ (Ramanathan et al. 2011). This study also tried to understand the effect of different growing temperatures (25, 20, 15, and 4°C) on the Ag nanoparticles synthesis and found temperature-dependent trends on Ag nanoparticles synthesis rate, with the fastest Ag^+ ions reduction at 25°C and slowest at 4°C. The TEM analysis demonstrated the production of spherical shaped Ag nanoparticles with a size range of 2–5 nm with few Ag nanoplates of size 100–150 nm at an optimal growth temperature of 20°C. At 25°C, a mixture of triangular and hexagonal nanoplates (50–150 nm) along with spherical nanoparticles was observed. At 15°C, a combination of nanoplates (50–150 nm) and spherical particles were found with lower proportions of spherical particles (70–100 nm), while at 4°C, nanoplates were produced with very few spherical nanoparticles. The results suggested that *Morganella psychroolerans* has the potential to tune shape anisotropy of Ag nanoparticles at different temperatures. Cell-free supernatant of psychrophilic bacteria, such as *Phaeocystis antratica*, *Pseudomonas proteolytica*, *Pseudomonas meridiana*, *Arthobacter kerguelensis*, and *Arthobacter gangotriensis*, were exploited for the biosynthesis of Ag nanoparticles (Shivaji et al. 2011). The average size of synthesized Ag nanoparticles varied from 6–12 nm when studied using TEM. Cell-free supernatants were only able to synthesize nanoparticles in the presence of light. This finding is the opposite of previous results found by Saifuddin et al. 2009, which reported that the biosynthesis of Ag nanoparticles using cell-free supernatant of *B. subtilis* would occur only under dark conditions (Saifuddin et al. 2009). The cell-free supernatant of *P. antratica* and *A. kerguelensis* showed rapid biosynthesis of Ag nanoparticles after 2 hours of incubation. The nanoparticles produced using cell-free supernatant of *P. antratica* were stable for up to 8 months when stored in the dark. The ability of the cell-free supernatant of *A. kerguelensis* was independent of synthesis temperature and pH. On the other hand, the stability of Ag nanoparticles was compromised, irrespective of growth temperature when culture pH was 7. These results demonstrate that cell-free supernatant can play a role in the biosynthesis of Ag nanoparticles and that production of nanoparticles varies with bacterial species, and different biomolecules might be involved in the synthesis.

In conclusion, biomolecules of thermophilic and psychrophilic microorganisms are of importance to industries due to their stability at extreme temperatures in some industrial processes. These biomolecules at the industrial level can assist in reducing the requirement for different chemicals used in the synthesis of nanoparticles to produce biocompatible materials via an eco-friendly method. In the case of thermophilic microorganisms, they were demonstrated to have a superior biosynthetic ability for yielding industrially relevant nanoparticles, such as Au, Ag, Se, Pd, Cu. Palladium and copper biosynthesis by thermophilic bacteria are of great importance, since, chemical synthesis of Pd and Cu nanoparticles are expensive and not eco-friendly. Also, biosynthesis of CeO_2 and Gd_2O_3 nanoparticles using thermophilic fungus constitute interesting examples of eco-friendly biosynthesis. Tables 10.2 and 10.3 provide some examples of metal nanoparticle synthesized by thermophilic and psychrophilic microorganisms, respectively. Unlike halophiles, the exploitation of algae from hot or cold environments have not been described yet. The major downside associated with the thermophilic and psychrophilic biosynthesis of metal nanoparticles is the poor understanding of biosynthesis mechanisms and biomolecules involved in it. Research addressing the mechanisms of biosynthesis, enzyme involvement, and biochemistry could assist in the extensive use of thermophiles and psychrophiles in industry.

Table 10.3 Nanoparticle biosynthesis by psychrophiles

Psychrophiles	*NP*	*Site of Synthesis*	*Temp* (°C)	*pH*	*Size* (nm)	*Shape*	*References*
			Bacteria				
Morganella psychrotolerans	Ag	Extracellular	4	6.8	2–5	Spherical	Ramanathan et al. 2011
Phaeocystis antratica	Ag	Extracellular	25/35	7	12.2 ± 5.7	Spherical	Shivaji et al. 2011
Pseudomonas proteolytica	Ag	Extracellular	8, 22, 30	5,7,10	6.9 ± 2.5	Spherical	Shivaji et al. 2011
Pseudomonas meridiana	Ag	Extracellular	8, 22, 30	5,7,10	6.2 ± 2.4	Spherical	Shivaji et al. 2011
Arthobacter kerguelensis	Ag	Extracellular	25/35	7	8.1 ± 5.9	Spherical	Shivaji et al. 2011
Arthobacter gangotriensis	Ag	Extracellular	8, 22, 30	5,7,10	6.5 ± 3.2	Spherical	Shivaji et al. 2011

NP: Nanoparticle

NANOPARTICLE BIOSYNTHESIS BY ACIDOPHILES AND ALKALIPHILES

Extreme organisms that thrive at acidic pH (below 3) and alkaline pH (above 9) are called acidophiles and alkaliphiles, respectively (Rothschild and Mancinelli 2001). Highly acidic habitats, such as mine drainage, solfataric fields, acid thermal hot springs, coal spoils, and bioreactors are favorable for the growth of acidophiles. These habitats feature low pH, temperature ranges of 25 to 90°C, and pressure of up to 5 MPa, low salinity, some heavy metals in either aerobic or anaerobic environments (Seckbach and Libby 1970, Hallberg and Lindström 1994, He et al. 2004). Acidophiles use a variety of pH homeostatic mechanisms that involve restriction of proton influx into the cytoplasm to help maintain the intracellular pH. Acidophiles are mostly distributed in the bacterial and archaeal domain and contribute to numerous biogeochemical cycles, including the iron and sulfur cycles (Druschel et al. 2004). One significant role of acidophiles lies in their biotechnological application of reducing metal pollutants. These can also be a source of acid-stable enzymes and other catalysts (Van Den Burg 2003, Golyshina and Timmis 2005). Alkaliphiles are commonly isolated from environments, such as garden soils, and are also found to proliferate in alkali thermal hot springs, shallow hydrothermal systems, sewage, and hypersaline soda lakes. These habitats contain a wide range of temperatures and usually feature high pH and moderate to high concentrations of dissolved salts (Xu et al. 1999, Hoover et al. 2003, Ma et al. 2004, Kanekar et al. 2012). Since alkaliphiles live in a low H^+ environment, the challenge for these organisms is to continuously neutralize the cytoplasm and encourage H^+ influx to drive ATP synthesis (Krulwich et al. 2009, Mesbah et al. 2009). Alkaliphiles play essential roles in industrial applications, for example, in biological detergent containing alkaline cellulases or alkaline proteases. The enzymes and other metabolites from acidophiles and alkaliphiles have been exploited for the biosynthesis of different nanoparticles. In this section, we will discuss various metal nanoparticles biosynthesis by acidophiles and alkaliphiles.

Biosynthesis of Nanoparticles by Acidophilic Bacteria

Lactobacillus acidophilus, a gram-positive bacteria typically found in the human gastrointestinal (GI tract), known to possess both reducing and capping agents, is reported to synthesize silver nanoparticles by reducing silver ions into elemental silver (Rajesh et al. 2015). *L. acidophilus* reacts with the culture media supplemented with 1 mM $AgNO_3$ for 24 hours at 37°C to yield Ag nanoparticles. The TEM analysis depicted the biosynthesis of spherical Ag nanoparticles of 4–50 nm size ranges. The silver nanoparticles produced by this microorganism are dispersed and stable over time. *Lactobacillus acidophilus* has also been reported to biosynthesize Se nanoparticles extracellularly (Rajasree and Gayathri 2015). *Lactobacillus acidophilus* was incubated with media containing 4 mM of sodium selenite at 37°C to biosynthesize elemental Se nanoparticles of spherical shape with particle size in a range of 40–60 nm. FTIR study revealed the O–H stretching mode and N–H stretch in amine groups, which can be implied for the presence of proteins on the surface of the Se nanoparticles.

In the case of other acidophilic microorganisms, *Arthobacter nitroguajacolicus*, isolated from a gold mine in Iran, was reported to synthesize gold nanoparticles extracellularly and intracellularly (Dehnad et al. 2015). *A. nitroguajacolicus* culture medium reacted with 1 mM $HAuCl_4$ for 24 hours to yield spherical Au nanoparticles of average particle size of 40 nm. A chemolithotroph, acidophilic, and aerobic bacterium, *Acidithiobacillus ferrooxidans*, uses ferrous or reduced inorganic sulfur compounds as energy source. This property of bacteria was exploited to synthesize electron-dense magnetite particles (Fe_3O_4), which are biocompatible, and demonstrated no toxicity when *in vitro* cytotoxic and genotoxic tests were performed (Yan et al. 2012). FTIR study also revealed the presence of amide bonds which can be implied for the presence of proteins on the surface of Fe_3O_4 nanoparticles; this capping rendered the particles' biocompatibility. The synthesis of Ag, Au, Se, Cd, Fe based nanoparticles are not limited to the bacteria described in this section, and other examples of bacteria synthesizing these nanoparticles are presented in Table 10.4.

Table 10.4 Nanoparticles biosynthesis by acidophiles.

Acidophiles	*NP*	*Mechanism*	*Temp* (°C)	*pH*	*Size* (nm)	*Shape*	*References*
Bacteria							
Lactobacillus acidophilus	Ag		37		4–50	Spherical	Rajesh et al. 2015
Lactobacillus acidophilus strain 01	Ag	Intracellular	30		45–60	Spherical	Namasivayam et al. 2010
Lactobacillus acidophilus	Se	Extracellular	35		20–150	Spherical	Rajasree and Gayathri 2015
Pilimelia columellifera SL19	Ag		27	4	12.7	Spherical	Golińska et al. 2016
Pilimelia columellifera SL24	Ag		27	4	15.9	Spherical	Golińska et al. 2016
Arthobacter nitroguajacolicus	Au	Intra/Extracellular	27	5.6	40	Spherical	Dehnad et al. 2015
Actinobacteria C9	Ag	Extracellular	27	7	8–60	Spherical	Anasane et al. 2016
Actinobacteria SF23	Ag	Extracellular	27	7	4–36	Spherical	Anasane et al. 2016
Streptacidiphilus sp. strain CGG11n	Ag	Extracellular	26	2–4	16	Spherical	Railean-Plugaru et al. 2016
Fungi							
Verticillium sp.	Au	Intracellular			20	Spherical, Triangular, Hexagonal	Mukherjee et al. 2001a
Verticillium sp.	Ag	Extracellular followed by precipitation onto the cells	25	5–6	25	Spherical	Mukherjee et al. 2001b
Bipolaris nodulosa	Ag	Extracellular			10–60	Spherical, Triangular, Hexagonal	Saha et al. 2010
Verticillium luteoalbum	Au	Intracellular	28	5–6	<10 nm	Spherical, Triangular, Hexagonal	Gericke and Pinches 2006

Pichia jadinii	Au	Intracellular	28	5–6	100 nm	Spherical, Triangular, Hexagonal	Gericke and Pinches 2006
Fusarium oxysporum	Au	Extracellular			20–40	Spherical and Triangular	Mukherjee et al. 2002
Fusarium oxysporum	ZrO_2	Extracellular	27	3.6	3–11	Quasi-Spherical	Bansal et al. 2004
Fusarium oxysporum	Au-Ag alloy	Extracellular			8–14		Senapati et al. 2005
Fusarium oxysporum	$BaTiO_3$	Extracellular	27	7	4–5	Quasi-Spherical	Bansal et al. 2006
Penicillium spp.	Ag	Intracellular	28		20–35	Spherical	Zhang et al. 2009
Penicillium purpurogenum	Ag	Extracellular	35	6.8	8–10	Spherical	Nayak et al. 2011
Penicillium chrysogenum	Au	Intracellular	27–29	5–5.6	5–100	Spherical, Triangular, Rod Shape	Sheikhloo and Salouti 2011
Penicillium citrinum	Ag	Extracellular	28		109	Spherical	Honary et al. 2013
Penicillium aurantiogriseum	CuO	Extracellular	28	5–9	130	Spherical	Honary et al. 2012
Penicillium waksmanii	CuO	Extracellular	28	5–9	110	Spherical	Honary et al. 2012
Penicillium citrinum	CuO	Extracellular	28	5–9	158	Spherical	Honary et al. 2012
Aspergillus flavus	Ag	Extracellular	37	6.2	8.92	Spherical	Vigneshwaran et al. 2007
Aspergillus terreus	Au	Intracellular	32		186	Spherical/ irregular	Baskar et al. 2014
Aspergillus spp.	Pb	Extracellular			5–20	Spherical	Pavani et al. 2012

NP: Nanoparticle

Biosynthesis of Nanoparticles by Acidophilic Fungi

Fungi seem to be a more viable microorganism for the synthesis of nanoparticles at an industrial scale, since, a fungal mycelial mesh can withstand flow pressure, agitation, and other conditions in bioreactors or chambers, where plant-based material and bacteria fail. *Verticillium* sp. an acidophilic filamentous fungi commonly found in decaying vegetation and soil, has been described for the biosynthesis of Ag nanoparticles. This *Verticillium* sp., when challenged with 0.1 mM $AgNO_3$ aqueous solution, produced Ag nanoparticles intracellularly of average particle size of 25 nm (Sastry et al. 2003). The mechanism of intracellular Ag nanoparticles synthesis is not fully known; however, the authors speculated that the steps for Ag nanoparticles synthesis involved trapping Ag^+ on the fungal surface via electrostatic interaction between Ag^+ and negatively charged carboxylate groups, followed by reduction of Ag^+ ions with enzymes present in the cell wall, and leading to the formation of Ag nanoparticles. This *Verticillium* sp. has also been reported to synthesize Au nanoparticles intra and extracellularly with mostly spherical, triangular, and hexagonal shapes (Mukherjee et al. 2001a).

In addition to the intracellular production of nanoparticles, fungi are exceptional for the extracellular production of nanoparticles. The extracellular secretion of reductive proteins, which can be easily handled, makes fungus an attractive option for cost-effective extracellular biosynthesis of nanoparticles. *Fusarium oxysporum* is the only fungus explored for the biosynthesis of a vast array of nanoparticles, such as Ag, Au, Au-Ag alloy, SiO_2, Ti, Zr, and other nanoparticles extracellularly (Mukherjee et al. 2002, Bansal et al. 2004, Senapati et al. 2005, Bansal et al. 2006). *F. oxysporum*, when exposed to an equimolar solution of $HAuCl_4$ and $AgNO_3$ demonstrated the formation of highly stable Au-Ag alloy nanoparticles (Senapati et al. 2005). Zirconia (ZrO_2) nanoparticles have also been reported to be produced extracellularly by *F. oxysporum* when challenged with aqueous ZrF_6^{2-} (Bansal et al. 2004). The size of ZrO_2 nanoparticles biosynthesized by *F. oxysporum* was found to be quasi-spherical with a size range of 3–11 nm. The protein of MW 24–28 kDa was found to be responsible for the formation of zirconia nanoparticles, as suggested by the PAGE analysis. Acidophilic fungi, *Aspergillus* spp., have been reported to produce intracellular and extracellular lead (Pb) nanoparticles (Pavani et al. 2012). The nanoparticles ranged between 5–20 nm were found on the on the cell surface. Also, microparticles of size ranging from 1.7–5.8 μm were observed extracellularly in media. Other acidophilic fungi have also been reported to produce nanoparticles, such as Ag, Au, and Cu, and are reported in Table 10.4.

Biosynthesis of Nanoparticles by Alkaliphile Bacteria

Spirulina platensis, a free-floating cyanobacterium naturally found in tropical and subtropical lakes with high pH and high concentrations of carbonate and bicarbonate, has been exploited for the biosynthesis of gold and silver nanoparticles. *S. platensis* produced spherical nanoparticles of a size range of 5–20 nm (Kalabegishvili et al. 2013). The same cyanobacteria, *S. platensis*, was also successful in biosynthesizing Au nanoparticles extracellularly when incubated with culture media supplemented by 1 mM $HAuCl_4$, yielding spherical Au nanoparticles of particle size ranging from 5 to 40 nm (Kalabegishvili et al. 2013). Eco-friendly biosynthesis of Se nanoparticles was conducted by using the alkaliphilic bacterium *Pseudomonas alcaliphila* isolated from the seawater near Hokkaido, Japan (Zhang et al. 2011b). Before the Se nanoparticles biosynthesis, *P. alcaliphilia* was cultivated with and without poly(vinylpyrrolidone) PVP for 24 hours. In the culture without PVP, a concentration of 100 mM of sodium selenite pentahydrate was added to the bacterial culture media, and displayed a time-dependent color change, indicating the synthesis of nanoparticles. Initially, the solution was light gray, which after 6 hours of incubation, became red in color, and the intensity of the red color increased overtime for up to 48 hours. Few spherical Se nanoparticles were observed after the first 6 hours of incubation with selenium salt; with increment in incubation time to 12 hours, spherical nanoparticles of 50–200 nm in size were observed (Zhang et al. 2011b). After 24 hours, spherical Se nanoparticles of 500 nm diameter were observed. When PVP was added to the media, the diameter of Selenium nanospheres changed from 20 nm at the early stages to 200 nm at the final stages after 24 hours. The Se nanoparticles were found to be capped by PVP and were stable and uniform for more than a month. However, biosynthesized nanoparticles without PVP aggregated and aged at room temperature within 10–20 days (Zhang et al. 2011b). This time-dependent size variation of Se nanoparticles can be an asset for the industrial production of Se nanoparticles, where different sized particles can be generated by controlling reaction time. Other alkaliphile bacteria have also been reported to biosynthesize metal nanoparticles. The nanoparticles mostly obtained by extracellular biosynthesis were Ag and Au, and they are shown in Table 10.5. There are also reports of the existence of alkaliphilic fungi and algae. However, the literature lacks information regarding biosynthesis of metal-based nanoparticles using them.

Table 10.5 Nanoparticles biosynthesis by alkaliphiles

Alkaliphiles	*NP*	*Mechanism*	*Temp* (°C)	*pH*	*Size* (nm)	*Morphology*	*References*
			Bacteria				
Spirulina platensis	Au	Extracellular	25	5–8	5–40	Spherical	Kalabegishvili et al. 2013
Nostoc sp.	Ag	Extracellular	25	7	16.33	Spherical	Ahmed et al. 2015
Bacillus licheniformis	Au	Extracellular	37		38	Spherical	Singh et al. 2014
Bacillus licheniformis	Ag	Extracellular	37	7	50	Spherical	Kalimuthu et al. 2008
Bacillus licheniformis	CdS	Extracellular	37		35	Triangle	Shivashankarappa and Sanjay 2015
Pseudomonas alcaliphila	Se	Extracellular	28	7.3	50–500	Spherical	Zhang et al. 2011b

NP: Nanoparticle

Since acidophilic and alkaliphilic microorganisms can produce biomolecules, which are stable at very low and high pH, their application is more significant in industrial biotechnology, and is not limited to nanoparticle biosynthesis. Due to unique biomolecules present and enzymatic activities, acidophilic and alkaliphilic microorganisms manage to biosynthesize novel nanoparticles, such as Pb nanoparticles and ZrO_2 nanoparticles, whose salts, otherwise, are toxic to various organisms. Clear understanding and elucidation of biosynthetic mechanisms of these acidophiles and alkaliphiles have been the bottleneck for taking this biochemical synthesis to industrial scale.

APPLICATIONS OF BIO-NANOPARTICLES

Bio-nanoparticles for Biomedical Applications

During the last decade, vast developments have been done in exploiting the power of nanotechnology in various fields of biomedicine. The most critical use of nanoparticles in biomedicine has been for disease diagnosis and treatment. Application, however, is not limited only to disease diagnosis and treatment, but for imaging, improving pharmacotherapies for complex medical cases, antibacterial activities, drug delivery, and cancer immunotherapy.

In most studies, Ag nanoparticles have shown their antimicrobial properties against a vast array of microorganisms. For example, Ag nanoparticles that are biosynthesized have demonstrated antibacterial activity against gram-positive and gram-negative pathogens, such as *E. coli, Bacillus subtilis, Staphylococcus aureus, Micrococcus luteus, Candida albicans, and Candida krusei* (Shahverdi et al. 2007, Ruparelia et al. 2008, Kathiresan et al. 2010, Saha et al. 2010, Suresh et al. 2010, Kumar and Mamidyala 2011). Apart from bearing antibacterial property, Ag nanoparticles are also known to have antifungal properties against *Aspergillus fumigatus, Candida albicans, Candida krusei, Fusarium* sp., *Mucor indicus, Humicola insolens,* and *Trichoderma reesei* (Panáček et al. 2009, Kumar and Mamidyala 2011, Vivek et al. 2011, Rajeshkumar et al. 2014). The cytotoxicity evaluation of Ag nanoparticles provided mixed results, since it has been shown to be toxic to some cell lines and not others (Akter et al. 2018). Dose-dependent cytotoxicity has also been reported for various normal and cancer cells (Syed et al. 2013, Karlsson et al. 2015, Raj et al. 2017). This cytotoxic property of Ag nanoparticles has also been suggested to be useful for the treatment of cancer. Ag nanoparticles have been reported to have toxicity toward H4IIE-luc rat hepatoma cells that are genetically modified cells for fast and unlimited growth (Botha et al. 2019). The variation of particle size influences the cytotoxicity of Ag nanoparticles. The effect of Ag nanoparticles on cell viability is demonstrated via the lactate dehydrogenase (LDH) assay and by the generation of reactive oxygen species (ROS) in a size-dependent manner in different cell lines (Akter et al. 2018). However, there are also reports of little or negligible toxicity to HepG2 cell lines when Ag nanoparticles were coated with polysaccharides (Kawata et al. 2009, Travan et al. 2009). The basic toxicity mechanism of nano-particles to cell line via ROS production is demonstrated in Figure 10.5.

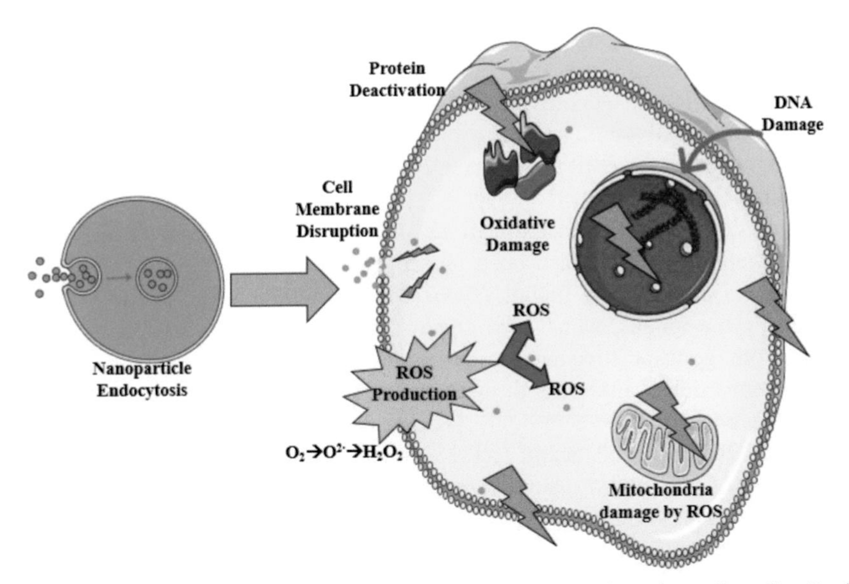

Figure 10.5 Basic mechanisms of toxicity of nanoparticles in eukaryotic cells. Toxic mechanisms of engineered nanoparticles involve the generation of intracellular and extracellular reactive oxygen species (ROS). Intracellular ROS is produced when nanosized particles penetrate and disrupt the cell membrane and mitochondria. Superoxide radicals from oxygen produce hydrogen peroxide, which deactivates proteins, enzymes, damages the cell membranes, and nucleic acids, causing oxidative damage and cell apoptosis.

Not only Ag, but Au, Se, and magnetite nanoparticles synthesized using bacteria have been used for biomedical applications. In the case of Au nanoparticles, these have also been shown to have antibacterial properties against gram-negative and gram-positive bacteria, such as *E. coli*, *K. pneumoniae*, MRSA, *S. aureus*, and *Pseudomonas aeruginosa* (Shamaila et al. 2016, Fan et al. 2019). Biosynthesized Au nanoparticles have the advantage of being capped with some biomolecules that can act as a substrate for functionalization and catalytic reactions. Biomolecules on the nanoparticle surface make them a better candidate for biosensing and drug delivery, such as drug delivery to the brain (Velasco-Aguirre et al. 2015). A study also reported the possibility to control protein orientation by tuning the surface chemistry of Au nanoparticles to deliver drugs more efficiently to the brain to treat Alzheimer's disease (Lin et al. 2015). Moreover, Au nanoparticles capped with biomolecules ensure cell endocytosis, reduce the level of reactive oxygen species (ROS) production and toxicity when used with the human leukemia cell lines (Connor et al. 2005). Recently, greater attention has been paid to the biomedical application of Se nanoparticles, where they can be used as antibacterial and for drug delivery. A study reported the successful antibacterial activity of Se nanoparticles against *Staphylococcus aureus* (MSSA and MRSA) and demonstrated the dose-dependent toxicity towards Caco-2 and human dermal fibroblasts cell lines (Nguyen et al. 2017, Huang et al. 2019). In the case of biosynthesized magnetite (Fe_3O_4) nanoparticles, they have also been reported to present strong antibacterial activity against both gram-positive and gram-negative bacteria (Prabhu et al. 2015). Biosynthesized Fe_3O_4 are biocompatible and demonstrated no toxicity when *in vitro* cytotoxic and genotoxic tests were performed (Yan et al. 2012). These nanoparticles can also be used for a biomedical imagery system called magnetic resonance tomography (MRT) for visualization of a cross-section of human tissue and organs (Blaney 2007).

In addition to biomedical applications of bacterial biosynthesized nanoparticles, fungal nanoparticles have also been investigated. For instance, the cerium oxide nanoparticles produced by fungus *Humicola* sp. are naturally capped by proteins secreted by the fungus, which also demonstrated significant antibacterial activity against both gram-positive and gram-negative bacteria (Karlsson et al. 2015). CeO_2 nanoparticles antibacterial activity is not much understood, but these were believed to penetrate through the cell membrane and deactivate enzymes, generating ROS and cell death. CeO_2 nanoparticles are found to be more toxic compared to CeO_2 microparticles when tested with IMR 32 cell lines (Kumari et al. 2014). The selective killing of cancer cells using CeO_2 nanoparticles has also been demonstrated with radiation. It is believed that the superoxide radical generation capacity of CeO_2 nanoparticles in acidic environments, commonly observed in cancer cells, further enhances the nanoparticle toxicity against cancer cells (Wason et al. 2013). Gadolinium oxide (Gd_2O_3) nanoparticles are other metal-based nanoparticles produced by *Humicola* sp., with huge applications in cancer detection technique of magnetic resonance imaging (MRI). Gd_2O_3 nanoparticles can also be applied for drug delivery and biomedical imaging (Khan et al. 2014). Gd_2O_3 nanoparticles also demonstrated to have good antibacterial activity with no significant toxicity toward HaCaT, DU145, and NCTC1469 cell lines and human red blood cells (RBCs) (Aashima et al. 2018, Miao et al. 2018). However, the accumulation of Gd_2O_3 nanoparticles in the human brain has been the bottleneck for its applications (Miao et al. 2018).

Bio-nanoparticles in Environmental Remediation

Possible ways of cleaning environmental systems using bio-nanoparticles are adsorption and degradation and dehalogenation of contaminants. Adsorption is an attractive method for the removal of contaminants, as it is known to be highly efficient, minimizes the production of chemical sludge, and does not require highly skilled human resources to run the process. Removal of heavy metals and textile dyes has been studied via adsorption using bio-nanoparticles. For instance, heavy metal chromium Cr(VI) has been reported to be reduced by the use of magnetite nanoparticle Fe_3O_4 (Simeonidis et al. 2016). The Fe(II) in magnetite can initiate the reduction of Cr(VI) to Cr(III), which results in reduced toxicity and formation of inner-sphere surface complexes at the surface of iron oxide due to the chelation of Cr(III) and –OH groups (Kendelewicz et al. 2000, Wang et al. 2018). Also, Fe_2O_3 demonstrated the synergistic adsorptive effects with other nanomaterials, such as graphene and multi-walled carbon nanotubes for the removal of toxic heavy metals, such as Cd(II), Pb(II), and Cu(II) (Fialova et al. 2014).

Palladium bio-nanoparticles have also attracted interest for application in remediation. Pd is one of the primary catalysts in chemistry for dehalogenation, reduction, hydrogenation, and C–C bond forming reactions under normal conditions (Mabbett et al. 2004, Windt et al. 2005, Wood et al. 2010). Since large scale Pd nanoparticle synthesis is not environmentally friendly and sustainable, the commercialization of bio-Pd focused dehalogenation of contaminants in wastewater and groundwater and soil remediation turns out to be the best possible option. Polychlorinated biphenyls (PCB) degradation using bio-Pd-catalyzed dehalogenation has been demonstrated, which presented faster dehalogenation of PCB in the presence of bio-Pd (Windt et al. 2005). Chlorocyclohexanes (HCH) and chlorobenzenes contaminated groundwater was treated using bio-Pd in a fluidized bed reactor, which showed superior treatment compared to the use of activated carbon filters (Hennebel et al. 2010). Although bio-nanoparticles have been studied for site remediation, they are still not being commercialized due to potential hazardous effects on the environment.

Bio-nanoparticles for Water Treatment

Bio-nanoparticles have a wide range of applications in the field of water and wastewater treatment, which ranges from adsorption of chemical dyes to disinfection of waters. For instance, nanofibers of microcrystalline cellulose incorporated with bio-Fe_2O_3 nanoparticles have been utilized for the removal of cationic dyes from water (Fard et al. 2018). Magnetite nanoparticles are also reported to be applicable in high gradient magnetic separation (HGMS) of flocculating particles (Blaney 2007). Bio-nanoparticles of zinc oxide (ZnO nanoparticles) have also been demonstrated to successfully adsorb anionic dyes, such as azo dyes, methyl orange (MO), and amaranth from aqueous solution (Zafar et al. 2019).

In the case of disinfection for water treatment, silver nanoparticles have been extensively investigated for coating membranes to prevent biofouling, which is a significant problem in membrane filtration processes. Silver nanoparticles possess large surface to volume ratio; hence it is more bioactive against microorganisms (Dankovich and Gray 2011, Carpenter et al. 2015). For instance, silver nanoparticles coated PVDF ultrafiltration membranes prevented organic and microbial fouling, improved hydrophobicity, which led to reduced contact angle and increased permeate flux (Li et al. 2013). Apart from using silver nanoparticles in membrane systems for minimizing biofouling, silver nanoparticles can also be incorporated or coated on beads, paper filters and ceramic filters to develop column filtration systems for inactivation of microorganisms (Mthombeni et al. 2012). For example, paper filters were coated with silver nanoparticles to inactivate microorganisms by percolating from the filter during filtration. Leaching of silver ions and their antimicrobial capabilities were investigated (Dankovich and Gray 2011). Results demonstrated the release of 0.1 ppm of silver nanoparticles and inactivation of 3 to 6 logs of *E. coli* and *E. faecalis*, respectively. These results show that silver nanoparticles are effective against hazardous microbes in drinking water. Although silver nanoparticles are already commercialized for home-water purification systems, the silver nanoparticles used in these commercial systems are chemically synthesized.

Bio-nanoparticles for Renewable Energy Production

Nanoparticles are also studied for supercapacitors, fuel cells, and batteries to assist in sustainable energy production. Various metal oxides, such as Fe_2O_4, Fe_3O_4, MnO_2, SnO_2 were studied to improve the electrode's electrochemical performance (Wu et al. 2010). Among these, Fe_3O_4 is considered a promising electrode material, as it has high theoretical capacity and low processing cost. However, these nanoparticles aggregate easily, possess poor conductivity, and show considerable volume variation, which limits the performance (Wan et al. 2015). Hybrid Fe_3O_4/BCN has tried to solve the issues mentioned above and produce working electrodes in lithium-ion batteries without current metal collectors, conducting additives, or binders (Wan et al. 2015). Other biogenic nanoparticles, such as platinum (Pt), palladium (Pd), and titanium oxide (TiO_2) have also been successfully applied to energy-related applications, such as photocatalytic hydrogen evolution from pure water, hydrogen production from fuel cells by using nanoparticle-based electrodes, and metal waste biorefining (Yong et al. 2007, Yong et al. 2010, Wu et al. 2012, Schröfel et al. 2014). Palladium (Pd) nanoparticles are of great importance in chemical and energy production industries. Pd nanoparticles can be used in hydrogen sensing due to their ability to absorb about a thousand times the volume of hydrogen with respect to its volume. Hydrogen is an important gas in chemical industries and its leakage can cause a massive explosion, but hydrogen sensing using Pd nanoparticles can prevent accidents.

CONCLUSION AND PROSPECTS

Bacteria, fungi, and algae can be exploited in nanotechnology, especially in the development of nanoparticles. The rich microbial diversity points to their innate potential for biofactories of nanoparticles. Several articles have reported nanoparticles biosynthesis by different types of microorganisms, but none of them have elucidated the pathways involved in the nanoparticle biosynthesis. The delineation of specific genes and how they are activated, as well as the characterization of enzymes involved in the biosynthesis of nanoparticles are required for the complete knowledge of the underlying mechanisms. Biomolecular and molecular mechanisms of nanoparticles synthesis need to be better understood to improve the synthesis rate and monodispersity of the nanoparticles. There is also a need to understand the mechanisms of the synthesis process to be able to develop large scale synthesis processes. Apart from having complex biosynthetic mechanisms and undeciphered process control methods, nanoparticles produced have massive applications in the environmental system from remediation of metal ions to water treatment via membrane and adsorption systems to renewable energy generation. A clear understanding of the mechanisms and pathways of nanoparticle synthesis could make the application of bionanoparticles more feasible for different fields in science and engineering.

ACKNOWLEDGMENTS

This work was supported by the following funding agencies: NSF BEINM Grant Number: 1705511; NSF CHE Grant Number: 1904472; the USDA National Institute of Food and Agriculture, AFRI Project No. 2018-67022-27969, and the Welch foundation award number (E-2011-20190330). The findings achieved herein are solely the responsibility of the authors.

References

Aashima, S.K. Pandey, S. Singh and S.K. Mehta. 2018. Biocompatible gadolinium oxide nanoparticles as efficient agent against pathogenic bacteria. J. Colloid Interface Sci. 529: 496–504.

Abboud, Y., T. Saffaj, A. Chagraoui, A. El Bouari, K. Brouzi, O. Tanane, et al. 2014. Biosynthesis, characterization and antimicrobial activity of copper oxide nanoparticles (CONPs) produced using brown alga extract (*Bifurcaria bifurcata*). Appl. Nanosci. 4: 571–576.

Agnihotri, M., S. Joshi, A.R. Kumar, S. Zinjarde and S. Kulkarni. 2009. Biosynthesis of gold nanoparticles by the tropical marine yeast *Yarrowia lipolytica* NCIM 3589. Mater. Lett. 63: 1231–1234.

Ahmad, A., S. Senapati, M.I. Khan, R. Kumar and M. Sastry. 2003. Extracellular biosynthesis of monodisperse gold nanoparticles by a novel extremophilic actinomycete, *Thermomonospora* sp. Langmuir. 19: 3550–3553.

Ahmed, E.A., E.H.A. Hafez, A.F.M. Ismail, S.M. Elsonbaty, H.S. Abbas and R.A.S.E. Din. 2015. Biosynthesis of Silver Nanoparticles By *Spirulina platensis* & *Nostoc* sp. Glo. Adv. Res. J. Microbiol.. 4: 036–049.

Akter, M., M.T. Sikder, M.M. Rahman, A.K.M.A. Ullah, K.F.B. Hossain, S. Banik, et al. 2018. A systematic review on silver nanoparticles-induced cytotoxicity: Physicochemical properties and perspectives. J. Adv. Res. 9: 1–16.

Anasane, N., P. Golińska, M. Wypij, D. Rathod, H. Dahm and M. Rai. 2016. Acidophilic actinobacteria synthesised silver nanoparticles showed remarkable activity against fungi-causing superficial mycoses in humans. Mycoses. 59: 157–166.

Azizi, S., M.B. Ahmad, F. Namvar and R. Mohamad. 2014. Green biosynthesis and characterization of zinc oxide nanoparticles using brown marine macroalga Sargassum muticum aqueous extract. Mater. Lett. 116: 275–277.

Bansal, V., D. Rautaray, A. Ahmad and M. Sastry. 2004. Biosynthesis of zirconia nanoparticles using the fungus Fusarium oxysporum. J. Mater. Chem. 14: 3303–3305.

Bansal, V., P. Poddar, A. Ahmad and M. Sastry. 2006. Room-temperature biosynthesis of ferroelectric barium titanate nanoparticles. J. Am. Chem. Soc. 128: 11958–11963.

Baskar, G., B.P. Vasanthi, M.V. Kumar and T. Dilliganesh. 2014. Characterization of intracellular gold nanoparticles synthesized by biomass of *Aspergillus terreus*. Acta Metall. Sin-Engl. 27: 569–572.

Behra, R., L. Sigg, M.J.D. Clift, F. Herzog, M. Minghetti, B. Johnston, et al. 2013. Bioavailability of silver nanoparticles and ions: From a chemical and biochemical perspective. J. R. Soc. Interface. 10: 20130396–20130396.

Blaney, L. 2007. Magnetite (Fe_3O_4): Properties, Synthesis, and Applications. Lehigh Rev. 15: 5.

Botha, T.L., E.E. Elemike, S. Horn, D.C. Onwudiwe, J.P. Giesy and V. Wepener. 2019. Cytotoxicity of Ag, Au and Ag-Au bimetallic nanoparticles prepared using golden rod (*Solidago canadensis*) plant extract. Sci. Rep. 9: 4169.

Carpenter, A.W., C.-F. de Lannoy and M.R. Wiesner. 2015. Cellulose nanomaterials in water treatment technologies. Environ. Sci. Technol. 49: 5277–5287.

Colombo, S., G. Toietta, L. Zecca, M. Vanoni and P. Tortora. 1995. Algae mediated green fabrication of silver nanoparticles and examination of its antifungal activity against clinical pathogens. J. Bacteriol. 177: 5561–5566.

Connor, E.E., J. Mwamuka, A. Gole, C.J. Murphy and M.D. Wyatt. 2005. Gold nanoparticles are taken up by human cells but do not cause acute cytotoxicity. Small 1: 325–327.

Correa-Llantén, D.N., S.A. Muñoz-Ibacache, M.E. Castro, P.A. Muñoz and J.M. Blamey. 2013. Gold nanoparticles synthesized by *Geobacillus* sp. strain ID17 a thermophilic bacterium isolated from Deception Island, Antarctica. Microb. Cell Fact. 12: 75–75.

Correa-Llantén, D.N., S.A. Muñoz-Ibacache, M. Maire and J.M. Blamey. 2014. Enzyme involvement in the biosynthesis of selenium nanoparticles by *Geobacillus wiegelii* strain GWE1 isolated from a drying oven. Int J. Bioengg. Life Sci. 8: 637–641.

D'Amico, S., T. Collins, J.-C. Marx, G. Feller, C. Gerday and C. Gerday. 2006. Psychrophilic microorganisms: Challenges for life. EMBO Rep. 7: 385–389.

Dahoumane, S.A., M. Mechouet, K. Wijesekera, C.D.M. Filipe, C. Sicard, D.A. Bazylinski et al. 2017. Algae-mediated biosynthesis of inorganic nanomaterials as a promising route in nanobiotechnology – A review. Green Chem. 19: 552–587.

Dameron, C.T., R.N. Reese, R.K. Mehra, A.R. Kortan, P.J. Carroll, M.L. Steigerwald, et al. 1989. Biosynthesis of cadmium sulphide quantum semiconductor crystallites. Nature. 338: 596–597.

Dankovich, T.A. and D.G. Gray. 2011. Bactericidal paper impregnated with silver nanoparticles for point-of-use water treatment. Environ. Sci. Technol. 45: 1992–1998.

Daraio, C. and S. Jin. 2012. Synthesis and patterning methods for nanostructures useful for biological applications. pp. 27–44. *In:* G.A. Silva and V. Parpura [eds]. Nanotechnology for Biology and Medicine: At the Building Block Level. Springer, New York.

Davis, T.A., B. Volesky and A. Mucci. 2003. A review of the biochemistry of heavy metal biosorption by brown algae. Water Res. 37: 4311–4330.

Dehnad, A., J. Hamedi, F. Derakhshan-Khadivi and R. Abuşov. 2015. Green synthesis of gold nanoparticles by a metal resistant *Arthrobacter nitroguajacolicus* isolated from gold mine. IEEE Trans. Nanobiosci. 14: 393–396.

Druschel, G.K., B.J. Baker, T.M. Gihring and J.F. Banfield. 2004. Acid mine drainage biogeochemistry at Iron Mountain, California. Geochem. Trans. 5: 13–13.

El-Rafie, H.M., M.H. El-Rafie and M.K. Zahran. 2013. Green synthesis of silver nanoparticles using polysaccharides extracted from marine macro algae. Carbohydr. Polym. 96: 403–410.

Erasmus, M., E.D. Cason, J. van Marwijk, E. Botes, M. Gericke and E. van Heerden. 2014. Gold nanoparticle synthesis using the thermophilic bacterium *Thermus scotoductus* SA-01 and the purification and characterization of its unusual gold reducing protein. Gold Bull. 47: 245–253.

Fan, Y., A.C. Pauer, A.A. Gonzales and H. Fenniri. 2019. Enhanced antibiotic activity of ampicillin conjugated to gold nanoparticles on PEGylated rosette nanotubes. Int. J. Nanomedicine 14: 7281–7289.

Fard, G.C., M. Mirjalili and F. Najafi. 2018. Preparation of nano-cellulose/A-Fe_2O_3 hybrid nanofiber for the cationic dyes removal: Optimization characterization, kinetic, isotherm and error analysis. Bulg. Chem. Commun. 50: 251–261.

Fialova, D., M. Kremplova, L. Melichar, P. Kopel, D. Hynek, V. Adam, et al. 2014. Interaction of heavy metal ions with carbon and iron based particles. Materials (Basel). 7: 2242–2256.

Fujio, Y. and S. Kume. 1991. Isolation and identification of thermophilic bacteria from sewage sludge compost. J. Ferment. Bioeng. 72: 334–337.

Gabani, P. and O.V. Singh. 2013. Radiation-resistant extremophiles and their potential in biotechnology and therapeutics. Appl. Microbiol. Biotechnol. 97: 993–1004.

Gericke, M. and A. Pinches. 2006. Microbial production of gold nanoparticles. Gold Bull. 39: 22–28.

Golińska, P., M. Wypij, D. Rathod, S. Tikar, H. Dahm and M. Rai. 2016. Synthesis of silver nanoparticles from two acidophilic strains of *Pilimelia columellifera* subsp. pallida and their antibacterial activities. J. Basic Microbiol. 56: 541–556.

Golyshina, O.V. and K.N. Timmis. 2005. Ferroplasma and relatives, recently discovered cell wall-lacking archaea making a living in extremely acid, heavy metal-rich environments. Environ Microbiol. 7: 1277–1288.

Goodman, C.M., C.D. McCusker, T. Yilmaz and V.M. Rotello. 2004. Toxicity of gold nanoparticles functionalized with cationic and anionic side chains. Bioconjug. Chem. 15: 897–900.

Hallberg, K.B. and E.B. Lindström. 1994. Characterization of *Thiobacillus caldus* sp. nov., a moderately thermophilic acidophile. Microbiology 140: 3451–3456.

Hatchett, D.W. and M. Josowicz. 2008. Composites of intrinsically conducting polymers as sensing nanomaterials. Chem. Rev. 108: 746–769.

He, Z.-G., H. Zhong and Y. Li. 2004. *Acidianus tengchongensis* sp. nov., a new species of acidothermophilic archaeon isolated from an acidothermal spring. Curr. Microbiol. 48: 159–163.

He, S., Z. Guo, Y. Zhang, S. Zhang, J. Wang and N. Gu. 2007. Biosynthesis of gold nanoparticles using the bacteria *Rhodopseudomonas capsulata*. Mater. Lett. 61: 3984–3987.

Hennebel, T., S. De Corte, L. Vanhaecke, K. Vanherck, I. Forrez, B. De Gusseme, et al. 2010. Removal of diatrizoate with catalytically active membranes incorporating microbially produced palladium nanoparticles. Water Res. 44: 1498–1506.

Holt, K.B. and A.J. Bard. 2005. Interaction of Silver(I) Ions with the respiratory chain of *Escherichia coli*: An electrochemical and scanning electrochemical microscopy study of the antimicrobial mechanism of micromolar Ag+. Biochemistry. 44: 13214–13223.

Honary, S., H. Barabadi, E. Gharaeifathabad and F. Naghibi. 2012. Green synthesis of copper oxide nanoparticles using *Penicillium aurantiogrseum*, *Penicillium citrinum* and *Penicillium waksmanii*. Dig. J. Nanomater. Biostruct. 7: 999–1005.

Honary, S., H. Barabadi, E. Gharaei-Fathabad and F. Naghibi. 2013. Green synthesis of silver nanoparticles induced by the fungus *Penicillium citrinum*. Trop. J. Pharm. Res. 12: 7–11.

Hoover, R.B., E.V. Pikuta, A.K. Bej, D. Marsic, W.B. Whitman, J. Tang, et al. 2003. *Spirochaeta americana* sp. nov., a new haloalkaliphilic, obligately anaerobic spirochaete isolated from soda Mono Lake in California. Int. J. Syst. Evol. Microbiol. 53: 815–821.

Huang, T., J.A. Holden, D.E. Heath, N.M. O'Brien-Simpson and A.J. O'Connor. 2019. Engineering highly effective antimicrobial selenium nanoparticles through control of particle size. Nanoscale. 11: 14937–14951.

Hulkoti, N.I. and T.C. Taranath. 2014. Biosynthesis of nanoparticles using microbes—A review. Colloids Surf. B Biointerfaces. 121: 474–483.

Hussein, A.H., B.K. Lisowska and D.J. Leak. 2015. The genus geobacillus and their biotechnological potential. Adv. Appl. Microbiol. 92: 1–48.

Jang, G.G., C.B. Jacobs, R.G. Gresback, I.N. Ivanov, I.I.I.H.M. Meyer, M. Kidder, et al. 2015. Size tunable elemental copper nanoparticles: Extracellular synthesis by thermoanaerobic bacteria and capping molecules. J. Mater. Chem. C. 3: 644–650.

Jeong, S., K. Woo, D. Kim, S. Lim, J. S. Kim, H. Shin, et al. 2008. Controlling the thickness of the surface oxide layer on cu nanoparticles for the fabrication of conductive structures by ink-jet printing. Adv. Funct. Mater. 18: 679–686.

Juibari, M.M., S. Abbasalizadeh, G.S. Jouzani and M. Noruzi. 2011. Intensified biosynthesis of silver nanoparticles using a native extremophilic *Ureibacillus thermosphaericus* strain. Mater. Lett. 65: 1014–1017.

Kalabegishvili, T., I. Murusidze, E. Kirkesali, A. Rcheulishvili, E. Ginturi, N. Kuchava, et al. 2013. Gold and silver nanoparticles in spirulina platensis biomass for medical application. Ecol. Chem. Eng. S. 20: 621–631.

Kalimuthu, K., R. Suresh Babu, D. Venkataraman, M. Bilal and S. Gurunathan. 2008. Biosynthesis of silver nanocrystals by *Bacillus licheniformis*. Colloids Surf. B Biointerfaces. 65: 150–153.

Kanekar, P.P., S.P. Kanekar, A.S. Kelkar and P.K. Dhakephalkar. 2012. Halophiles – taxonomy, diversity, physiology and applications. pp. 1–34. *In:* T. Satyanarayana and B.N. Johri [eds]. Microorganisms in Environmental Management: Microbes and Environment. Springer, Netherlands.

Karlsson, H.L., M.S. Toprak and B. Fadeel. 2015. Toxicity of metal and metal oxide nanoparticles. pp. 75–112. *In:* G.F. Nordberg, B.A. Fowler and M. Nordberg [eds]. Handbook on the Toxicology of Metals, 4th Ed. Academic Press.

Kathiresan, K., S. Manivannan, M.A. Nabeel and B. Dhivya. 2009. Studies on silver nanoparticles synthesized by a marine fungus, *Penicillium fellutanum* isolated from coastal mangrove sediment. Colloids Surf. B Biointerfaces. 71: 133–137.

Kathiresan, K., N.M. Alikunhi, S. Pathmanaban, A. Nabikhan and S. Kandasamy. 2010. Analysis of antimicrobial silver nanoparticles synthesized by coastal strains of *Escherichia coli* and *Aspergillus niger*. Can. J. Microbiol. 56: 1050–1059.

Kawata, K., M. Osawa and S. Okabe. 2009. *In vitro* toxicity of silver nanoparticles at noncytotoxic doses to HepG2 human hepatoma cells. Environ. Sci. Technol. 43: 6046–6051.

Kendelewicz, T., P. Liu, C.S. Doyle and G.E. Brown. 2000. Spectroscopic study of the reaction of aqueous Cr(VI) with Fe_3O_4 (111) surfaces. Surf. Sci. 469: 144–163.

Kessi, J. and K.W. Hanselmann. 2004. Similarities between the abiotic reduction of selenite with glutathione and the dissimilatory reaction mediated by *Rhodospirillum rubrum* and *Escherichia coli*. J. Biol. Chem. 279: 50662–50669.

Kessi, J. 2006. Enzymic systems proposed to be involved in the dissimilatory reduction of selenite in the purple non-sulfur bacteria *Rhodospirillum rubrum* and *Rhodobacter capsulatus*. Microbiology. 152: 731–43.

Khan, S.A. and A. Ahmad. 2013. Fungus mediated synthesis of biomedically important cerium oxide nanoparticles. Mater. Res. Bull. 48: 4134–4138.

Khan, S.A., S. Gambhir and A. Ahmad. 2014. Extracellular biosynthesis of gadolinium oxide (Gd_2O_3) nanoparticles, their biodistribution and bioconjugation with the chemically modified anticancer drug taxol. Beilstein J. Nanotechnol. 5: 249–257.

Kowshik, M., N. Deshmukh, W. Vogel, J. Urban, S.K. Kulkarni and K.M. Paknikar. 2002. Microbial synthesis of semiconductor CdS nanoparticles, their characterization, and their use in the fabrication of an ideal diode. Biotechnol. Bioeng. 78: 583–588.

Krulwich, T.A., D.B. Hicks and M. Ito. 2009. Cation/proton antiporter complements of bacteria: Why so large and diverse? Mol. Microbiol. 74: 257–260.

Kumar, C.G. and S.K. Mamidyala. 2011. Extracellular synthesis of silver nanoparticles using culture supernatant of *Pseudomonas aeruginosa*. Colloids Surf. B Biointerfaces. 84: 462–466.

Kumari, M., S.P. Singh, S. Chinde, M.F. Rahman, M. Mahboob and P. Grover. 2014. Toxicity study of cerium oxide nanoparticles in human neuroblastoma cells. Int. J. Toxicol. 33: 86–97.

Li, X., R. Pang, J. Li, X. Sun, J. Shen, W. Han, et al. 2013. *In situ* formation of Ag nanoparticles in PVDF ultrafiltration membrane to mitigate organic and bacterial fouling. Desalination. 324: 48–56.

Liang, X.J., X. Li, H. Xu, Z.-S. Chen and G. Chen. 2011. Biosynthesis of nanoparticles by microorganisms and their applications. J. Nanomater. 2011.

Liu, H., Y. Ye, J. Chen, D. Lin, Z. Jiang, Z. Liu, et al. 2012. *In situ* photoreduced silver nanoparticles on cysteine: An insight into the origin of chirality. Chem. Eur. 18: 8037–8041.

Lin, W., T. Insley, M.D. Tuttle, L. Zhu, D.A. Berthold, P. Král, et al. 2015. Control of protein orientation on gold nanoparticles. J. Phys. Chem. C. 119: 21035–21043.

Ma, Y., Y. Xue, W.D. Grant, N.C. Collins, A.W. Duckworth, R.P. van Steenbergen, et al. 2004. *Alkalimonas amylolytica* gen. nov., sp. nov., and *Alkalimonas delamerensis* gen. nov., sp. nov., novel alkaliphilic bacteria from soda lakes in China and East Africa. Extremophiles. 8: 193–200.

Mabbett, A.N., P. Yong, J.P.G. Farr and L.E. Macaskie. 2004. Reduction of Cr(VI) by "palladized" biomass of *Desulfovibrio desulfuricans* ATCC 29577. Biotechnol. Bioeng. 87: 104–109.

Mann, S. 1993. Molecular tectonics in biomineralization and biomimetic materials chemistry. Nature. 365: 499–505.

Mata, Y.N., M.L. Blázquez, A. Ballester, F. González and J.A. Muñoz. 2008. Characterization of the biosorption of cadmium, lead and copper with the brown alga *Fucus vesiculosus*. J. Hazard. Mater. 158: 316–323.

Mata, Y.N., E. Torres, M.L. Blázquez, A. Ballester, F. González and J.A. Muñoz. 2009. Gold(III) biosorption and bioreduction with the brown alga *Fucus vesiculosus*. J. Hazard. Mater. 166: 612–618.

Mehta, D. and T. Satyanarayana. 2013. Diversity of hot environments and thermophilic microbes. pp. 3–60. *In:* T. Satyanarayana, J. Littlechild and Y. Kawarabayasi [eds]. Thermophilic Microbes in Environmental and Industrial Biotechnology: Biotechnology of Thermophiles. Springer, Netherlands.

Mesbah, N.M., G.M. Cook and J. Wiegel. 2009. The halophilic alkalithermophile *Natranaerobius thermophilus* adapts to multiple environmental extremes using a large repertoire of Na(K)/H antiporters. Mol. Microbiol. 74: 270–281.

Miao, X., S.L. Ho, T. Tegafaw, H. Cha, Y. Chang, I.T. Oh, et al. 2018. Stable and non-toxic ultrasmall gadolinium oxide nanoparticle colloids (coating material = polyacrylic acid) as high-performance T1 magnetic resonance imaging contrast agents. RSC Adv. 8: 3189–3197.

Mishra, R.R., S. Prajapati, J. Das, T.K. Dangar, N. Das and H. Thatoi. 2011. Reduction of selenite to red elemental selenium by moderately halotolerant *Bacillus megaterium* strains isolated from Bhitarkanika mangrove soil and characterization of reduced product. Chemosphere. 84: 1231–1237.

Mohammed Fayaz, A., M. Girilal, M. Rahman, R. Venkatesan and P.T. Kalaichelvan. 2011. Biosynthesis of silver and gold nanoparticles using thermophilic bacterium *Geobacillus stearothermophilus*. Process Biochem. 46: 1958–1962.

Molnár, Z., V. Bódai, G. Szakacs, B. Erdélyi, Z. Fogarassy, G. Sáfrán, et al. 2018. Green synthesis of gold nanoparticles by thermophilic filamentous fungi. Sci. Rep. 8: 3943.

Moon, J.-W., I.N. Ivanov, C.E. Duty, L.J. Love, A.J. Rondinone, W. Wang, et al. 2013. Scalable economic extracellular synthesis of CdS nanostructured particles by a non-pathogenic thermophile. J. Ind. Microbiol. Biotechnol. 40: 1263–1271.

Mthombeni, N.H., L. Mpenyana-Monyatsi, M.S. Onyango and M.N.B. Momba. 2012. Breakthrough analysis for water disinfection using silver nanoparticles coated resin beads in fixed-bed column. J. Hazard. Mater. 217–218: 133–140.

Mukherjee, P., A. Ahmad, D. Mandal, S. Senapati, S.R. Sainkar, M.I. Khan, et al. 2001a. Bioreduction of $AuCl_4^-$ ions by the fungus, *Verticillium* sp. and surface trapping of the gold nanoparticles formed. Angew. Chem. 40: 3585–3588.

Mukherjee, P., A. Ahmad, D. Mandal, S. Senapati, S.R. Sainkar, M.I. Khan, et al. 2001b. Fungus-mediated synthesis of silver nanoparticles and their immobilization in the mycelial matrix: A novel biological approach to nanoparticle synthesis. Nano Lett. 1: 515–519.

Mukherjee, P., S. Senapati, D. Mandal, A. Ahmad, M.I. Khan, R. Kumar, et al. 2002. Extracellular synthesis of gold nanoparticles by the fungus *Fusarium oxysporum*. ChemBioChem. 3: 461–463.

Muthukannan, R. and B. Karuppiah. 2011. Rapid synthesis and characterization of silver nano particles by novel *Pseudomonas* sp.“ram bt - 1”. J. Ecobiotechnol. 3: 24–28.

Namasivayam, S.K.R., E.K. Gnanendra and R. Reepika. 2010. Synthesis of silver nanoparticles by *Lactobaciluus acidophilus* 01 strain and evaluation of its *in vitro* genomic DNA toxicity. Nano-Micro Lett. 2: 160–163.

Nayak, R.R., N. Pradhan, D. Behera, K.M. Pradhan, S. Mishra, L.B. Sukla, et al. 2011. Green synthesis of silver nanoparticle by *Penicillium purpurogenum* NPMF: The process and optimization. J. Nanopart. Res. 13: 3129–3137.

Neal, A.L., L.K. Clough, T.D. Perkins, B.J. Little and T.S. Magnuson. 2004. *In situ* measurement of Fe(III) reduction activity of *Geobacter pelophilus* by simultaneous in situ RT-PCR and XPS analysis. FEMS Microbiol. Ecol. 49: 163–169.

Nguyen, T.H.D., B. Vardhanabhuti, M. Lin and A. Mustapha. 2017. Antibacterial properties of selenium nanoparticles and their toxicity to Caco-2 cells. Food Control. 77: 17–24.

Ollivier, B., P. Caumette, J.L. Garcia and R.A. Mah. 1994. Anaerobic bacteria from hypersaline environments. Microbiol. Rev. 58: 27–38.

Oren, A. 2013. Life at high salt concentrations, intracellular KCl concentrations, and acidic proteomes. Front. Microbiol. 4: 315–315.

Ovais, M., A.T. Khalil, M. Ayaz, I. Ahmad, S.K. Nethi and S. Mukherjee. 2018. Biosynthesis of metal nanoparticles via microbial enzymes: A mechanistic approach. Int. J. Mol. Sci. 19: 4100.

Oza, G., S. Pandey, R. Shah and M. Sharon. 2012. A mechanistic approach for biological fabrication of crystalline gold nanoparticles using marine algae, *Sargassum wightii*. Eur. J. Exp. Biol. 2: 505–512.

Panáček, A., M. Kolář, R. Večeřová, R. Prucek, J. Soukupová, V. Kryštof, et al. 2009. Antifungal activity of silver nanoparticles against *Candida* spp. Biomaterials. 30: 6333–6340.

Pavani, K.V., N.S. Kumar and B.B. Sangameswaran. 2012. Synthesis of lead nanoparticles by *Aspergillus* species. Pol. J. Microbiol. 61: 61–63.

Pawar, V., A. Shinde, A.R. Kumar, S. Zinjarde and S. Gosavi. 2012. Tropical marine microbe mediated synthesis of cadmium nanostructures. Sci. Adv. Mater. 4: 135–142.

Pető, G., G.L. Molnár, Z. Pászti, O. Geszti, A. Beck and L. Guczi. 2002. Electronic structure of gold nanoparticles deposited on SiOx/Si(100). Mater. Sci. Eng. C. 19: 95–99.

Pimprikar, P.S., S.S. Joshi, A.R. Kumar, S.S. Zinjarde and S.K. Kulkarni. 2009. Influence of biomass and gold salt concentration on nanoparticle synthesis by the tropical marine yeast Yarrowia lipolytica NCIM 3589. Colloids Surf. B Biointerfaces. 74: 309–316.

Poli, A., G. Anzelmo and B. Nicolaus. 2010. Bacterial exopolysaccharides from extreme marine habitats: Production, characterization and biological activities. Mar. Drugs. 8: 1779–1802.

Prabhu, Y.T., K.V. Rao, B.S. Kumari, V.S.S. Kumar and T. Pavani. 2015. Synthesis of Fe_3O_4 nanoparticles and its antibacterial application. Int. Nano Lett. 5: 85–92.

Prasad, R., R. Pandey and I. Barman. 2016. Engineering tailored nanoparticles with microbes: Quo vadis? Wiley Interdiscip. Rev. Nanomed. Nanobiotechnol. 8: 316–330.

Railean-Plugaru, V., P. Pomastowski, M. Wypij, M. Szultka-Mlynska, K. Rafinska, P. Golinska, et al. 2016. Study of silver nanoparticles synthesized by acidophilic strain of *Actinobacteria* isolated from the of *Picea sitchensis* forest soil. J. Appl. Microbiol. 120: 1250–1263.

Raj, A., P. Shah and N. Agrawal. 2017. Dose-dependent effect of silver nanoparticles (AgNPs) on fertility and survival of *Drosophila*: An *in-vivo* study. PloS One. 12: e0178051–e0178051.

Rajasree, R. and S. Gayathri. 2015. Extracellular synthesis of selenium nanoparticle using some species of lactobacillus. Indian J. Geo-Mar. Sci. 43: 766–775.

Rajesh, S., V. Dharanishanthi and A.V. Kanna. 2015. Antibacterial mechanism of biogenic silver nanoparticles of *Lactobacillus acidophilus*. J. Exp. Nanosci. 10: 1143–1152.

Rajeshkumar, S., C. Malarkodi, K. Paulkumar, M. Vanaja, G. Gnanajobitha and G. Annadurai. 2014. Algae mediated green fabrication of silver nanoparticles and examination of its antifungal activity against clinical pathogens. Int. J. Met. 2014.

Ramanathan, R., A.P. O'Mullane, R.Y. Parikh, P.M. Smooker, S.K. Bhargava and V. Bansal. 2011. Bacterial kinetics-controlled shape-directed biosynthesis of silver nanoplates using *Morganella psychrotolerans*. Langmuir. 27: 714–719.

Rampelotto, P.H. 2013. Extremophiles and extreme environments. Life (Basel) 3: 482–485.

Romera, E., F. González, A. Ballester, M.L. Blázquez and J.A. Muñoz. 2007. Comparative study of biosorption of heavy metals using different types of algae. Biores. Technol. 98: 3344–3353.

Rothschild, L.J. and R.L. Mancinelli. 2001. Life in extreme environments. Nature. 409: 1092–1101.

Ruparelia, J.P., A.K. Chatterjee, S.P. Duttagupta and S. Mukherji. 2008. Strain specificity in antimicrobial activity of silver and copper nanoparticles. Acta Biomater. 4: 707–716.

Saha, S., J. Sarkar, D. Chattopadhyay, S. Patra, A. Chakraborty and K. Acharya. 2010. Production of silver nanoparticles by a phytopathogenic fungus *Bipolaris nodulosa* and its antimicrobial activity. Dig. J. Nanomater. Biostructures. 5: 887–895.

Saifuddin, N., C.W. Wong and A.A.N. Yasumira. 2009. Rapid biosynthesis of silver nanoparticles using culture supernatant of bacteria with microwave irradiation. J. Chem. 6: 61–70.

Sarikaya, M., C. Tamerler, A.K.Y. Jen, K. Schulten and F. Baneyx. 2003. Molecular biomimetics: Nanotechnology through biology. Nat. Mater. 2: 577.

Sarmiento, F., R. Peralta and J.M. Blamey. 2015. Cold and hot extremozymes: Industrial relevance and current trends. Front. Bioeng. Biotechnol. 3: 148–148.

Sastry, M., A. Ahmad, M.I. Khan and R. Kumar. 2003. Biosynthesis of metal nanoparticles using fungi and actinomycete. Curr. Sci. 85: 162–170.

Schröfel, A., G. Kratošová, I. Šafařík, M. Šafaříková, I. Raška and L.M. Shor. 2014. Applications of biosynthesized metallic nanoparticles – A review. Acta Biomater. 10: 4023–4042.

Seckbach, J. and W.F. Libby. 1970. Vegetative life on venus? Or investigations with algae which grow under pure CO_2 in hot acid media at elevated pressures. Space Life Sci. 2: 121–143.

Senapati, S., A. Ahmad, M.I. Khan, M. Sastry and R. Kumar. 2005. Extracellular biosynthesis of bimetallic Au–Ag alloy nanoparticles. Small. 1: 517–520.

Seshadri, S., A. Prakash and M. Kowshik. 2012. Biosynthesis of silver nanoparticles by marine bacterium, *Idiomarina* sp. PR58-8. Bull. Mater. Sci. 35: 1201–1205.

Shah, R., G. Oza, S. Pandey and M. Sharon. 2012. Biogenic fabrication of gold nanoparticles using *Halomonas salina*. J. Microbiol. Biotechnol. Res. 2(4).

Shahverdi, A.R., A. Fakhimi, H.R. Shahverdi and S. Minaian. 2007. Synthesis and effect of silver nanoparticles on the antibacterial activity of different antibiotics against *Staphylococcus aureus* and *Escherichia coli*. Nanomedicine. 3: 168–171.

Shamaila, S., N. Zafar, S. Riaz, R. Sharif, J. Nazir and S. Naseem. 2016. Gold nanoparticles: An efficient antimicrobial agent against enteric bacterial human pathogen. Nanomaterials (Basel). 6: 71.

Shankar, P.D., S. Shobana, I. Karuppusamy, A. Pugazhendhi, V.S. Ramkumar, S. Arvindnarayan, et al. 2016. A review on the biosynthesis of metallic nanoparticles (gold and silver) using bio-components of microalgae: Formation mechanism and applications. Enzyme Microb. Technol. 95: 28–44.

Sheikhloo, Z. and M. Salouti. 2011. Intracellular biosynthesis of gold nanoparticles by the fungus Penicillium chrysogenum. Int. J. Nanosci. Nanotechnol. 7: 102–105.

Shen, N., X.-Y. Xia, Y. Chen, H. Zheng, Y.-C. Zhong and R.J. Zeng. 2015. Palladium nanoparticles produced and dispersed by *Caldicellulosiruptor saccharolyticus* enhance the degradation of contaminants in water. RSC Adv. 5: 15559–15565.

Shivaji, S., S. Madhu and S. Singh. 2011. Extracellular synthesis of antibacterial silver nanoparticles using psychrophilic bacteria. Process Biochem. 46: 1800–1807.

Shivashankarappa, A. and K.R. Sanjay. 2015. Study on biological synthesis of cadmium sulfide nanoparticles by *Bacillus licheniformis* and its antimicrobial properties against food borne pathogens. Nanoscience and Nanotechnology Research. 3: 6–15.

Show, S., A. Tamang, T. Chowdhury, D. Mandal and B. Chattopadhyay. 2015. Bacterial (BKH1) assisted silica nanoparticles from silica rich substrates: A facile and green approach for biotechnological applications. Colloids Surf. B Biointerfaces. 126: 245–250.

Siddiqi, K.S., A. Husen and R.A.K. Rao. 2018. A review on biosynthesis of silver nanoparticles and their biocidal properties. J. Nanobiotechnol. 16: 14–14.

Simeonidis, K., S. Mourdikoudis, E. Kaprara, M. Mitrakas and L. Polavarapu. 2016. Inorganic engineered nanoparticles in drinking water treatment: A critical review. Environ. Sci.: Water Res. Technol. 2: 43–70.

Singaravelu, G., J.S. Arockiamary, V.G. Kumar and K. Govindaraju. 2007. A novel extracellular synthesis of monodisperse gold nanoparticles using marine alga, *Sargassum wightii* Greville. Colloids Surf. B Biointerfaces. 57: 97–101.

Singh, S., A.S. Vidyarthi, V.K. Nigam and A. Dev. 2014. Extracellular facile biosynthesis, characterization and stability of gold nanoparticles by *Bacillus licheniformis*. Artif. Cells Nanomed. Biotechnol. 42: 6–12.

Suresh, A.K., D.A. Pelletier, W. Wang, J.-W. Moon, B. Gu, N.P. Mortensen, et al. 2010. Silver nanocrystallites: Biofabrication using *Shewanella oneidensis*, and an evaluation of their comparative toxicity on gram-negative and gram-positive bacteria. Environ. Sci. Technol. 44: 5210–5215.

Syed, A., S. Saraswati, G.C. Kundu and A. Ahmad. 2013. Biological synthesis of silver nanoparticles using the fungus *Humicola* sp. and evaluation of their cytoxicity using normal and cancer cell lines. Spectrochim Acta A Mol. Biomol. Spectrosc. 114: 144–147.

Thakkar, K.N., S.S. Mhatre and R.Y. Parikh. 2010. Biological synthesis of metallic nanoparticles. Nanomedicine. 6: 257–262.

Tian, B. and Y. Hua. 2010. Carotenoid biosynthesis in extremophilic Deinococcus-Thermus bacteria. Trends Microbiol. 18: 512–520.

Tiquia-Arashiro, S.M. 2014. Biotechnological applications of thermophilic carboxydotrophs. pp. 29–101. *In*: S.M. Tiquia-Arashiro [ed.]. Thermophilic Carboxydotrophs and their Applications in Biotechnology. Springer International Publishing.

Travan, A., C. Pelillo, I. Donati, E. Marsich, M. Benincasa, T. Scarpa, et al. 2009. Non-cytotoxic silver nanoparticle-polysaccharide nanocomposites with antimicrobial activity. Biomacromolecules. 10: 1429–1435.

Turner, P., G. Mamo and E.N. Karlsson. 2007. Potential and utilization of thermophiles and thermostable enzymes in biorefining. Microb. Cell Fact. 6: 9–9.

Van Den Burg, B. 2003. Extremophiles as a source for novel enzymes. Curr. Opin. Microbiol. 6: 213–218.

Velasco-Aguirre, C., F. Morales, E. Gallardo-Toledo, S. Guerrero, E. Giralt, E. Araya, et al. 2015. Peptides and proteins used to enhance gold nanoparticle delivery to the brain: Preclinical approaches. Int. J. Nanomedicine. 10: 4919–4936.

Vigneshwaran, N., N.M. Ashtaputre, P.V. Varadarajan, R.P. Nachane, K.M. Paralikar and R.H. Balasubramanya. 2007. Biological synthesis of silver nanoparticles using the fungus *Aspergillus flavus*. Mater. Lett. 61: 1413–1418.

Vivek, M., P.S. Kumar, S. Steffi and S. Sudha. 2011. Biogenic silver nanoparticles by *Gelidiella acerosa* extract and their antifungal effects. Avicenna J. Med. Biotechnol. 3: 143–148.

Wan, Y., Z. Yang, G. Xiong, R. Guo, Z. Liu and H. Luo. 2015. Anchoring Fe_3O_4 nanoparticles on three-dimensional carbon nanofibers toward flexible high-performance anodes for lithium-ion batteries. J. Power Sources. 294: 414–419.

Wang, L., Y. Sun, Z. Li, A. Wu and G. Wei. 2016. Bottom-up synthesis and sensor applications of biomimetic nanostructures. Materials (Basel). 9: 53.

Wang, M., J. Tian, M. Xiang and X. Liu. 2017. Living strategy of cold-adapted fungi with the reference to several representative species. Mycology. 8: 178–188.

Wang, C., H. Liu, Z. Liu, Y. Gao, B. Wu and H. Xu. 2018. Fe(3)O(4) nanoparticle-coated mushroom source biomaterial for Cr(VI) polluted liquid treatment and mechanism research. R Soc. Open Sci. 5: 171776–171776.

Wason, M.S., J. Colon, S. Das, S. Seal, J. Turkson, J. Zhao, et al. 2013. Sensitization of pancreatic cancer cells to radiation by cerium oxide nanoparticle-induced ROS production. Nanomedicine: Nanotechnology, Biology and Medicine. 9(4): 558–569.

Windt, W.D., P. Aelterman and W. Verstraete. 2005. Bioreductive deposition of palladium (0) nanoparticles on *Shewanella oneidensis* with catalytic activity towards reductive dechlorination of polychlorinated biphenyls. Environ. Microbiol. 7: 314–325.

Wood, J., L. Bodenes, J. Bennett, K. Deplanche and L.E. Macaskie. 2010. Hydrogenation of 2-Butyne-1,4-diol using novel bio-palladium catalysts. Ind. Eng. Chem. Res. 49: 980–988.

Wu, Z.S., W. Ren, D.W. Wang, F. Li, B. Liu and H. M. Cheng. 2010. High-Energy MnO_2 Nanowire/Graphene and Graphene Asymmetric Electrochemical Capacitors. ACS Nano. 4: 5835–5842.

Wu, X., Q. Song, L. Jia, Q. Li, C. Yang and L. Lin. 2012. Pd-Gardenia-TiO_2 as a photocatalyst for H_2 evolution from pure water. Int. J. Hydrog. Energy. 37: 109–114.

Xu, Y., P. Zhou and X. Tian. 1999. Characterization of two novel haloalkaliphilic archaea *Natronorubrum bangense* gen. nov., sp. nov. and *Natronorubrum tibetense* gen. nov., sp. nov. Int. J. Syst. Evol. Microbiol. 49: 261–266.

Yan, L., X. Yue, S. Zhang, P. Chen, Z. Xu, Y. Li, et al. 2012. Biocompatibility evaluation of magnetosomes formed by *Acidithiobacillus ferrooxidans*. Mater. Sci. Eng. C. 32: 1802–1807.

Yong, P., M. Paterson-Beedle, I.P. Mikheenko and L.E. Macaskie. 2007. From bio-mineralisation to fuel cells: Biomanufacture of Pt and Pd nanocrystals for fuel cell electrode catalyst. Biotechnol. Lett. 29: 539–544.

Yong, P., I.P. Mikheenko, K. Deplanche, M.D. Redwood and L.E. Macaskie. 2010. Biorefining of precious metals from wastes: An answer to manufacturing of cheap nanocatalysts for fuel cells and power generation via an integrated biorefinery? Biotechnol. Lett. 32: 1821–1828.

Zafar, M.N., Q. Dar, F. Nawaz, M.N. Zafar, M. Iqbal and M.F. Nazar. 2019. Effective adsorptive removal of azo dyes over spherical ZnO nanoparticles. J. Mater. Res. Tech. 8: 713–725.

Zhang, X., X. He, K. Wang, Y. Wang, H. Li and W. Tan. 2009. Biosynthesis of size-controlled gold nanoparticles using fungus, *Penicillium* sp. J. Nanosci. Nanotechnol. 9: 5738–5744.

Zhang, X., S. Yan, R.D. Tyagi and R.Y. Surampalli. 2011a. Synthesis of nanoparticles by microorganisms and their application in enhancing microbiological reaction rates. Chemosphere. 82: 489–494.

Zhang, W., Z. Chen, H. Liu, L. Zhang, P. Gao and D. Li. 2011b. Biosynthesis and structural characteristics of selenium nanoparticles by *Pseudomonas alcaliphila*. Colloids Surf. B Biointerfaces. 88: 196–201.

Zhang, J., J. Wanner and O.V. Singh 2015. Extremophiles and biosynthesis of nanoparticles current and future perspectives. pp. 101–121. *In:* O.V. Singh [ed.]. Bio-Nanoparticles: Biosynthesis and Sustainable Biotechnological Implications. Wiley-Blackwell.

11

Synthesis and Biological Applications of Greener Nanoparticles

Roland M. Miller[1], Francis J. Osonga[2] and Omowunmi A. Sadik[2*]

[1]Department of Chemistry
Center for Research in Advanced Sensing Technologies & Environmental Sustainability (CREATES)
State University of New York at Binghamton
P.O. Box 6000, Binghamton, NY, 13902, USA
Email: rmiller8@binghamton.edu

[2]Department of Chemistry and Environmental Science
New Jersey Institute of Technology, University Heights, 151 Warren Street
Newark, NJ, 07102, USA
Tel: (973) 596-2833; Fax: (973) 596-3586
Email: fosonga1@binghamton.edu; sadik@njit.edu

INTRODUCTION

A nanoparticle is a microscopic particle that has at least one dimension on the nanoscale, which is below 100 nm. One of the many unique properties that nanoparticles have is their large surface area-to-volume ratio. Nanoparticles have been used in a myriad of biological and other applications such as antimicrobial, anti-inflammation, anticancer, and many others. The demand for new antimicrobial treatments had led to their increasing use.

*For Correspondence: sadik@njit.edu

The use of "greener" approaches for the synthesis of nanoparticles has recently become of interest to scientists. Natural products have been studied extensively to synthesize nanoparticles. One major group of natural compounds that have been studied for synthesis of nanoparticles are flavonoids. Flavonoids are polyphenolic compounds that are abundant in the plant kingdom. Some of the compounds that will be discussed in this chapter are 2-(3,4-dihydroxyphenyl)-3,5,7-trihydroxy-4H-chromen-4-one (quercetin), 5,7-dihydroxy-2-(4-hydroxyphenyl)-4H-chromen-4-one (apigenin), and 2-(3,4-dihydroxyphenyl)-5,7-dihydroxy-4H-chromen-4-one (luteolin). Quercetin and apigenin possess important biological properties, such as antiviral, anticancer, and antimicrobial (Daglia 2012, Mandalari et al. 2007, Wu et al. 2017). Luteolin has also been shown to possess anticancer properties (Lin et al. 2008). We have reported the modification of these compounds for their use as reducing and stabilizing agents for nanoparticle synthesis. We have also studied their antimicrobial properties, as well as other applications (Osonga et al. 2016a, b, 2017, 2018a, b, 2019a, b).

The chapter presents the applications of the greener synthesized nanoparticles that have been synthesized in our lab. In particular, this chapter will focus on the "greener" synthesis of metal-based nanoparticles, specifically gold and silver nanoparticles, with their applications. Applications of these nanoparticles will target their interactions with microbes, including some catalytic and sensing properties. The chapter will conclude with a section on the current commercial synthesis of nanoparticles.

Sustainable Nanotechnology

The term sustainable nanotechnology could be broken down into two words: sustainable and nanotechnology. In order for something to be considered sustainable, it must replenish itself without becoming depleted or damaged permanently. Sustainability of the various global resources has been a main concern in the last few decades. Sustainability has major relevance to climate change. John Elkington developed the term Triple Bottom Line (TBL) to account for the three pillars of sustainability, including environment, society, and economy (Elkington 1998). The three pillars are often referred to as the three Ps, meaning people, planet, and profits (Slaper and Hall 2013), and are not easily measurable. Figure 11.1 shows the pillars and some measurements associated with each pillar (Slaper and Hall 2013). If one of these pillars is not followed, the system for sustainability will fail. This system can and has been used in a variety of organizations, including businesses, governments, and even local communities (Slaper and Hall 2013).

Michael Faraday was the first person to describe the properties of nanoparticles (Faraday 1857). However, Richard Feynman gave the first talk on nanotechnology, describing molecular machines, which could be built atomically in 1959 (Singh et al. 2010). The term nanotechnology was first coined in 1974. The first paper on nanotechnology was published in 1981, titled "An approach to the development of general capabilities for molecular manipulation." Since then, the development of instruments, such as Transmission Electron Microscopy (TEM), Atomic Force

Microscopy (AFM), Dynamic Light Scattering (DLS), and other instrumentation have helped in the analysis and characterization of nanoparticles (Singh et al. 2010). Nanotechnology has become one of the most researched areas in all areas of science.

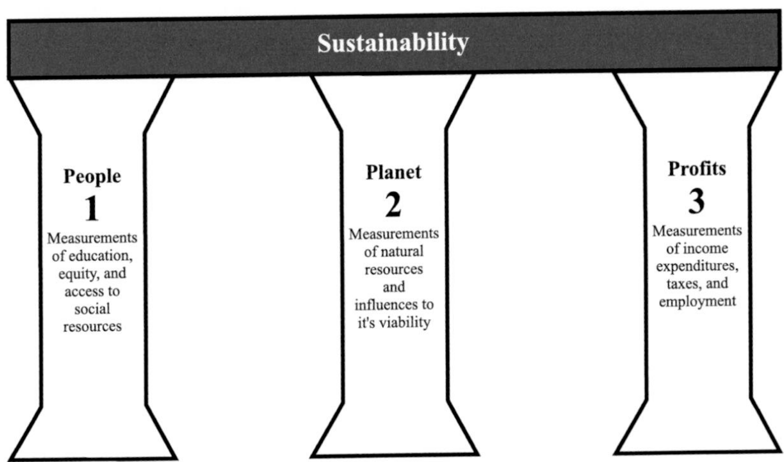

Figure 11.1 Three pillars that make up sustainability with the measurements of each pillar.

Various groups and organizations have been formed to focus on and inform others about nanotechnology. For example, the Sustainable Nanotechnology Organization (SNO) was created in 2011. This organization has attracted the attention of hundreds to thousands of chemists, biologists, physicists, and material scientists around the world. SNO focuses on three specific areas: research, responsibility, and education. Their goal is to raise awareness about nanotechnology, and make it an academic discipline similar to chemistry, biology, biomedical science, and other disciplines. SNO believes that nanotechnology is still treated as a secondary topic, the foundations are not laid out within disciplines, students do not see the connections between nanotechnology and other sciences, and nanotechnology can be a unifying science with sustainable approaches and concepts (Education 2020).

Nanoparticles can be divided into two main types: organic and inorganic nanoparticles. Organic nanoparticles include carbon-based nanoparticles and inorganic nanoparticles include noble metal nanoparticles and semiconductor nanoparticles. Generally, the synthesis of nanoparticles can be divided into two methods: top-down and bottom-up. In the top-down approach, larger nanoparticles are used to direct the assembly of other nanoparticles, whereas the bottom-up approach builds up towards larger structures starting at the molecular level and ideally include the ability to control the resulting structures (Singh et al. 2010).

The most widely used methods for nanoparticle synthesis are wet-chemical procedures (Thakkar et al. 2010). In our lab, we use a chemical reduction method for the synthesis of our nanoparticles. In this synthesis, three components are fundamental: a metal ion precursor, a reducing agent, and a capping agent. The metal ion precursor is the metal ion that will be reduced during the synthesis to form the nanoparticle (gold and silver). The reducing agent is the compound that will reduce the metal ion precursor to form the nanoparticle. The capping agents stabilize the nanoparticle while also preventing their conglomeration.

The synthesis of nanoparticles should embrace the concept of sustainable nanotechnology, which involves the nanoscale control of synthesis and processing of matter without footprints that give rise to environmental degradation. Thus, there is a search for synthetic methods that utilize fewer amounts of materials, energy, and water, while reducing or replacing the need for organic solvents (Sadik et al. 2014). Hence, safer approaches in nanosynthesis have been directed towards the use of plants that contain functional compounds that facilitate the reaction (Nadagouda et al. 2014). Sustainable nanotechnology seeks to integrate the principles of green chemistry in the synthesis of nanoparticles by using a suitable solvent, nontoxic reducing, capping, or stabilizing agent with less expenditure of energy (Osonga et al. 2018a, b).

Green Chemistry Perspective for Green Synthesis

Green chemistry is defined as the "design of chemical products and processes to reduce or eliminate the use and generation of hazardous substances" (Anstas and Williamson 1996, 1998). Green chemistry was first mentioned nearly 20 years ago (Anastas and Eghbali 2010). Since then, the concept has made dramatic impact in almost all industries, including aerospace, household products, and pharmaceutical (Anastas and Eghbali 2010). Even the universities and governmental funding agencies have adopted this concept (Anastas and Eghbali 2010). Green chemistry is still an ongoing concept, especially in the area of research.

There are 12 principles of green chemistry. These are (i) prevention, (ii) atom economy, (iii) less hazardous chemical synthesis, (iv) designing safer chemicals, (v) safer solvents and auxiliaries, (vi) design for energy efficiency, (vii) use of renewable feedstock, (viii) reduce derivatives, (ix) catalysis, (x) design for degradation, (xi) real-time analysis for pollution prevention, and (xii) inherently safer chemistry for accident prevention. A description of each of the principles is shown in Figure 11.2.

Since the "creation" of green chemistry, there have been many changes to human health, the environment, and the economy and business sectors. For human health, there have been advances in cleaner air, water, food, and less exposure to toxic chemicals. The environment has also seen improvements in the number of compounds that have ended up there. This, in turn, leads to plants and animals suffering less from the toxic chemicals once released in the environment. Green chemistry has also lowered the potential of global warming, ozone depletion, and smog formation. This concept has also led to less usage of landfills. The elimination of unnecessary synthetic steps, higher yields for chemical reactions, reduction of waste, and improved competitiveness of chemical manufacturers and their customers have influenced the economy and the business of companies (EPA 2019).

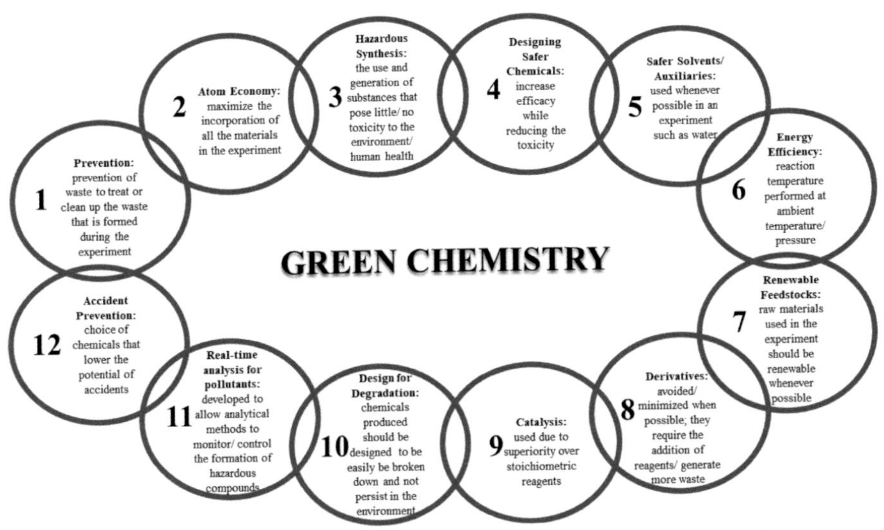

Figure 11.2 Twelve principles of green chemistry with brief descriptions.

Green Synthesis and its Significance

Green and sustainable nanotechnology are mainly focused on the application of the principles of green chemistry and sustainability. The development of nanotechnology should be coupled with the evaluation of societal, environmental, and economic impacts, along with posing the basis for comprehensive sustainability assessment of different nanoparticles with the same functionality. Sustainable nanotechnology relates to the research and development of nanomaterials that have economic and societal benefits while, at the same time, minimizing negative environmental impact (Cinelli et al. 2016). This is an emerging focus in the field of nanomaterials and nano-enabled products. Over the past decade, the perspective that “nano is dangerous” has shifted to “nano can be made safe” by employing engineering principles to their safe design, which have positively impacted nano-manufacturing and Nano-Environmental Health & Safety (Grassian et al. 2016). There has been a multidisciplinary effort with basic, mechanistic, functional, and computational study designs. The sustainable nanotechnology community is faced with concerns about the increasing knowledge gap between these new material designs and their safety. The major challenges in nanosynthesis are the design, synthesis, and assembly of well-controlled anisotropic nanoparticles. However, it is essential to note that industrial techniques for the production of metal nanoparticles often require catalytic temperatures above 200°C and long reaction times exceeding ten hours. In contrast, green synthesis requires only ambient conditions and minutes to create stable nanoparticles. The significance of this work is that it promotes the concept of sustainable nanosynthesis.

The use of natural products is one of the ways to implement sustainable nanosynthesis. Flavonoids, such as quercetin, apigenin, and luteolin, have been studied in our lab for over a decade (Zhou et al. 2007). We have also studied the interactions of flavonoids with heavy metals, such as lead and hexavalent chromium (Okello et al. 2012, 2016). Mwilu et al. then demonstrated how a totally phosphorylated quercetin (QPP) compound could act as a substrate for alkaline phosphatase in enzyme-linked immunosorbent assay (Mwilu et al. 2014). These studies led to the synthesis of quercetin diphosphate (QDP), quercetin monophosphate (QP), as well as apigenin triphosphate (ATRP) (Osonga et al. 2017).

The modification of flavonoids is not a new concept, though. There have been many derivatives made with quercetin, including esters, phenylisocyanated, methylated, sulfated, glutathionated, and glucuronidated derivatives and mono, di-, and mixed conjugates (Wei et al. 2014, Gao et al. 2009). Wei et al. investigated *in vitro* studies using phosphate esters of flavonoids as pancreatic cholesterol esterase and acetylcholinesterase inhibitors (Wei et al. 2014). The results from the study showed high potency for the target in comparison to the parent compounds (Wei et al. 2014). Why should these naturally occurring flavonoids be modified? The process of modification of their structures increases their bioavailability (Rice-Evans et al. 1996). This can result in the synthesis of new compounds that are superior to or as good as the parent compounds *in vitro* (Osonga et al. 2017).

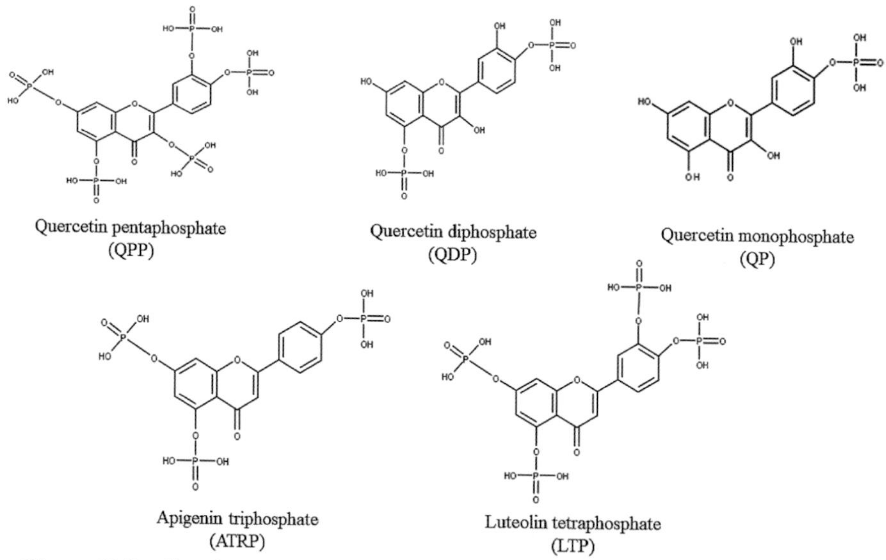

Quercetin pentaphosphate (QPP)

Quercetin diphosphate (QDP)

Quercetin monophosphate (QP)

Apigenin triphosphate (ATRP)

Luteolin tetraphosphate (LTP)

Figure 11.3 Chemical structure of modified flavonoids used for nanoparticle synthesis.

Flavonoids suffer from poor oral bioavailability due to their limited water solubility, and hence the solubility of quercetin and other polyphenols could be increased through formulation or modification of their chemical structure. The sequential synthesis was conducted by protecting the selected hydroxyl group on quercetin, then installing phosphate groups, and followed by deprotection with the results of the synthesis yielding QDP and QP (Osonga et al. 2017). We have successfully conducted studies on the total modification of all the hydroxyl groups on the flavonoids and produced modified flavonoids, namely QPP, ATRP, and luteolin tetraphosphate (LTP). The solubility studies on the modified phosphorylated derivatives in water revealed that there was an increased solubility in water over the parent molecules. For example, QPP demonstrated 84000-fold enhancement in solubility over the unmodified parent quercetin, while the solubility of LTP and apigenin triphosphate (ATRP) in water increased 2970 and 3660-fold over their parent molecules, luteolin and apigenin, respectively (Osonga et al. 2017).

The compounds that have been synthesized in our lab, QPP, QDP, QP, ATRP, as well as LTP, have been used to synthesize a variety of nanoparticles using water as a solvent (Figure 11.3). The increase in the number of phosphate groups helped to increase the solubility of the compounds in water (Osonga et al. 2017). The resulting water-soluble molecules allow for a "greener" approach for nanoparticle synthesis because the synthetic conditions could be water-based and at ambient temperatures.

Biomedical Applications of Gold and Silver Nanoparticles

The use of nanoparticles in biomedical applications is an ever-growing research topic. Nanoparticles exhibit various chemical and physical characteristics that differ compared to small molecules or bulk materials due to their shapes and sizes (Elahi et al. 2018). Inorganic nanoparticles, such as gold and silver have been exploited in different applications, such as nanomedicine and biomedical devices (Haider and Kang 2015). The two most commonly studied inorganic nanoparticles are gold and silver. Each metal has unique properties that it can be used for in their applications. Nanoparticles can be divided up into different classes based on their dimensions: one-dimensional nanoparticle, such as nanorods, nanotubes, and nanowires, two-dimensional, such as nanoplates, truncated nanotriangles, and nanopentagons, three-dimensional, such as nanodumbbells, nanostars, nanocubes, and nanospheres (Elahi et al. 2018). The other shapes are often superior in comparison to spherical-shaped nanoparticles (Elahi et al. 2018). Some of the abovementioned shapes can be seen in Figure 11.4.

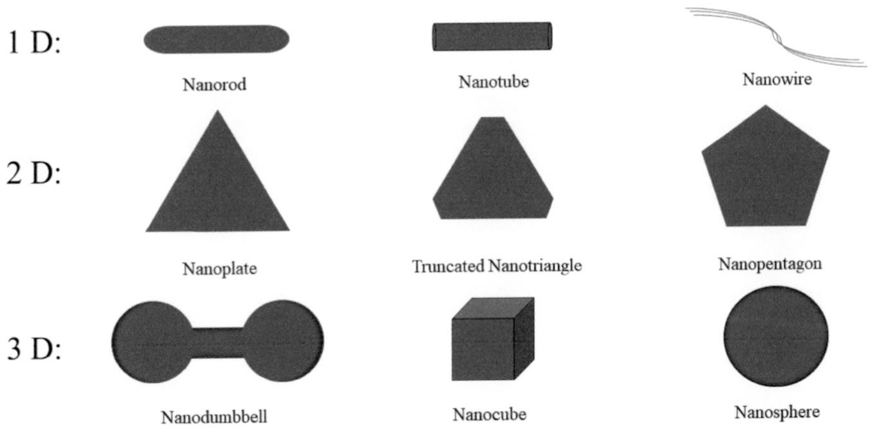

Figure 11.4 Dimensions and shape of various nanoparticles.

Gold Nanoparticles in the Biomedical Field

Gold (Au) nanoparticles have been used in a myriad of applications, including therapeutics, detection, drug delivery, sensing, imaging, and catalysts (Elahi et al. 2018). Au nanoparticles, like other nanoparticles, have large surface-to-volume ratios, but have a few unique characteristics, such as high biocompatibility and low toxicity, which makes them an excellent choice for the biomedical applications.

Gold nanorods are one of the most studied shapes of gold nanoparticles due to their anisotropic shape as well as displaying two separate Surface Plasmon Resonance (SPR) bands (Elahi et al. 2018). One of the bands is in the visible range while the second band is found in the near-infrared (IR) region. Having this property makes nanorods a great target for photochemical therapy *in vivo* (Elahi et al. 2018). Spherical, triangular, cubic, and hexagonal Au nanoparticles have all shown to have antimicrobial properties (Elahi et al. 2018). Au nanoparticles can easily bind to thiol groups and amino acids containing thiols, such as cysteine, which make these nanoparticles accessible for sensors.

Silver Nanoparticles in the Biomedical Field

Silver (Ag) nanoparticles are known for their antimicrobial properties. Silver's antibacterial properties date all the way back to Hippocrates' treatment of ulcers (Chaloupka et al. 2010). Ag nanoparticles can be used in biomedical imaging using surface-enhanced Raman scattering and sensing in single-step immunoassays (Chaloupka et al. 2010). Ag nanoparticles have been impregnated into external ventricular drain catheters for humans. This study showed that the impregnated Ag nanoparticles were beneficial in stopping catheter-associated ventriculitis in human patients *in vivo* (Lackner et al. 2008). The use of Ag nanoparticles also includes applications in bone cement, wound dressings, and anti-inflammatory treatment (Chaloupka et al. 2010). In fact, Ag nanoparticles are used in commercial products, such as food packaging and soaps (Fabrega et al. 2011).

Greener Synthesis of Nanoparticles

The use of greener synthesis for nanoparticles has become a wide area of research. In our lab, we have synthesized a variety of modified flavonoids that have been used to synthesize nanoparticles. In the next section, we will discuss the synthesis of Au nanoparticles and Ag nanoparticles using the flavonoids from our lab, the use of a polymer (polyamic acid) that has been studied in our lab for a few decades, and a photochemical synthesis of nanoparticles.

PROCESS FOR BIOSYNTHESIS

Water-based Synthesis

Flavonoid Derived-Gold Nanoparticles

The modified flavonoids, such as QPP, ATRP, LTP, and quercetin sulfonic acid (QSA) have been used to synthesize Au nanoparticles (Osonga et al. 2016b, 2018b).

Our synthetic approach adopted a one-pot, water-soluble synthesis method which can be reproduced easily without any negative environmental impact whatsoever, and hence allowing this green synthesis to be utilized by the industry as an innovation to the current method of nanoparticle production. The key reasons are that quercetin and apigenin derived flavonoids are far less expensive, nontoxic, water-soluble, and more efficient at reduction, stabilization, and capping than industrially used organic solvents and surfactants. These water-soluble, phosphorylated flavonoids were utilized both as reducing agent and stabilizer. The synthesis was achieved at room temperature using water as a solvent and required no capping agent.

Synthesis using QPP, QSA, and ATRP. The use of these compounds in water with gold ions formed nanoparticles without the addition of any other extraneous compounds, unlike conventional nanosynthesis. This proved that the flavonoids acted as reducing as well as capping agents for the nanoparticles. The schematic for the synthesis of gold nanoparticles using QDP is shown in Figure 11.5. The Au nanoparticles were observed with UV-Vis spectroscopy, TEM, surface area electron diffraction (SAED), energy dispersive absorption spectroscopy (EDS), and X-ray diffraction (XRD). The starting solution of the flavonoids was yellow; then with the addition of the gold (III) ions, the solution turned to a red wine color, which is a signature characteristic of the formation of Au nanoparticles (Figure 11.6A). The Au nanoparticles had an absorbance near 540 nm (Osonga et al. 2016b, 2018b). The temperature of the reaction was also varied from room temperature up to 90°C. It was observed that at higher temperatures, the rate of reaction for the reduction of the gold (III) ions increased. The concentrations of the gold ions and the derivatized flavonoids were also varied.

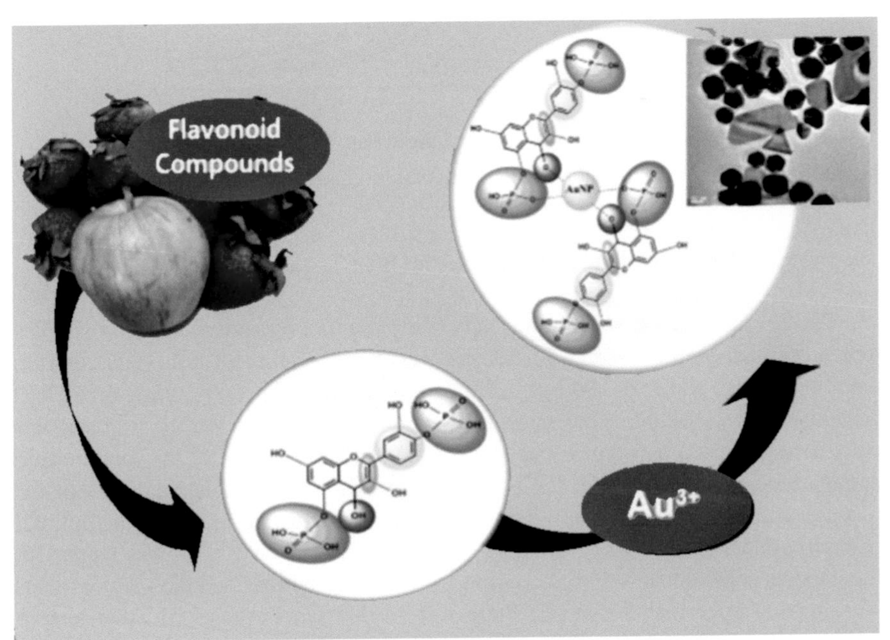

Figure 11.5 Formation of Au nanoparticles using QDP as the reducing and capping agent.

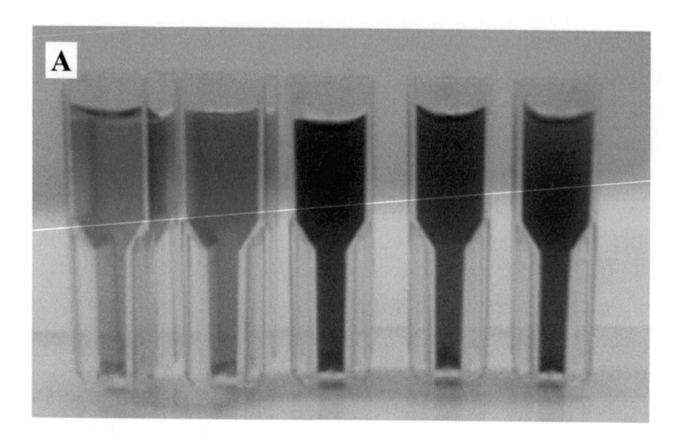

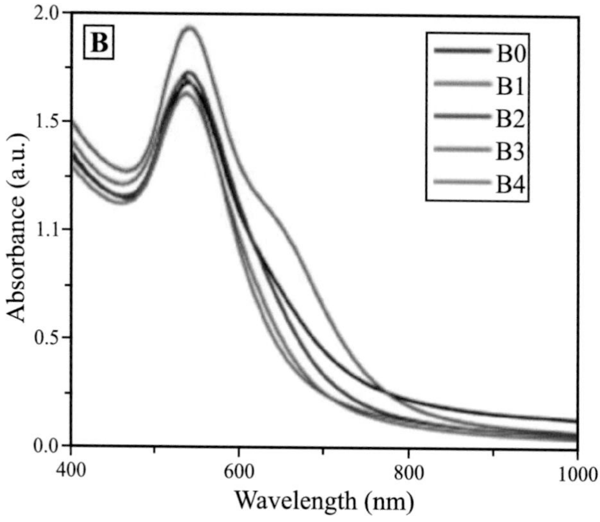

Figure 11.6 (A) Pictorial color depicted formation of gold nanoparticles when 4.5 mM gold (III) chloride were reacted with 5 mM of QDP at room temperature in a molar ratio of 1:1 (B0), 1:2 (B1), 1.5:1 (B2), 2:1 (B3), and 2:3 (B4), respectively. (B) UV-Vis spectra showing the formation of gold nanoparticles for B0, B1, B2, B3, and B4, respectively.

The design criterion within this study is the use of a one-pot method. In simpler terms, only nano-pure water, 150–500 μL of 3.5 mM–5 mM of gold (III) chloride metal precursor, and 150–1000 μL of one chosen flavonoid (QPP, QSA, or ATRP) were used for synthesis. The reagents were allowed to react for 10 minutes and were then analyzed by a color change and UV-Vis absorbance at the known gold nanoparticle SPR wavelength (530 nm) (Figure 11.6B). The concentrations of the metal precursors and the chosen flavonoids were manipulated to determine optimal reaction efficiency. For QPP, QSA, and ATRP, as the concentration was increased, the nanoparticle concentration increased as well. The synthesized Au nanoparticles were mostly spherical. However, there were other nanoparticles of other shapes that were formed, including triangular, cubical, and hexagonal nanoparticles. The sizes of these particles ranged from 2–45 nm. EDS confirmed the presence of the gold nanoparticles. SAED and XRD both showed patterns corresponding to (111), (200), (220), and (311) planes, demonstrating the crystallinity of these particles as face-centered cubic (Osonga et al. 2016b). This work featured the reduction, stabilization, and capping of metal precursor gold (III) chloride into a myriad of gold nanoparticle structures. Using the quercetin derived flavonoids, quercetin pentaphosphate (QPP), quercetin sulfonic acid (QSA), and apigenin derived flavonoid, apigenin triphosphate (ATRP), the data shows that each green flavonoid was able to perform all three (reducing, stabilizing, and capping) responsibilities on gold ions to successfully produce anisotropic gold nanocubes, hexagons, and nanoprisms under ambient temperatures without the need for any catalyst and within a five minute reaction time (Figure 11.7).

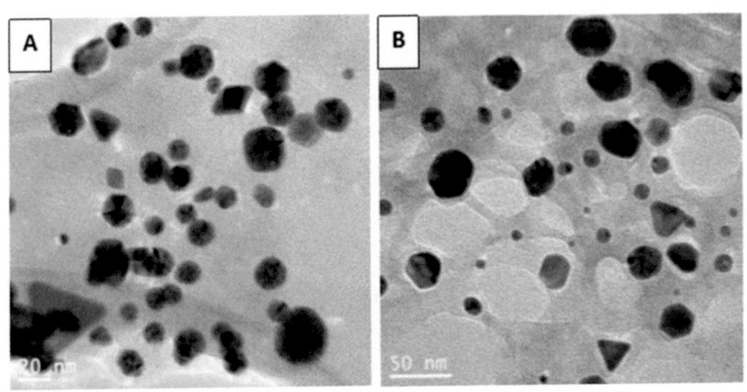

Figure 11.7 TEM micrographs of gold nanoparticles synthesized using (A) QPP (B) ATRP, with average size of 16 nm and 15 nm, respectively.

Synthesis using LTP. For LTP, when the concentration was increased, the results indicated an increase in nanoparticle formation. Gold nanocubes were formed in this synthesis, even at room temperature. The range for these nanoparticles was much wider than the previous study ranging from one to 1000 nm in size. The SAED and XRD studies demonstrated the same characteristics as the previous study of face-centered cubic structures (Osonga et al. 2018b). The synthesized phosphorylated derivative was subsequently explored as reducing agents for the sustainable synthesis of gold nanoparticles. The role of the flavonoid derivative as reducing, capping, and stabilizing agent was demonstrated in the synthesis of anisotropic gold nanoparticles (Osonga et al. 2018b).

Hydrogen tetrachloroaureate was reacted with different concentrations of the water-soluble flavonoid derivative at room temperature (25°C). The synthesis of anisotropic nanoparticles usually requires high temperatures and shape directing agents, such as Cetyl Trimethyl Ammonium Bromide (CTAB). LTP was reacted with gold and silver metal precursors, in separate trials, yielding both anisotropic cuboidal and isotropic spherical metal-based nanoparticles. Size distributions of the synthesized metal-based nanoparticles ranged between 5–15 nm in diameter. Transmission Electron Micrograph (TEM) images showed the formation of cuboidal Au nanoparticles, which were able to maintain stability for eight months in solution without degradation.

In this work, the quantitative metrics, such as lattice constant were examined to compare the properties of the Au nanocubes made by our water-based greener synthesis method and those made by other methods. The TEM images of Au nanocubes exhibited truncated edges that were found to agree with Au nanocubes using organic reducing, capping, and stabilizing agents (Osonga et al. 2018b). The XRD pattern indicated four peaks at $2\theta = 38.19°$, $44.56°$, $64.75°$, and $77.85°$, which were assigned to (111), (200), (220), and (311) planes of the face-centered cubic (fcc) of Au nanoparticles, respectively. This was a confirmation of the formation of elemental Au (JCPDS 04–0784). Based on the TEM images and XRD results, 3D models of the gold nanocubes were developed using Vesta software. The lattice constant of the SAED pattern of the Au nanocubes was calculated to be 4.052 Å, which is in agreement with that for fcc gold of 4.078 Å (JCPDS 04–0784), and comparable to that for Au nanocubes in the literature of 4.068 Å. This confirmed that LTP was used as the reducing and capping agent, using water as the solvent for Au nanoparticles (Osonga et al. 2018b).

Flavonoid Derived-Silver Nanoparticles

The modified flavonoids, QPP, QDP, QSA, ATRP, and LTP, have also been used to synthesize Ag nanoparticles (Osonga et al. 2016a, 2018a, b). It has been previously reported that flavonoids can reduce Ag^+ to Ag^0 as well as enhance the stability of nanoparticles (Ling et al. 2009). The Ag nanoparticles were characterized using the previously discussed methods for the Au nanoparticles. Typically, UV-Vis spectroscopy is the first method that is performed in our lab to observe characteristic SPR peaks for the nanoparticles. QPP, ATRP, and QSA Ag nanoparticles were found to have SPR peaks at 430, 399, and 404 nm (Osonga et al. 2016a). LTP Ag nanoparticles had a redshift with an SPR peak at 445 nm in solution (Osonga et al. 2018b). QDP Ag nanoparticles were also observed to have an SPR peak ranging from 438 to 453 nm (Osonga et al. 2018a).

Synthesis using QPP, QSA, and ATRP. For QPP, QSA, and ATRP, the sizes ranged from five to 80 nm for all compounds with mostly spherical to hexagonal shapes. EDS spectra were recorded to confirm the presence of the Ag in the nanoparticles. XRD was performed to show the crystallinity of the Ag nanoparticles. The presence of (111) and (200) planes confirmed the face-centered cubic structure of the nanoparticles. The XRD was also used to calculate the average size of the nanoparticles, which was calculated to be 22 nm, which was in good agreement with the TEM data. SAED confirmed the presence of the crystallinity with strong planes for the (111) and (200) reflection planes (Osonga et al. 2016a).

Synthesis using LTP and QDP. LTP was also used to synthesize Ag nanoparticles (Osonga et al. 2018b). The solution was the first yellow, and then when the silver ions were added, the solution turned to a yellow brown, which is characteristic of Ag nanoparticles. The TEM for these Ag nanoparticles showed the nanoparticles were spherical with an average size of 12–13 nm. It was also observed that when the concentration of LTP was increased, the number of nanoparticles formed also increased. When the concentration of the silver ions was increased, the nanoparticle size was found to increase. These were found to be more quasi-spherical in shape.

The XRD for the Ag nanoparticles gave diffraction peaks at 38.19°, 44.31°, 64.54°, and 77.55°, corresponding to (111), (200), (220), and (311) planes showing crystallinity. EDS was used to confirm the presence of Ag in the nanoparticles (Osonga et al. 2018b).

QDP was used to synthesize Ag nanoparticles (Osonga et al. 2018a). The UV-Vis data showed an SPR peak for these ranging from 438–453 nm. One of the solutions showed a second SPR peak at 584 nm. This demonstrates that the nanoparticles are anisotropic. This was then confirmed with TEM. The size of the particles ranged from five to 80 nm. The nanoparticles were mostly spherical; however, in two instances, the formation of nano triangles, truncated nano triangles, and hexagons were formed. EDS was again used to confirm the presence of the silver in the nanoparticles (Osonga et al. 2018a).

Polyamic Acid Synthesis of Gold and Silver Nanoparticles

Polymers, specifically polyamic acid (PAA), have also been studied along with flavonoids (Andreescu et al. 2005, Kariuki et al. 2015, 2016, 2017). PAA membranes are electroactive, conductive, biodegradable, and also contain amide and carboxyl functional groups (Kariuki et al. 2015). The applications for these nanoparticles will be discussed in the applications section below.

PAA is synthesized using two different compounds: an aniline component and a dianhydride component. The first polymer that will be discussed was synthesized using 4,4-oxydianline (ODA) and pyromellitic dianhydride (PMDA) (Kariuki et al. 2015). The ODA was dissolved in dimethylformamide, and the PMDA was dissolved in dimethylacetamide. These solvents are not eco-friendly since they are organics. However, the synthesis of the nanoparticles does not require any more solvents. Au and Ag nanoparticles were synthesized with the polymer. The nanoparticles were characterized by UV-Vis, TEM, EDS, and XRD (Kariuki et al. 2015).

The Au nanoparticles were observed to have an SPR band around 510–525 nm. The TEM showed the nanoparticles were elliptical with an average size of approximately 10 nm. EDS confirmed the presence of gold in the nanoparticle solution. XRD was performed on the nanoparticles, as well. The diffraction pattern gave peaks at 38.12°, 43.07°, 64.47°, and 77.22° corresponding to (111), (200), (220), and (311) planes of face-centered gold. The XRD was also used to calculate the average size and was calculated to be 9.7±0.5 nm. This result matched the TEM results for the size of the nanoparticles (Kariuki et al. 2015).

The Ag nanoparticles were observed to have an SPR band centered ~440 nm. The TEM demonstrated the Ag nanoparticles were spherical with an average size close to 4 nm. EDS also confirmed the presence of silver in the solution. The XRD pattern showed four peaks at 38.12°, 44.07°, 64.27°, and 77.22°. These also correspond to (111), (200), (220), and (311) planes of face-centered cubic silver. Using the XRD data, the average size of the nanoparticles was calculated to be 4.2±0.5 nm, which is in good agreement with the TEM results (Kariuki et al. 2015).

Our lab has also used *para*-phenylenediamine (PDA) to synthesize PAA with PDA and PMDA instead of ODA, as previously mentioned (Kariuki et al. 2016).

This new polymer was also used to synthesize Ag nanoparticles. The SPR peak for these nanoparticles was observed around 430–500 nm. The band was very broad, being about 80 nm wide, signaling a wide range of sizes. The shape of these particles was found to be quasi-spherical with TEM. The size distribution was found to have a majority size of about 6–10 nm. EDS confirmed the presence of the Ag in the nanoparticle solution. XRD has also been used to analyze the crystallinity of the nanoparticles. The only peak that was observed that was from the polymer was found at 37.9°. This peak corresponds to the (111) plane of Ag in a face-centered cubic structure. The application for this polymer and Ag nanoparticles will be discussed later on (Kariuki et al. 2016).

A water-soluble polymer (WSP) was also synthesized with the addition of triethylamine after the addition of PDA and PMDA. This polymer led to the formation of gold nanoflowers. The SPR peaks for the nanoparticles ranged from 450–750 nm. EDS showed the presence of gold in the nanoparticle solution. The XRD pattern showed planes of (111), (200), (220), and (311), indicating a face-centered cubic structure. Temperature effects were also studied. The results suggested that at high temperatures, the reduction rate was accelerated and destroyed the anisotropic character of the Au nanoparticles. The applications for these nanoparticles will be discussed below (Kariuki et al. 2017).

Photochemical Synthesis

Photochemical synthesis of nanoparticles has become a method of interest due to not having to use a reducing agent during synthesis. The use of plant extracts is commonly used to photochemically synthesize silver nanoparticles (Verma et al. 2016, Kumar et al. 2016). Other molecules, such as citric acid and dopamine hydrochloride, Triton X-100, and poly (vinyl pyrrolidone) have also been used to photochemically synthesize gold nanoparticles (Yang et al. 2007, Pal 2004, Dong and Zhou 2007, Mallick et al. 2001, Wang et al. 2008).

In our lab, we have synthesized gold nanoplates photochemically using QDP (Osonga et al. 2019a). It was confirmed that the presence of sunlight was needed by leaving a sample in the dark when no color change was observed even after ten days of incubation. UV-Vis spectroscopy, TEM, EDS, SAED, and XRD were used to characterize the nanoparticles. Mie theory was also calculated for computational studies of the nanoparticles (Osonga et al. 2019a). We successfully demonstrated the photochemical synthesis of gold nanoplates in the presence of sunlight using QDP. The nanoplate sizes ranged from 10–200 nm, and all synthesis was carried out in the water.

The UV-Vis spectra for these nanoparticles gave SPR peaks ranging from 550 to 760 nm. The nanoparticles were observed to form within two minutes of being in direct sunlight. After 90 minutes of exposure, two SPR peaks were present. EDS was later used to confirm the presence of the gold in the nanoparticle solution. TEM was performed to look at the shape and size of the nanoparticles. It was confirmed that the formation of gold nanoplates was formed. The average edge length was found to be 73 to 1200 nm. Large nanoprisms were also observed with TEM. XRD showed four peaks at 38.29°, 44.51°, 64.80°, and 77.70°. These peaks correspond with (111), (200), (220), and (311) planes that show the gold to be in face-centered cubic orientation. The intensities of the peaks showed that the nanoplates had a majority of (111) facets. The crystal lattice was calculated using SAED. The lattice constant was calculated to be 4.060 Å. These nanoparticles were shown to be stable up to one year (Osonga et al. 2019a). These nanoparticles were used as a catalyst for the reduction of methylene blue, but the results will be discussed later on.

BIOLOGICAL APPLICATIONS OF NANOPARTICLES

Applications of Silver Nanoparticles

Most of the nanoparticles that have been synthesized in our lab were explored for their antimicrobial properties. The QPP, QSA, and ATRP silver nanoparticles were tested for their antimicrobial and cytotoxicity properties *in vitro* (Osonga et al. 2016a). The microbes that were tested against were *Staphylococcus epidermidis*, *Escherichia coli*, *Citrobacter freundii* for the antimicrobial tests, and IEC6 (non-cancerous immortalized rabbit intestinal cell line) for the cytotoxicity (Osonga et al. 2016a). The synthesized nanosilver exhibits good antibacterial properties (Osonga et al. 2016a).

For *E. coli*, it was observed that 200 ng/mL did not show much inhibition, whereas 800 ng/mL and 2 μg/mL Ag nanoparticles displayed strong inhibition effects. After 24-hour incubation, agar plates treated with 8 and 20 μg/mL Ag nanoparticles with 10^7 colony-forming unit (CFU) in 20 μL of *E. coli*, did not show any visible colonies even after 72-hour incubation. *C. freundii* was analyzed at 10^4 and 10^5 CFU in 40 μL. The 200 ng/mL Ag nanoparticles did not display any toxicity. However, 2 and 3 μg/mL Ag nanoparticles did not fully inhibit the growth of *C. freundii*. The inhibition was calculated to be over 99% and less than 50% at concentrations of 10^4 and 10^5 CFU, respectively. A 72-hour incubation test was performed using 5 and 8 μg/mL with 10^6 CFU in 100 μL, and there was no visible formation of growth. For *S. epidermidis*, 10^4 and 10^5 CFU in 20 μL concentrations were used. 800 ng/mL Ag nanoparticles did not show any inhibition, while 2 and 3 μg/mL Ag nanoparticles displayed no colony formation after 24-hour incubation. However, after 72-hour incubation, the 2 and 3 μg/mL Ag nanoparticles showed minimal inhibition, and even lower for 10^5 CFU. The Ag nanoparticles were found to possess bacteriostatic and bactericidal effects. The 5 μg/mL Ag nanoparticles were used at the lowest concentration for cytotoxicity testing, because it was the highest concentration that showed the best inhibition of bacterial growth. The IEC-6 cells were used at a concentration of 10^4 cells/mL. The 5 μg/mL Ag nanoparticles gave less than a 10% toxicity against the IEC-6 cells (Osonga et al. 2016a).

QDP Ag nanoparticles were studied for their inhibition and the expression of virulence genes of *E. coli* SM10, along with their antifungal and antibacterial properties for *Enterobacter aerogenes*, *S. epidermidis*, *C. freundii*, *Trichaptum biforme*, and *Aspergillus nidulans*. QDP Ag nanoparticles were shown to decrease swarming motility by over 60% against *E. coli*. Swarming motility is a bacterium's ability to move, in this case on the agar, and allow to it to form colonies. QDP Ag nanoparticles were also shown to downregulate *fliC* which is a flagellar gene. This corresponds to the swarming motility decrease observed. The anisotropic QDP Ag nanoparticles showed strong antibacterial properties against *E. coli*, *S. epidermidis*, and *C. freundii* at 10 and 100 μM. The anisotropic QDP Ag nanoparticles were also tested for their antifungal properties, and the results showed that the nanoparticles worked better against *A. nidulans* compared to *T. bioforme*. *T. bioforme* was not eliminated at 100 μM, unlike *A. nidulans*. It was also shown that the anisotropic Ag nanoparticles displayed the best results for all of the tests in this study (Osonga et al. 2018a).

The results obtained from this study are very promising, and future research will be based on comparison with commercial drugs to investigate whether flavonoid derived Ag nanoparticles could be used as an alternative to antibiotic-based approaches to control pathogenic bacteria. This could have significant potential in the treatment of diseases due to acquired antibiotic-resistant bacteria that are posing major public health concerns. The anisotropic silver nanoparticles analyzed in this work demonstrated significant bactericidal activity against all strains tested. In particular, the anisotropic nature of the synthesized nanoparticles presents a unique antibacterial property, compared to spherical nanoparticles. Thus, the phosphorylated derivatives described above may overcome the limitations of the parent Quercetin. These limitations include low bioavailability, poor aqueous solubility, as well as rapid body clearance, fast metabolism, and enzymatic degradation. This work has demonstrated antibacterial activity of anisotropic silver nanoparticles using *E. coli* as the model organism. This work also extensively studied the *in vitro* effects of the nanoparticles on swarming motility, proteins, genes, and enzymes in virulence of bacteria (Osonga et al. 2018a).

Ag nanoparticles were also synthesized using two different forms of PAA (Kariuki et al. 2015, 2016). The nanoparticles were tested for their antimicrobial, catalytic properties, and for the development of an electrochemical sensor for nitrobenzene. The Ag nanoparticles were tested against gram-positive bacteria, *Listeria monocytogenes* and *S. epidermidis*, gram-negative bacteria, *E. coli,* and *Pseudomonas aeruginosa*, and cytotoxicity against Caco-2 (colon cancer cells) and IEC-6 cell lines (Kariuki et al. 2015). It was observed that the Ag nanoparticles showed antibacterial properties for all of the bacteria that were tested. The results showed that the Ag nanoparticles either decreased the colony number or decreased the size of the colonies that were formed. The cytotoxicity experiments demonstrated that there was a change in the growth pattern when the Caco-2 cells were treated with the nanoparticles. There was no change for the IEC-6 cells after treatment with the nanoparticles. Three different concentrations of nanoparticles were tested for cytotoxicity against IEC-6 cells (5, 15, 30 μg/mL).

The different concentrations gave relative cell viabilities of 92.5%, 87.5%, and 72.5% after treatment, respectively. The same Ag nanoparticles were tested for their catalytic activity against methylene blue using sodium borohydride as the reducing agent. Methylene blue is an organic dye that is released from various industries that cause diarrhea, abdominal pain, nausea, vomiting, headache, fever, and dizziness in humans if consumed (Jha and Shimpi 2018). The rate constant was calculated to be 1.09×10^{-2}/s using first-order kinetics (Kariuki et al. 2015). The Ag nanoparticles that were used for the sensor of nitrobenzene were embedded into the PAA membrane and then placed on top of a glassy carbon electrode (Kariuki et al. 2016). These nanoparticles were in the size range from 6 to 10 nm. The detection limit for the sensor was found to be 1.68 μM, with a sensitivity of 7.88 μA/μM. The linear range was also found to be 10–600 μM. This sensor had good comparisons to previously developed sensors for nitrobenzene in the literature (Kariuki et al. 2016).

Applications of Gold Nanoparticles

The gold nanoparticles in our lab have been tested for their catalytic properties, Surface Enhanced Raman Spectroscopy (SERS) enhancement, antimicrobial, and cytotoxicity properties (Kariuki et al. 2015, 2017, Osonga et al. 2016b, 2019a). PAA was used to synthesize Au nanoparticles as well as QDP (Kariuki et al. 2015, Osonga et al. 2019a). The PAA-Au nanoparticles were tested for their catalytic properties for the reduction of 4-nitrophenol in the presence of sodium borohydride. 4-nitrophenol is another pollutant that is released from various industries that causes kidney and liver damage, nausea, cyanosis, and headaches. It can be reduced to 4-aminophenol that is biologically active and is an important intermediate in drugs, such as paracetamol and phenacetin (Jha and Shimpi 2018). These Au nanoparticles were approximately 10 nm and quasi-spherical in shape. The rate constant was calculated to be 5.2×10^{-3}/s. The reaction without the addition of the Au nanoparticles displayed no change even after 60 minutes, whereas with the nanoparticles the reaction appeared to be completed after 500 seconds (Kariuki et al. 2015). Osonga et al. photochemically synthesized gold nanoplates using QDP with size lengths of 73–1200 nm (Osonga et al. 2019a). In this study, the catalytic activity of gold nanoplates was tested for the reduction of methylene blue in the presence of sodium borohydride. For comparison, the gold nanoplates and gold nanospheres were tested. The rate constants were calculated to be 3.44×10^{-2} and 1.11×10^{-2}/s, respectively. The results demonstrated that the nanoplates were more than three times higher than the nanospheres (Osonga et al. 2019a). Both sets of Au nanoparticles show promising results for the use of the reduction of pollutants in waterways.

The WSP was also used to synthesize Au nanoparticles (Kariuki et al. 2017). These nanoparticles were in the shape of nanoflowers. They were then tested for SERS properties. SERS is an important spectroscopic method used in surface science, biophysics, and analytical chemistry. This is due to its high sensitivity and selectivity for adsorbates. In this study, 4-mercaptobenzioc acid was used as a SERS probe due to its structure containing a thiol group on one end and carboxylic acid group on the other end that allows strong interactions with metal surfaces. The Raman spectrum with the Au nanoparticles and 4-mercaptobenzioc acid was significantly different compared to just the 4-mercaptobenzoic acid alone. Some of the peaks became weaker, some completely disappeared, and some peaks also shifted and broadened with respect to 4-mercaptobenzoic acid alone. The analytical enhancement factor was also intensified in the order of 10^{23} for one of the peaks (1092/cm). These results showed that these nanoparticles could be used as biolabeling, biosensing, imaging, and many other applications (Kariuki et al. 2017).

The previously mention flavonoids, QPP, QSA, and ATRP, were also used to synthesize Au nanoparticles (Osonga et al. 2016b). These Au nanoparticles were spherical, triangular, cubicle, hexagonal, and rectangular in shape. These Au nanoparticles were tested for their antimicrobial and cytotoxicity properties. *E. coli*, *S. epidermidis*, *C. fruendii*, and IEC-6 cell lines were used as model organisms. 800 ng/mL to 3 μg/mL Au nanoparticles showed strong inhibitory effects against *E. coli* at a concentration of 10^4 and 10^5 CFU after 24-hour incubation. For *S. epidermidis*, no antibacterial effects were observed until the Au nanoparticles were at a concentration of 2 or 3 μg/mL for 24-hour incubation. At a concentration of 3 μg/mL, there was no colony formation after 72-hour incubation. The Au nanoparticles did not show any toxicity towards *C. freundii* until a concentration of 2 or 3 μg/mL, depending on the concentration of bacteria, 10^4 or 10^5 CFU, respectively. However, only at a concentration of 5 μg/mL was there no meaningful growth after 72-hour incubation for any of the bacteria. The 5 μg/mL Au nanoparticles were used to test for their cytotoxicity against IEC-6 cells. The evaluation was found that there was less than a 10% decrease at the concentration of 5 μg/mL. These results show that these Au nanoparticles show modest cytotoxicity against non-cancerous cells (Osonga et al. 2016b).

Commercial Synthesis of Nanoparticles

Current industrial procedures requiring the production of mass quantities of nanoparticles with certain morphology are performed by using harmful organic solvents to successfully achieve the reduction, stabilization, and capping of nanoparticles. This type of synthesis can be broken down into two different methods, including physical and chemical synthesis. Physical methods are expensive, whereas chemical synthesis involves the use of toxic materials (typically organic solvents) and generates hazardous by-products (Osonga et al. 2016a). Reducing agents commonly used throughout nanoparticle industries are borohydrides, citrates, and ascorbic acids. Sodium borohydride and hydrazine are typically used as reducing agents for Au and Ag nanoparticle formation. Sodium borohydride can create caustic salt and flammable gases, which are environmental hazards (Agnihotri et al. 2013, Christy and Umadevi 2012, Nadagouda et al. 2014). Sodium borohydride and sodium citrate have been used for the reducing agents, and they are typically paired with a capping agent, to prevent agglomeration (Elia et al. 2014). Capping agents commonly used include thiols, phosphines, phenanthroline, and chiral diphosphite. Although all of the stated chemical reagents are not particularly environmentally friendly, the use of reducing agents, such as hydrazine used in industries, is considered the most harmful to biological systems. Surfactants, such as Polyvinylpyrrolidone (PVP), are commonly used in large quantities to limit nanoparticle growth as well. Surfactants have also been used to synthesize gold and other metal-based nanoparticles (Sharma et al. 2012). However, these procedures use toxic organic solvents due to the capping agents being hydrophobic, causing environmental issues/concerns as well as possible human harm (Osonga et al. 2018b). These reducing, stabilizing, and capping agents, commonly used to synthesize nanoparticles, require high temperatures and create massive amounts of sludge wastes that are simply discarded into environmental habitats, wreaking havoc on biological systems. Other issues with traditional methods include the requirement of intricate preparation, expensive materials and equipment, high temperatures, and long reaction times (Osonga et al. 2016b).

The use of green or "greener" chemistry and methods for nanoparticle synthesis eliminates the use of chemicals that pose harm to humans as well as the environment. Greener nanosynthesis utilizes fewer amounts of materials, water, and energy; while reducing or replacing the need for organic solvents. The green synthesis outperforms current industrial methods of mass synthesis of silver and gold nanoparticles, and ventures into environmentally sustainable territories of chemistry that is now necessary for modern research. This work also provides insights into the mechanism of flavonoid-based nanoparticle synthesis while eliminating the use of hazardous and toxic organic solvents and adopting the use of water as a solvent.

ACKNOWLEDGMENTS

The authors acknowledge the National Science Foundation Grant # IOS-1543944 and Bill & Melinda Gates Foundation for funding.

References

Agnihotri, S., S. Mukherji and S. Mukherji. 2013. Immobilized silver nanoparticles enhance contact killing and show the highest efficacy: Elucidation of the mechanism of bactericidal action of silver. Nanoscale. 5: 7328–7340.

Anastas, P.T. and T.C. Williamson. 1996. Green Chemistry: Designing Chemistry for the Environment. American Chemical Series Books, Washington, D.C.

Anastas, P.T. and T.C. Williamson. 1998. Green Chemistry: Theory and Practice. Oxford University Press, New York.

Anastas, P. and N. Eghbali. 2010. Green chemistry: Principles and practices. Chem. Soc. Rev. 39: 301–312.

Andreescu, D., A.K. Wanekaya, O.A. Sadik and J. Wang. 2005. Nanostructured polyamic acid membranes as novel electrode materials. Langmuir. 21: 6891–6899.

Benefits of Green Chemistry. 2019. EPA. Washington, USA.

Chaloupka, K., Y. Malam and A.M. Seifalian. 2010. Nanosilver as a new generation of nanoproducts in biomedical applications. Trends Biotechnol. 28: 580–588.

Christy, A.J. and M. Umadevi. 2012. Synthesis and characterization of monodispersed silver nanoparticles. Adv. Nat. Sci. Nanotechnol. 3: 035013.

Cinelli, M., S.R. Coles, O. Sadik, B. Karn and K. Kirwan. 2016. A framework of criteria for the sustainability assessment of nanoproducts. J. Cleaner Prod. 126: 277–287.

Daglia, M. 2012. Polyphenols as antimicrobial agents. Curr. Opin. Biotechnol. 23: 174–181.

Dong, S.A. and S.-P. Zhou. 2007. Photochemical synthesis of colloidal gold nanoparticles. Mater. Sci. Eng. B. 140: 153–159.

Education, 2020. Sustainable Nanotechnology Organization (SNO). http://www.susnano.org/outreach-education.html.

Elahi, N., M. Kamali and M.H. Baghersad. 2018. Recent biomedical applications of gold nanoparticles: A review. Talanta. 184: 537–556.

Elia, P., R. Zach, S. Hazan, S. Kolusheva, Z. Porat and Y. Zeiri. 2014. Green synthesis of gold nanoparticles using plant extracts as reducing agents. Int. J. Nanomed. 9: 4007–4021.

Elkington, J. 1998. Cannibals with Forks: The Triple Bottom Line of 21st Century Business. New Society Publishers.

Fabrega, J. and S.N. Luoma, C.R. Tyler, T.S. Galloway and J.R. Lead. 2011. Silver nanoparticles: Behavior and effects in the aquatic environment. Environ. Int. 37: 517–531.

Faraday, M. 1857. Experimental relations of gold (and other metals) to light. Philosophical Trans. R. Soc. London. 147: 145–181.

Gao, Q., G. Lian and F. Lin. 2009. The first total synthesis of 7-O-β-d-glucopyranosyl-4'-O-α-l-rhamnopyranosyl apigenin via a hexanoyl ester-based protection strategy. Carbohydr. Res. 344: 511–515.

Grassian, V.H., A.J. Haes, I.A. Mudunkotuwa, P. Demokritou, A.B. Kane, C.J. Murphy, et al. 2016. NanoEHS—defining fundamental science needs: No easy feat when the simple itself is complex. Environ. Sci.: Nano. 3: 15–27.

Haider, A. and I.K. Kang. 2015. Preparation of silver nanoparticles and their industrial and biomedical applications: A comprehensive review. Adv. Mater. Sci. Eng. 2015.

Jha, M. and N.G. Shimpi. 2018. Spherical nanosilver: Bio-inspired green synthesis, characterizations, and catalytic applications. Nano-Structures & Nano-Objects. 16: 234–249.

Kariuki, V.M., I. Yazgan, A. Akgul, A. Kowal, M. Parlinska and O.A. Sadik. 2015. Synthesis and catalytic, antimicrobial and cytotoxicity evaluation of gold and silver nanoparticles using biodegradable, π-conjugated polyamic acid. Environ. Sci.: Nano. 2: 518–527.

Kariuki, V.M., S.A. Fasih-Ahmad, F.J. Osonga and O.A. Sadik. 2016. An electrochemical sensor for nitrobenzene using π-conjugated polymer-embedded nanosilver. Analyst. 141: 2259–2269.

Kariuki, V.M., J.C. Hoffmeier, I. Yazgan and O.A. Sadik. 2017. Seedless synthesis and SERS characterization of multi-branched gold nanoflowers using water-soluble polymers. Nanoscale. 9: 8330–8340.

Kumar, V., D.K. Singh, S. Mohan and S.H. Hasan. 2016. Photo-induced biosynthesis of silver nanoparticles using aqueous extract of *Erigeron bonariensis* and its catalytic activity against Acridine Orange. J. Photochem. Photobiol., B. 155: 39–50.

Lackner, P., R. Beer, G. Broessner, R. Helbok, K. Galiano, C. Pleifer, et al. 2008. Efficacy of silver nanoparticles-impregnated external ventricular drain catheters in patients with acute occlusive hydrocephalus. Neurocrit. Care. 8: 360–365.

Lin, Y., R. Shi, X. Wang and H.-M. Shen. 2008. Luteolin, a flavonoid with potential for cancer prevention and therapy. Curr. Cancer Drug Targets. 8: 634–646.

Ling, L.T., S.A. Yap, A.K. Radhakrishnan, T. Subrahmanian, H.M. Cheng and U.D. Palanisamy. 2009. Standardized *Mangifera indica* extract is an ideal antioxidant. Food Chem. 113: 1154–1159.

Mallick, K., Z.L. Wang and T. Pal. 2001. Seed-mediated successive growth of gold particles accomplished by UV irradiation: A photochemical approach for size-controlled synthesis. J. Photochem. Photobiol., A. 140: 75–80.

Mandalari, G., R.N. Bennet, G. Bisignano, D. Trombetta, A. Saija, C.B. Faulds, et al. 2007. Antimicrobial activity of flavonoids extracted from bergamot (*Citrus bergamia* Risso) peel, a byproduct of the essential oil industry. J. Appl. Microbiol. 103: 2056–2064.

Mwilu, S.K., V.A. Okello, F.J. Osonga, S. Miller and O.A. Sadik. 2014. A new substrate for alkaline phosphatase based on quercetin pentaphosphate. Analyst. 139: 5472–5481.

Nadagouda, M.N., N. Iyanna, J. Lalley, C. Han, D.D. Dionysiou and R.S. Varma. 2014. Synthesis of silver and gold nanoparticles using antioxidants from blackberry, blueberry, pomegranate, and turmeric extracts. ACS Sustain. Chem. Eng. 2: 1717–1723.

Okello, V.A., S. Mwilu, N. Noah, A. Zhou, J. Chong, M.T. Knipfing, et al. 2012. Reduction of hexavalent chromium using naturally-derived flavonoids. Environ. Sci. Technol. 46: 10743–10751.

Okello, V.A., F.J. Osonga, M.T. Knipfing, V. Bushlyar and O.A. Sadik. 2016. Reactivity, characterization of reaction products, and immobilization of lead in water and sediments using quercetin pentaphosphate. Environ. Sci. Process Impacts. 18: 306–313.

Osonga, F.J., V.M. Karuki, I. Yazgan, A. Jimenez, D. Luther, J. Schulte, et al. 2016a. Synthesis and antibacterial characterization of sustainable nanosilver using naturally-derived macromolecules. Sci. Total Environ. 563–564: 977–986.

Osonga, F.J., I. Yazgan, V.M. Kariuki, D. Luther, A. Jimenez, P. Le, et al. 2016b. Greener synthesis and characterization, antimicrobial, and cytotoxicity studies of gold nanoparticles of novel shapes and sizes. RSC Adv. 6: 2302–2313.

Osonga, F.J., J.O. Onyango, S.K. Mwilu, N.M. Noah, J. Schulte, M. An, et al. 2017. Synthesis and characterization of novel flavonoid derivatives via sequential phosphorylation of quercetin. Tetrahedron Lett. 58: 1474–1479.

Osonga, F.J., A. Akgul, I. Yazgan, A. Akgul, R. Ontman, V.M. Kariuki, et al. 2018a. Flavonoid-derived anisotropic silver nanoparticles inhibit growth and change the expression of virulence genes in *Escherichia coli* SM10. RSC Adv. 8: 4649–4661.

Osonga, F.J., P. Le, D. Luther, L. Sakhaee and O.A. Sadik. 2018b. Water-based synthesis of gold and silver nanoparticles with cuboidal and spherical shapes using luteolin tetraphosphate at room temperature. Environ. Sci. Nano. 5: 917–932.

Osonga, F.J., V.M. Kariuki, V.M. Wambua, S. Kalra, B. Nweke, R.M. Miller, et al. 2019a. Photochemical synthesis and catalytic applications of gold nanoplates fabricated using quercetin diphosphate macromolecules. ACS Omega 4: 6511–6520.

Osonga, F.J., A. Akgul, R.M. Miller, G.B. Eshun, I. Yazgan, A. Akgul, et al. 2019b. Antimicrobial activity of a new class of phosphorylated and modified flavonoids. ACS Omega 4: 12865–12871.

Pal, A. 2004. Photochemical synthesis of gold nanoparticles via controlled nucleation using a bioactive molecule. Mater. Lett. 58: 529–534.

Rice-Evans, C.A., N.J. Miller and G. Paganga. 1996. Structure-antioxidant activity relationships of flavonoids and phenolic acids. Free Radix Biol. Med. 20: 933–956.

Sadik, O.A., N. Du, V. Kariuki, V. Okello and V. Bushlyar. 2014. Current and emerging technologies for the characterization of nanomaterials. ACS Sustain. Chem. Eng. 2: 1707–1716.

Sharma, R.K., S. Gulati and S. Mehta. 2012. Preparation of gold nanoparticles using tea: A green chemistry experiment. J. Chem. Educ. 89: 1316–1318.

Singh, M., S. Manikandan and A.K. Kumaraguru. 2010. Nanoparticles: A new technology with wide applications. Res. J. Nanosci. Nanotechnol. 1: 1–11.

Slaper, T.F. and T.J. Hall. 2013. The Triple Bottom Line: What is it and how does it work? Indiana Business Review (IBR).

Thakkar, K.N., S.S. Mhatre and R.Y. Parikh. 2010. Biological synthesis of metallic nanoparticles. Nanomedicine. 6: 257–262.

Verma, D.K., S.H. Hasan and R.M. Banik. 2016. Photo-catalyzed and phyto-mediated rapid green synthesis of silver nanoparticles using herbal extract of *Salvinia molesta* and its antimicrobial efficacy. J. Photochem. Photobiol., B. 155: 51–59.

Wang, L., G. Wei, C. Guo, L. Sun, Y. Sun, Y. Song, et al. 2008. Photochemical synthesis and self-assembly of gold nanoparticles. Colloids Surf. A: Physiochem. Eng. Aspects 312: 148–153.

Wei, Y., A.Y. Peng, B. Wang, L. Ma, G. Peng, Y. Du, et al. 2014. Synthesis and biological evaluation of phosphorylated flavonoids as potent and selective inhibitors of cholesterol esterase. Eur. J. Med. Chem. 74: 751–758.

Wu, T., H. Li, J. Chen, Y. Cao, W. Fu, P. Zhou, et al. 2017. Apigenin, a novel candidate involving herb-drug interaction (HDI), interacts with organic anion transporter 1 (OAT1). Pharmacol. Rep. 69: 1254–1262.

Yang, S., R. Zhang, Q. Wang, B. Ding and Y. Wang. 2007. Coral-shaped 3D assemblies of gold nuclei induced by UV irradiation and its disintegration. Colloids Surf. A: Physiochem. Eng. Aspects. 311: 174–179.

Zhou, A., S. Kikandi and O.A. Sadik. 2007. Electrochemical degradation of quercetin: Isolation and structural elucidation of the degradation products. Electrochem. Commun. 9: 2246–2255.

12

Liposomal Delivery: A Powerful Tool to Promote the Efficacy of Antimicrobial Agents

Mona I. Shaaban[1], Mohamed A. Shaker[2,3*] and Fatma M. Mady[4]

[1]Microbiology and Immunology Department
Faculty of Pharmacy, Mansoura University, PO Box 35516, Mansoura, Egypt
Tel: +201066944268; Fax: +20502200242
E-mail: mona_ibrahem@mans.edu.eg

[2]Pharmaceutics and Pharmaceutical Technology Department
College of Pharmacy, Taibah University
PO Box 30040, Al-Madinah Al-Munawarah, Saudi Arabia
Tel: +966541951635; Fax: +9661484618888
E-mail: mshaker@mun.ca

[3]Pharmaceutics Department
Faculty of Pharmacy, Helwan University, PO Box 11795, Cairo, Egypt

[4]Pharmaceutics Department
Faculty of Pharmacy, Minia University, PO Box 61519, Minia, Egypt
Tel: +966541453203; Fax: +961484618888
E-mail: madyfatma@gmail.com

INTRODUCTION

Infectious diseases have been occurring on the earth since the start of the creation and the earliest period of our life (Brier 2004, Sabbahy 2017). Their history can be traced back to the ancient Egyptians (Sabbahy 2017). Since then, the infectious

*For Correspondence: mshaker@mun.ca

diseases have become a major threat of human health, producing millions of cases of death annually (Brier 2004, Sabbahy 2017). Recently with the horrifying propagation of antimicrobial-resistant infections, the emerging multidrug-resistant/pan-resistant pathogens have spread rapidly (Dong et al. 2015, Moghadas-Sharif et al. 2015, Alhariri et al. 2017). This developed resistance is caused by the formation of degrading enzymes for antimicrobial, effective efflux of antimicrobials, and surface adsorption of microbes through the production of a bio-polymeric film (Yoneyama and Katsumata 2006). This biofilm is considered the major virulence that provides a solid firm barrier to even small molecule antimicrobial agents (Hoiby et al. 2010). It provides the protective and supportive medium for microbial adhesion and growth for developing life-threatening microbial strains (Tseng et al. 2013). This developing challenge constitutes a leading research concern on the therapeutic future of antimicrobials and, consequently, many investigators had exploited numerous pharmaceutical carriers and delivery systems to combat/destroy such resistant microbes (Hoiby et al. 2010, Dong et al. 2015, Moghadas-Sharif et al. 2015, Alhariri et al. 2017).

Among these delivery strategies/systems, liposomes are nano-sized carriers for the delivery of antimicrobials (Hoiby et al. 2010). Liposomes have been extensively used over the last decades and getting special attention in the formulation of numerous antimicrobial delivery systems for prevention/eradication of biofilm formation (Drulis-Kawa and Dorotkiewicz-Jach 2010). Liposomes are nano-sized spherical vesicles consisting of an aqueous core surrounded by thermodynamically stable spherical bilayers assembled from appropriate phospholipid molecules (Anwekar et al. 2011). As injectable, nontoxic, and biodegradable carriers, liposomes can easily incorporate/encapsulate the diversity of hydrophobic and/or hydrophilic antimicrobial molecules (Sharma and Sharma 1997, Drulis-Kawa and Dorotkiewicz-Jach 2010, Anwekar et al. 2011). During the assembly of phospholipids bilayers, hydrophilic antimicrobials can straightforwardly encapsulate in the aqueous core when dissolved in it, while hydrophobic antimicrobials easily incorporate within the phospholipids bilayers (Bakker-Woudenberg et al. 1994).

Comparing with other nano-carriers, liposomes offer many useful benefits in the antimicrobial delivery, including the ability to modify their physicochemical characters to accommodate a specific biological character (Bakker-Woudenberg et al. 1994, Sharma and Sharma 1997, Mozafari 2005, Drulis-Kawa and Dorotkiewicz-Jach 2010, Anwekar et al. 2011, Akbarzadeh et al. 2013), escape from the body's immune system (reticuloendothelial systems) via stealth liposomes (Drulis-Kawa and Dorotkiewicz-Jach 2010, Anwekar et al. 2011), target the loaded antimicrobial through the fusion of their phospholipid bilayers with another cellular membrane bilayers at the action site (Bakker-Woudenberg et al. 1994, Sharma and Sharma 1997, Akbarzadeh et al. 2013), alter their vesicle size/charge, phospholipid composition, number of assembled bilayers (uni- or multi-lamellae), and the surface decoration with conjugated ligands (peptide and/or polymer) (Gregoriadis 1995, Sharma and Sharma 1997).

As a consequence of such delivery benefits, various researchers have explored the application of liposomes in the delivery of numerous antimicrobials by striving to accomplish two main goals (Goyal et al. 2005, Mozafari 2005). The first is to provide the protection of antimicrobial from the anticipated hydrolysis inside the body, which is considered a cause for the destruction of the antimicrobials (Bakker-Woudenberg et al. 1994). Encapsulation of antimicrobials within liposomes is expected to increase their chemical/biological stability and protect them from early dilution, inactivation, and degradation inside the body circulation (Sachetelli et al. 2000, Halwani et al. 2008). The second is achieving targeted and sustained delivery of antimicrobials to their site of action after administration to the body. It is expected that controlling the release of antimicrobial will aid in minimizing the development of microbial resistance and prevent the biofilm formation (Robinson et al. 2001, Drulis-Kawa and Dorotkiewicz-Jach 2010).

The chapter focuses on liposomes as nano-carriers for improving the efficacy of antimicrobial and relating all the strategies applied in the formulation of various liposomes to the clinical practice and therapy of infectious diseases. The review also outlines the recent efforts employed to improve currently existing antimicrobials. Emphasis is on the understating of liposomal preparation and relating their formulation to the current methods for optimizing the treatment of infectious diseases.

LIPOSOMAL DELIVERY OF ANTIMICROBIAL AGENTS

Liposomal Nano-carriers as Antimicrobial Delivery Systems

Although the liposomal vesicles are varying in their surface charges, sizes, and numbers of membranes, the main components of bi-layered liposomes are phospholipids (Akbarzadeh et al. 2013). Phospholipids are obtained from different origins that might be natural, modified natural, or of synthetic origins (Akbarzadeh et al. 2013). Various phospholipids are used for the formation of liposomal vesicles, including but not restricted to phosphatidylcholine, phosphatidylethanolamine, phosphatidylserine, phosphatidylinositol, and phosphatidylglycerol (Mozafari 2005). Formulation of liposomes requires the use of one or more saturated phospholipids as a building unit for the bilayers; however, cholesterol or its derivatives are essentially incorporated for increasing the stability of the obtained vesicles (Wu et al. 2012). Cholesterol is essential for keeping the structural integrity and increasing the thermodynamic stability of the prepared liposomes (Laouini et al. 2012). Cholesterol is also necessary to diminish both the absolute mobility for the vesicles and the fluid permeability of the membrane, as well as to enhance the *in vivo* biological activity of the liposomes (Laouini et al. 2012). Recently, it is also revealed that liposomal stability may be improved by a surface coating of the vesicles with charged biopolymers (Gharib et al. 2012). This surface charged biopolymer imparts electrostatic repulsive force that diminishes the possibility of vesicle aggregation, especially, upon storage (Yang et al. 2015, Juang et al. 2016, Alhariri et al. 2017). Clearly, all the phospholipids used in liposomes are classified based on their charges into : Cationic phospholipids such as triester phosphatidylcholines, Anionic phospholipids such as diacetyl phosphate, phosphatidylserine, phosphatidylglycerol; Amphoteric (zwitterionic) phospholipids, and non-ionic phospholipids (Szoka Jr and Papahadjopoulos 1980).

The hydrodynamic diameter of the prepared liposomes is a critical parameter for estimating the plasma residence period of the circulating vesicles inside the living body (Liu et al. 2015, Saadat et al. 2016). The number of membrane bilayers (lamellae) considerably affects the encapsulation efficiency of the loaded antimicrobials (Ma et al. 2013, Colzi et al. 2015). Consequently, antimicrobial loaded liposomes are categorized based on the size and the number of lamellae into three main categories. The first is the multilamellar vesicles (MLVs), which have a diameter larger than 0.5 mm and their membrane consists of more than one lamella. The second is the large unilamellar vesicles (LUVs), which have a diameter equal or more than 0.1 mm and their membrane consists of just one lamella. The last one is the small unilamellar vesicles (SUVs), which have a diameter in the range of 0.02–0.1 mm and their membrane consists of just one lamella (Bordi et al. 2006). Nonetheless, the recent literature focuses on the formulation of "multivesicular vesicles (MVVs)", which are considered liposomes inside liposomes (Ebato et al. 2003, Grant et al. 2004). These liposomes are also known as double liposomes (DL) and intended to provide double protection to the antimicrobial drugs against various degradative enzymes (Katayama et al. 2003).

Preparation of Antimicrobial Liposomes

Several techniques have been reported on the preparation of liposomes, however, there are three main traditional strategies used for the preparation of antimicrobial liposomes. The distinction among these strategies is the way by which the phospholipid phase is dried from its solution in an organic solvent (usually chloroform), before being re-dispersed in an aqueous buffer/media (hydration step) (Mozafari 2005). The lipid-soluble antimicrobials are dissolved in the organic solvent with the phospholipid phase, and water-soluble antimicrobials are dissolved in the hydrating buffer/media (Drulis-Kawa and Dorotkiewicz-Jach 2010, Laouini et al. 2012). These are the specific methodologies to generally encapsulate the antimicrobials during the vesicle's assembly and liposome formation ["passive encapsulation"] (Anwekar et al. 2011, Akbarzadeh et al. 2013). With some other antimicrobials, the encapsulation step occurs after the formation of the vesicles ["active encapsulation"] (Anwekar et al. 2011, Akbarzadeh et al. 2013). Thus, the procedures are carried out to obtain the entire assembly of liposomal vesicles, which depend on either mechanical dispersion, solvent dispersion/injection, or supercritical fluid assisted methods (Szoka Jr and Papahadjopoulos 1980, Anwekar et al. 2011). In the following section, an individual description for each of the preparation methodologies will be briefly illustrated.

Mechanical Dispersion Methods

Many strategies used for preparing antimicrobial liposomes belong to the mechanical dispersion techniques, including but not restricted to both film hydration and reversed phase evaporation.

Film Hydration Method (Bangham Method). Thin film hydration is an early and commonly used method for preparation (Laouini et al. 2012). In this method, the lipid constituents (phospholipids and cholesterol) are first dissolved in the distinct organic solvent, followed by evaporation of the organic solvent by using a rotary evaporator. As a result, the lipid forms a thin film on the rotating flask's wall. This dried film is then subjected to the hyderation step through the addition of aqueous buffer/media with continued rotation using a rotary evaporator. This hydrating solution is warmed up to a temperature slightly higher than that of lipid phase glass transition temperature, to facilitate the assembly of spherical vesicles. This way of liposomes preparation is simple to handle and reproducible; in any case, it yields a population of heterogeneous MLVs with variable shapes and sizes. That can be subject to additional techniques, such as vesicle extrusion to produce LUVs or ultra-sonication to gives SUVs (Mui et al. 2003).

Reversed Phase Evaporation (RPE). Reversed phase evaporation is a preparative strategy mainly employed to prepare giant unilamellar liposomes (~20 μm diameters) possessing a large trapping volume (Wu et al. 2004, Gupta et al. 2017). Subsequently, such vesicles are able to encapsulate a greater amount of antimicrobial when compared to liposomes prepared by film hydration method with the same composition (Yang et al. 2015). In this technique, the phospholipid is self-assembled at the oil-water interface rather than hydrated from a dry lipid film (Wu et al. 2004, Yang et al. 2015, Gupta et al. 2017). The lipid constituents (phospholipids and cholesterol) are first dissolved in water immiscible solvent, usually in chloroform/ diethyl ether (1:1). This organic solution is emulsified with water to form simple w/o emulsion or multiple w/o/w emulsion, producing liposomes. The obtained emulsion is then subjected to centrifugation and/or organic solvent evaporation (Wu et al. 2004, Yang et al. 2015, Gupta et al. 2017). Hence, this method is also called "emulsification method" or "solvent evaporation method".

Solvent Injection/Dispersion Method

Solvent injection has been mainly developed to prepare small vesicles (less than 0.1 mm diameter) with a narrow size of distribution (Stano et al. 2004). Liposomes prepared by this method also show a high encapsulation efficiency without the need for further processing techniques, such as ultra-sonication or vesicle extrusion (Stano et al. 2004). Briefly, the lipid constituents (phospholipids and cholesterol) are first dissolved in distinct water miscible organic solvent (mostly ethanol) or water immiscible liquid (mostly ether). This lipid solution is injected by a syringe directly into a continuously stirred aqueous buffer/medium with a temperature slightly higher than that of lipid phase glass transition temperature. Small spherical vesicles with a narrow size distribution are then self-assembled in that aqueous buffer/media and simply collected by precipitation or centrifugation (Laouini et al. 2012). Using water insoluble solvent in this injection methodology facilitates its easy removal compared to using water soluble solvent. For example, ether is easily vaporized from its mixtures with water.

Supercritical Assisted (SuperLip) Method

Supercritical assisted liposome formulation is a rapid and efficient method recently used to improve antimicrobial encapsulation efficiency (Campardelli et al. 2016).

Regardless of the physicochemical nature of the candidate antimicrobial, this technique avoids the existing loss in antimicrobial content that occurs during the hydration procedure (Campardelli et al. 2018). In this technique, the solution of the antimicrobial agent is prepared and formulated as small droplets using the micrometric nozzle. These droplets are then coated with the assembled phospholipid bilayers using supercritical carbon dioxide (Campardelli et al. 2016, 2018). Several candidate antimicrobials, such as ofloxacin and ampicillin were formulated with the SuperLip technique (Campardelli et al. 2016, 2018).

Additionally Employed Techniques to Promote Formulation Efficiency

Ultrasonication Technique

Ultrasonication is probably the foremost among frequently utilized techniques to obtain SUVs. Herein, the prepared MLVs are subjected to ultrasonic waves with a frequency of more than 20 kHz. Liposomes are subjected to ultrasonication using an ultrasonic bath or an ultra-sonic probe either under normal or reduced pressure. This technique is very fast, nonetheless, it poses many limitations and drawbacks. These include the relatively low encapsulation efficiency, conceivable degradation of both phospholipid and encapsulated compounds, contamination with the probe metal impurities, and the coexistence of MLVs together with SUVs (Szoka Jr and Papahadjopoulos 1980).

Vesicle Extrusion Technique (VET)

Vesicle extrusion technique is a commonly applied technique for the formation of large liposomes (Nayar et al. 1989). Usually, it is employed to increase the entrapping volume and encapsulation efficiency of the loaded antimicrobials (Castoldi et al. 2017). This extrusion technique depends mainly on the repeated passage of the prepared liposomes through polycarbonate filters (of different pore sizes such as 1, 0.6, 0.4, and 0.1 μm) under pressure (Nayar et al. 1989, Juang et al. 2016, Castoldi et al. 2017).

Frozen and Thawed Multilamellar Vesicles

Freezing and thawing technique is commonly used to increase the antimicrobials encapsulation efficiency (Ma et al. 2013). Frozen and thawed multilamellar vesicles are prepared through the deep freezing of the prepared liposomes using a cryogenic liquid (liquid nitrogen) followed by defrosting in a heat-adjusted water bath (Ma et al. 2013, Colzi et al. 2015, Liu et al. 2015). The thawing water bath is adjusted at a temperature of ten degrees above the glass transition temperature of the phospholipids used in the initial preparation (Ma et al. 2013, Liu et al. 2015, Colzi et al. 2015).

Smart Formulations for Antimicrobial Liposomes

Stealth (Long Circulatory) Liposomes

Stealth liposomes are a special form of liposomes with an outer coat composed of synthetic polymers, such as polyethylene glycol (Allen et al. 1991). The presence of this outer coat disguises their detection by the immune system and avoids capture by the mononuclear phagocyte system, and hence prolongs their circulation time and enhances their effectiveness. The polyethyleneglycol (PEG) layer can be incorporated as a surface coating during liposomal preparation, or surface conjugation to the prepared liposomes via crosslinked lipid (Allen et al. 1991) or physically attached to the surface of the liposomal vesicle. This technique can be useful to enhance the affinity towards cancer cells (Hussein and Anderson 2004, Robert et al. 2004), infected tissues, and intracellular pathogens. For example, antibiotic ciprofloxacin encapsulated inside liposomes sheathed with polyoxyethylene glycol were successfully able to eradicate *Pseudomonas aeruginosa* (Bakker-Woudenberg et al. 2002), *Klebsiella pneumoniae* (Bakker-Woudenberg et al. 2001), and *Streptococcus pneumonia* (Ellbogen et al. 2003) in lung infected rat model. It was shown that the quinolone loaded stealth liposomes prolong the residence time in blood and lungs and lower the toxicity even at high doses. Similarly, prepared gentamicin stealth liposomes eradicate lung infection induced by *Klebsiella pneumoniae* in a rat model (Schiffelers et al. 2001). In intracellular pathogenic infection, isoniazid and rifampin stealth liposomes are more effective than free drugs against tuberculosis infected mice (when treated twice a week for six weeks). Also, liposome-encapsulated drugs eliminate mycobacteria from the liver and spleen more effectively than the free antimicrobial agents (Deol et al. 1997). Furthermore, intravenous co-administration of rifampicin and isoniazid as lung-specific stealth liposomes preparations to guinea pigs, prolonged the therapeutic levels of the antituberculosis drugs in plasma and organs up to seven days (Labana et al. 2002). This decrease in tuberculosis therapy dose regimen is associated with increased patient compliance and effective management of tuberculosis (TB). Favorably, the six weeks treatment course of a weekly dose significantly diminished the TB infection localized in the liver, spleen, and lungs, relative to the untreated guinea pig (Pandey et al. 2004).

Surface Charged Liposomes

Charging the liposomal surface also enhances the antimicrobial efficacy of the loaded drugs, which is mainly associated with the increase in the liposomal uptake by microbes. Cationic charged liposomes carry out ionic interaction with the negatively charged bacteria, facilitating the fusion with bacterial cell envelope, thus becoming more effective against bacterial infection. Ciprofloxacin-loaded cationic liposomes were more active in killing gram-negative isolates of *Klebsiella pneumoniae, Pseudomonas aeruginosa*, and *Escherichia coli* as compared with the free drug (Gubernator et al. 2007). In the same way, both meropenem and gentamicin loaded charged liposomes displayed enhancement in the antimicrobial action for beating gram-positive/gram-negative strains *(Pseudomonas aeruginosa, Klebsiella pneumoniae, Escherichia coli*, and *Staphylococcus aureus*)

(Drulis-Kawa et al. 2006a, b). Specifically, the enhancement in antimicrobial action was in the order of cationic liposomes > neutral liposomes > anionic liposomes > free antimicrobial (Drulis-Kawa et al. 2006a, b). Similarly, cationic ticarcillin-nanoliposomes exhibited higher encapsulation efficacies, elevated antipseudomonal potential, and a higher killing rate compared to the neutral and negatively charged preparations in the eradication of the pathogen from liver, spleen, and skin of the infected animals (Gharib et al. 2012). Furthermore, Gharib and colleagues prepared cationic nanoliposomal epigallocatechin gallate with high encapsulation efficiency and slow release of loaded antimicrobial agents (Gharib et al. 2013). Using skin infected burned mice models, these positively charged vesicles revealed the highest bactericidal efficacy and killing rate against *Staphylococcus aureus* strain (resistant to methicillin) compared to neutral and anionic nanoliposomes, as well as to the free antimicrobial agent. In the same instance, the cationic liposomes have more affinity to *Staphylococcus epidermidis* (Sanderson and Jones 1996) and *P. aeruginosa* (Dong et al. 2015) in the bacterial biofilm with disruption of the bacterial aggregates.

Immuno-liposomes

Interestingly, liposomes can also be used to target specific sites/organisms by attaching selective ligands to their surfaces. Various ligands can be used, such as complete specific monoclonal antibodies or antibody fragments (Manjappa et al. 2011). Such liposomes are called immuno-liposomes and were recently shown to be an effective carrier in targeting infected cells and microbial pathogens. In that approach, immuno-liposomes are highly effective against *Streptococcus oralis* (naming them anti-oralis liposomes), mainly by incorporating chlorhexidine and triclosan (Robinson et al. 2000, 1998). Maruyama et al. (1990) demonstrated the ability of these immuno-liposomes to bioaccumulate efficiently in a specific body organ and tissues in the experimental mice models. Immuno-liposomes were able to target the lung to deliver antimicrobial agents for the successful treatment of pneumonia. This targeting ability (lung bioaccumulation) is also enhanced by increasing the level of the conjugated protein to the used phospholipids. Also, the size of the prepared immuno-liposomes has a remarkable effect on antimicrobial targeting. Larger vesicles accumulate inside the lung more efficiently than small vesicles.

Surface Decorated Liposomes

Furthermore, decorating the outer liposome surface with lectins, lipids, polysaccharides, or peptides enables specific recognition of bacterial membranes and enhances the liposomal delivery to specific targets. Lectin decorated liposomes have been prepared as triclosan carriers, by Jones et al. (1993, 1994). In comparison with free triclosan, these liposomes exhibited significantly higher efficacy against skin biofilm-forming bacteria (such as *Proteus vulgaris* and *Staphylococcus. epidermidis*) and oral pathogens (such as *Streptococcus sanguis*) (Jones et al. 1993, 1994). The drug was adsorbed by the biofilm-associated bacteria and pierced this extracellular polymeric film and got directly distributed inside the bacteria (Jones et al. 1993, 1994). Likewise, metronidazole loaded lectin decorated liposomes were able to inhibit *Streptococcus mutans* biofilm maintained in the periodontal pocket in relation to metronidazole uncoated liposomes and the free drug (Vyas et al. 2001). In addition to that, the liposomal surface conjugated with a specific antigen was able to elicit the immune response. These liposomes were exploited in a unique vaccination approach. Liposomal vesicles could be attached to preparations with viral membranes and phosphatidylcholine to comprise the virosome. For example, hepatitis A virus vaccine, influenza virus vaccine, malaria vaccine, hepatitis B antigen (HBsAg) are considered safe, effective, and well-tolerated liposomal viral vaccines (Chang and Yeh 2012, López-Sagaseta et al. 2016).

Antimicrobial Liposomes

Liposomal β-lactams

Entrapment of ticarcillin in nanoliposomes was performed by the extrusion method, with lipid film of egg lecithin and cholesterol. The ticarcillin liposomes were then negatively or positively charged via the addition of dicetylphosphate and stearyl amine, respectively to the liposomal preparation. Cationic charged liposomal ticarcillin exhibited higher encapsulation efficacies (76%) compared to the encapsulation of anionic (43%) and neutral (55%) nano form, as measured by HPLC. Moreover, the cationic ticarcillin preparations were significantly more effective against *P. aeruginosa* with MIC 3.0 mg/L compared to the neutral, free, and anionic ticarcillin nanoliposome. Positive nanoliposome ticarcillin acted upon *Pseudomonas aeruginosa* with a killing rate of 100%. The *in vivo* animal study also showed that the animal survival rate was obviously correlated to the charge present on the surface of the obtained liposomes. Negatively, neutral, and positively charged ticarcillin liposomes showed 20, 60, and 100% effectiveness, respectively. Furthermore, the positively charged ticarcillin liposomes treated animals recovered with complete bacterial eradication in the liver, spleen, and skin. This is clearly explained by the presence of electrostatic interaction between cationic nanoliposome loaded-ticarcillin and the lipopolysaccharides of *P. aeruginosa* that facilitates the entrance of the ticarcillin through the bacterial outer membrane (Gharib et al. 2012). Liposomal meropenem was prepared by the thin phospholipid film hydration approach; the dry phospholipid layer was hydrated using an aqueous solution of meropenem (35 mg/mL) to prepare vesicles with the mean size 104–152 nm. Liposomal loaded meropenem showed activity against *P. aeruginosa*, *E. coli*, and *K. pneumonia* at or below the MICs of the free antibiotic (Gubernator et al. 2007). At the same instance, liposomal meropenem was also prepared using the thin lipid film method with the encapsulation efficiency of 1.16–3.7 percent. Cationic liposomes loaded meropenem revealed 2–4 folds lower MICs compared to meropenem encapsulated in the neutral and anionic formulations in muller Hinton broth medium. However, the antimicrobial activity of liposomes containing meropenem in serum medium was more than that of the free drug, which is related to the discontinuous fusion between the liposomal bilayers (phospholipids) and the cell membrane of bacteria. Therefore, a cationic liposomal formulation of meropenem could be more effective in topical preparations than the intravenous formulations (Drulis-Kawa et al. 2006a, b).

Liposomal Aminoglycosides

Liposomes entrapping aminoglycosides (such as gentamicin) improve the delivery of the drug to phagocytic cells and improve the intra-phagocytic eradication of various microbes, such as *Staphylococcus aureus* and *Brucella abortus* compared to free drug, as examined with bovine phagocytic cells. The antimicrobials have been successfully detected inside the phagocytic cells up to the third day of treatment with liposomal encapsulated aminoglycoside; however, the free aminoglycosides could not be detected on the third day of the treatment (Dees and Schultz 1990). Furthermore, Vitas and collaborators demonstrated that the cationic liposomes entrapped with gentamicin protect 70% of the murine acquired fatal infection with *Brucella abortus* strain (Vitas et al. 1997).

Using dehydration followed by rehydration of prepared liposomes, aminoglycosides (amikacin, gentamicin, and tobramycin) have been entrapped in liposomes (Kirby and Gregoriadis 1984). They used both cholesterol and dipalmitoylphosphatidylcholine (DPPC) in a molar ratio of 1:2 to prepare aminoglycoside encapsulated liposomes. These liposomes were able to reduce the MIC ($\leq$8 μg/mL) of *P. aeruginosa* isolates compared to the conventional drugs ($\geq$32 μg/mL). The liposomal preparation enhanced bacterial membrane fusion and improved antimicrobial uptake (Mugabe et al. 2006).

Sucrose coated lipid bilayers (DSPC and cholesterol) of the liposomal preparation improved the encapsulation efficiency of the loaded aminoglycosides. Nanosized liposomal aminoglycosides improve the entry of the antimicrobial agent inside the bacteria and improved the activity of the loaded aminoglycosides against resistant pathogens. For instance, tobramycin liposomes showed four-fold enhancement in the antimicrobial activity of loaded tobramycin against *Burkholderia cenocepacia* resistant strains during *in vitro* studies (Halwani et al. 2007). Alhariri and Omri (2013) also demonstrated the potentiated efficacy of tobramycin liposomes against *Pseudomonas aeruginosa* infected rat models. They enhanced the bacterial uptake through surface charging of the prepared liposomes by incorporating positively charged excipients, such as bismuth ethanedithiol. The formulated liposomes not only reduced bacterial count in the infected model rats, but also eliminated cellular communication and eliminated virulence factors, such as lipase, protease, and chitinase production. Similarly, liposomal N-acetylcysteine inhibited the pathogenicity and virulence of *Pseudomonas aeruginosa* (Hasanin and Omri 2014).

Gentamicin liposomes prepared from egg phosphatidylcholine, cholesterol, and oleic acid through the dehydration-rehydration technique, remarkably decrease the bactericidal gentamicin concentration against a resistant strain of *Staphylococcus aureus* to methicillin by 15-fold. Also, the combined liposomal preparation showed fractional inhibitory concentration index (FICI) of combination Gentamicin/oleic acid liposomes to be more effective in killing MRSA compared to the standard vancomycin (Atashbeyk et al. 2014). Furthermore, non-charged and negatively charged gentamicin liposomes showed enhanced antimicrobial activity against *examined strains of Klebsiella oxytoca and Pseudomonas aeruginosa*. The measured MIC (minimal inhibitory concentration) and MBC (minimal bactericidal concentration) revealed a significant lowering of the concentration compared to that of free gentamicin. In addition, gentamicin liposomes prevent and reduce the biofilm formation *in examined strains of Klebsiella oxytoca and Pseudomonas aeruginosa* (Alhariri et al. 2017).

Liposomal Colistin

Herein, the liposomal loading efficiency to colistin increased by using anionic lipid component, such as sodium cholesteryl sulfate, by raising the electrostatic attraction between colistin and the lipid membrane. These results highlight the suitability of applying an electrostatic attraction to entrap colistin in liposomes for pulmonary delivery for increasing colistin retention in the lungs. The entrapped colistin exhibited prolonged colistin retention in the lung, with less colistin being transferred to the bloodstream and kidneys, and the improved bio-distribution further resulted in the enhanced therapeutic effect in mice infected with a pulmonary isolate of *P. aeruginosa*, compared to the colistin solution (Li et al. 2017). The co-formulation of liposomal azithromycin/colistin elicited encapsulation efficiency of azithromycin > 98% and improved the release of azithromycin up to 30%. Hence, it enabled the administration of both antibiotics as inhalation therapy for treating multidrug resistant pathogens (Wallace et al. 2013). Polymyxin B liposomes prepared using DPPC and cholesterol also revealed the dramatic increase in antimicrobial efficacy and decrease in the adverse effect of polymyxin. This is related to the improved penetration of polymyxin B liposomes into resistant cells of *Pseudomonas aeruginosa* strains relative to the free polymyxin B (Alipour et al. 2008). *In vivo* activity of polymyxin B liposomes were also investigated by He and his research group (He et al. 2013). The liposomal polymyxin B was firstly prepared using modified reversed-phase evaporation procedure using DPPC and cholesterol as the phospholipid content. Liposomal polymyxin caused a significant lowering of the bacterial count in animal lungs infected with multi-drug resistant *Pseudomonas aeruginosa* with an elevated level of the drug into the animal lungs. In addition, intravenous administration of the liposomal polymyxin preparation prolonged the survival of infected mice (He et al. 2013).

Liposomal Macrolides

Azithromycin liposomes were also prepared, using dehydration-rehydration technique, to enhance the antimicrobial activity against *Pseudomonas aeruginosa*. The liposomal phospholipids were expected to interact with the bacterial membrane, and facilitate the introduction of azithromycin inside the bacterial cytoplasm (Solleti et al. 2014). Liposomal azithromycin preparation significantly eradicated the bacteria and successfully prevented bacterial biofilm formation. Meanwhile, azithromycin liposomes attenuate the bacterial ability for the production of different virulence and resistant factors with restriction of cellular motility. The recognized cytotoxicity and hemolytic side effect of azithromycin were also minimized. In another study done by Alhajlan et al. (2013), the clarithromycin liposomes with various charges have been formulated using the dehydration-rehydration technique. The entrapment efficiency of these preparations was up to 30% and 70% of the used drug during the preparation.

The formulation shows significant activity against resistant *P. aeruginosa* isolates. Moreover, the clarithromycin-entrapped liposomes exhibited a pronounced effect on pseudomonas virulence factors. Either charged or neutral preparations reduced elastase and protease production relative to that of free clarithromycin (Alhajlan et al. 2013). Recently, Li et al. (2017) were also able to formulate the clarithromycin liposomes, but with the use of ultra-sonic freeze spray drying technique. They use both mannitol and sucrose as cryoprotectants and co-cryoprotectants at concentrations of 15 and 5% w/v, respectively. The dry liposomal powder so obtained was in the form of a porous fluffy cake structure, constituting narrow size distribution microparticles with high clarithromycin recovery and uniform clarithromycin content (Li et al. 2017).

Liposomal Rifampicin

Rifampicin liposomes were also prepared using stearic acid, stearyl amine, and precirol. Using the solid lipid nano-formulation, 0.013% rifampin was mixed with the lipid content and sodium chloride solution (0.9% w/v) as an isotonic vehicle. The liposomal preparation had particle sizes 200–300 nm, and encapsulation efficiency of 70%. Free rifampin nanoform was significantly effective against mature *Staphylococcus epidermidis* compared to the free drug. Furthermore, rifampicin loaded formulations were effective in the delivery of the antibiotic to the bacterial interfaces, causing a remarkable reduction in viability of bacteria embedded in biofilm (Changsan et al. 2009). The same data was obtained by Moghadas-Sharif and collaborators, who reported the liposomal synthesis of anionic, cationic, and PEGylated rifampin liposomes using rehydration dehydration method with particle sizes 145, 134, and 142 nm, respectively. The encapsulation efficacy of rifampin was 60 percent. Cationic rifampin nanoliposome eradicated biofilm of *S. epidermidis* more efficiently compared to the anionic preparation, which could be attributed to the enhanced absorption of the cationic preparation due to electrostatic attraction to the negatively charged bacteria (Moghadas-Sharif et al. 2015).

Liposomal Polyene (Antifungal Agents)

Another application of nanoliposomes in antimicrobial administration is the liposomal polyenes. Amphotericin B belongs to the polyene class of fungicidal medications that has a broad spectrum effect against invasive fungal infection and Leishmania parasite. Amphotericin B works mainly on the fungus membrane through binding with its ergosterol molecules, resulting in cellular perforation, cellular electrolyte loss, and consequent fungus death. The initially marketed preparation was the formulation of amphotericin B (Amp) as a deoxycholate salt, however, nephrotoxicity and diffusion-related toxicity limited its administration for decades. However, liposomal amphotericin B formulation (LAmB) significantly reduced its toxicity and retained its potent antifungal effects (Wingard et al. 2000). Amphotericin B liposomes successfully demonstrate their efficacy as antifungal lock treatment against fungal biofilm produced by *Candida albicans* (Schinabeck et al. 2004).

Effective encapsulation of polyene macrolides, such as nystatin and amphotericin B was accompanied by the development of injectable dosage forms. Enhanced encapsulation of amphotericin B has been achieved through distearoylglycerophosphoethanolamine-polyethyleneglycol (DSPE-PEG) by hydration of aqueous sucrose solution (9% w/v). This preparation improves the physicochemical stability and residence time in plasma. Also, AmB-encapsulating PEG liposomes (PEG-L-AmB) were less toxic and more effective against aspergillus infection in the pulmonary murine model compared to that of commonly administered amphotericin B formulations. More improvement in the antifungal efficacy was shown by monoclonal antibodies decorated liposomes loaded with amphotericin B. These immuno-liposomes were designed by conjugating polyethylene glycol chain at its terminal with 34A antibodies (Moribe and Maruyama 2002). The nystatin entrapment efficiency of the dipalmitoylphosphatidylcholine liposomes has been increased through the incorporation of cholesterol (70%). Prepared liposomes show physical stability and elevated antifungal activity compared to the free colistin against *Candida albicans* (Saadat et al. 2016).

FUTURE PERSPECTIVES FOR ANTIMICROBIAL LIPOSOMES

The aforementioned research results have certainly established the case for antimicrobial liposomes to get clinical approval and be introduced in the market as a commercial pharmaceutical product in the near future. Formulation of antimicrobial agents as liposomes demonstrates superior efficacy in the treatment of resistant pathogens compared to the conventional forms of antimicrobial treatments. The ability of liposomes to attain longer residence time in plasma with limited uptake by the mononuclear phagocyte system opens up new areas of treatment for severe and chronic infections, with safety, efficacy, and accuracy. The liposomal formulation retains the activity of the loaded antimicrobial agent against degrading enzymes produced by resistant isolates. In addition, the sustained release of the antimicrobial agent provides prolonged activity and stability. Nonetheless, it cannot be overlooked that two remaining challenges have to be completely overcome, in order for these liposomal formulations to be clinically established and commercially available. The first is biocompatibility and long-term safety after systemic administration. Rigorous clinical assessment for the complete biological response through clinically relevant testing is still needed. The second is the prospect of mass production with minimal batch to batch variation in the pharmaceutical manufacturing pipelines. Research investigations to meet these two challenges are still ongoing (Deol et al. 1997, Schiffelers et al. 2001, Ebato et al. 2003, Chang and Yeh 2012, He et al. 2013). The high investment cost of generating new antimicrobials will impel the pharmaceutical companies to conduct clinical studies on the liposomal delivery for the existing antimicrobials in addressing the emerging problem of microbial resistance.

References

Akbarzadeh, A., R. Rezaei-Sadabady, S. Davaran, S.W. Joo, N. Zarghami, Y. Hanifehpour, et al. 2013. Liposome: Classification, preparation, and applications. Nanoscale Res. Lett. 8: 102.

Alhajlan, M., M. Alhariri and A. Omri. 2013. Efficacy and safety of liposomal clarithromycin and its effect on *Pseudomonas aeruginosa* virulence factors. Antimicrob. Agents Chemother. 57: 2694–2704.

Alhariri, M. and A. Omri. 2013. Efficacy of liposomal bismuth-ethanedithiol-loaded tobramycin after intratracheal administration in rats with pulmonary *Pseudomonas aeruginosa* infection. Antimicrob. Agents Chemother. 57: 569–578.

Alhariri, M., M.A. Majrashi, A.H. Bahkali, F.S. Almajed, A.O. Azghani, M.A. Khiyami, et al. 2017. Efficacy of neutral and negatively charged liposome-loaded gentamicin on planktonic bacteria and biofilm communities. Int. J. Nanomedicine. 12: 6949–6961.

Alipour, M., M. Halwani, A. Omri and Z.E. Suntres. 2008. Antimicrobial effectiveness of liposomal polymyxin B against resistant Gram-negative bacterial strains. Int. J. Pharm. 355: 293–298.

Allen, T., C. Hansen, F. Martin, C. Redemann and A. Yau-Young. 1991. Liposomes containing synthetic lipid derivatives of poly (ethylene glycol) show prolonged circulation half-lives *in vivo*. Biochim. Biophys. Acta. 1066: 29–36.

Anwekar, H., S. Patel and A. Singhai. 2011. Liposome-as drug carriers. Int. J. of Pharm. & Life Sci. (IJPLS). 2(7): 945–951.

Atashbeyk, D.G., B. Khameneh, M. Tafaghodi and B.S. Fazly Bazzaz. 2014. Eradication of methicillin-resistant *Staphylococcus aureus* infection by nanoliposomes loaded with gentamicin and oleic acid. Pharm. Biol. 52: 1423–1428.

Bakker-Woudenberg, I.A., G. Storm and M.C. Woodle. 1994. Liposomes in the treatment of infections. J. Drug Target 2: 363–371.

Bakker-Woudenberg, I.A., T. Marian, L. Guo, P. Working and J.W. Mouton. 2001. Improved efficacy of ciprofloxacin administered in polyethylene glycol-coated liposomes for treatment of *Klebsiella pneumoniae* pneumonia in rats. Antimicrob. Agents Chemother. 45: 1487–1492.

Bakker-Woudenberg, I.A., M.T. ten Kate, L. Guo, P. Working and J.W. Mouton. 2002. Ciprofloxacin in polyethylene glycol-coated liposomes: Efficacy in rat models of acute or chronic *Pseudomonas aeruginosa* infection. Antimicrob. Agents Chemother. 46: 2575–2581.

Bordi, F., C. Cametti and S. Sennato. 2006. Advances in PlanarLipid Bilayers and Liposomes. Academic Press, London.

Brier, B. 2004. Infectious diseases in ancient Egypt. Infect. Dis. Clin. North Am. 18: 17–27.

Campardelli, R., I.E. Santo, E.C. Albuquerque, S.V. de Melo, G. Della Porta and E. Reverchon. 2016. Efficient encapsulation of proteins in submicro liposomes using a supercritical fluid assisted continuous process. J. Supercrit. Fluids. 107: 163–169.

Campardelli, R., P. Trucillo and E. Reverchon. 2018. Supercritical assisted process for the efficient production of liposomes containing antibiotics for ocular delivery. J. CO2 Util. 25: 235–241.

Castoldi, A., C. Herr, J. Niederstraßer, H.I. Labouta, A. Melero, S. Gordon, et al. 2017. Calcifediol-loaded liposomes for local treatment of pulmonary bacterial infections. Eur. J. Pharm. Biopharm. 118: 62–67.

Chang, H.-I. and M.-K. Yeh. 2012. Clinical development of liposome-based drugs: Formulation, characterization and therapeutic efficacy. Int. J. Nanomedicine. 7: 49.

Changsan, N., H.K. Chan, F. Separovic and T. Srichana. 2009. Physicochemical characterization and stability of rifampicin liposome dry powder formulations for inhalation. J. Pharm. Sci. 98: 628–639.

Colzi, I., A.N. Troyan, B. Perito, E. Casalone, R. Romoli, G. Pieraccini, et al. 2015. Antibiotic delivery by liposomes from prokaryotic microorganisms: Similia cum similis works better. Eur. J. Pharm. Biopharm. 94: 411–418.

Dees, C. and R.D. Schultz. 1990. The mechanism of enhanced intraphagocytic killing of bacteria by liposomes containing antibiotics. Vet. Immunol. Immunopathol. 24: 135–146.

Deol, P., G. Khuller and K. Joshi. 1997. Therapeutic efficacies of isoniazid and rifampin encapsulated in lung-specific stealth liposomes against *Mycobacterium tuberculosis* infection induced in mice. Antimicrob. Agents Chemother. 41: 1211–1214.

Dong, D., N. Thomas, B. Thierry, S. Vreugde, C.A. Prestidge and P.-J. Wormald. 2015. Distribution and inhibition of liposomes on *Staphylococcus aureus* and *Pseudomonas aeruginosa* Biofilm. PLoS One. 10: e0131806.

Drulis-Kawa, Z., J. Gubernator, A. Dorotkiewicz-Jach, W. Doroszkiewicz and A. Kozubek. 2006a. A comparison of the *in vitro* antimicrobial activity of liposomes containing meropenem and gentamicin. Cell Mol. Biol. Lett. 11: 360.

Drulis-Kawa, Z., J. Gubernator, A. Dorotkiewicz-Jach, W. Doroszkiewicz and A. Kozubek. 2006b. *In vitro* antimicrobial activity of liposomal meropenem against *Pseudomonas aeruginosa* strains. Int. J. Pharm. 315: 59–66.

Drulis-Kawa, Z. and A. Dorotkiewicz-Jach. 2010. Liposomes as delivery systems for antibiotics. Int. J. Pharm. 387: 187–198.

Ebato, Y., Y. Kato, H. Onishi, T. Nagai and Y. Machida. 2003. *In vivo* efficacy of a novel double liposome as an oral dosage form of salmon calcitonin. Drug Dev. Res. 58: 253–257.

Ellbogen, M.H., K.M. Olsen, M.J. Gentry-Nielsen and L.C. Preheim. 2003. Efficacy of liposome-encapsulated ciprofloxacin compared with ciprofloxacin and ceftriaxone in a rat model of pneumococcal pneumonia. J. Antimicrob. Chemother. 51: 83–91.

Gharib, A., Z. Faezizadeh and M. Godarzee. 2012. *In vitro* and *in vivo* activities of ticarcillin-loaded nanoliposomes with different surface charges against *Pseudomonas aeruginosa* (ATCC 29248). DARU J. Pharm. Sci. 20: 41.

Gharib, A., Z. Faezizadeh and M. Godarzee. 2013. Therapeutic efficacy of epigallocatechin gallate-loaded nanoliposomes against burn wound infection by methicillin-resistant *Staphylococcus aureus*. Skin Pharmacol. Phys. 26:68–75.

Goyal, P., K. Goyal, S.V. Kumar, A. Singh, O. Katare and D.N. Mishra. 2005. Liposomal drug delivery systems–clinical applications. Acta Pharm. 55: 1–25.

Grant, G.J., Y. Barenholz, E.M. Bolotin, M. Bansinath, H. Turndorf, B. Piskoun, et al. 2004. A novel liposomal bupivacaine formulation to produce ultralong-acting analgesia. Anesthesiology. 101: 133–137.

Gregoriadis, G. 1995. Engineering liposomes for drug delivery: Progress and problems. Trends Biotechnol. 13: 527–537.

Gubernator, J., Z. Drulis-Kawa, A. Dorotkiewicz-Jach, W. Doroszkiewicz and A. Kozubek. 2007. *In vitro* antimicrobial activity of liposomes containing ciprofloxacin, meropenem and gentamicin against gram-negative clinical bacterial strains. Lett. Drug Des. Discov. 4: 297–304.

Gupta, P.V., A.M. Nirwane, T. Belubbi and M.S. Nagarsenker. 2017. Pulmonary delivery of synergistic combination of fluoroquinolone antibiotic complemented with proteolytic enzyme: A novel antimicrobial and antibiofilm strategy. Nanomedicine. 13: 2371–2384.

Halwani, M., C. Mugabe, A.O. Azghani, R.M. Lafrenie, A. Kumar and A. Omri. 2007. Bactericidal efficacy of liposomal aminoglycosides against *Burkholderia cenocepacia*. J. Antimicrob. Chemother. 60: 760–769.

Halwani, M., S. Blomme, Z.E. Suntres, M. Alipour, A.O. Azghani, A. Kumar, et al. 2008. Liposomal bismuth-ethanedithiol formulation enhances antimicrobial activity of tobramycin. Int. J. Pharm. 358: 278–284.

Hasanin, A. and A. Omri. 2014. Liposomal N-acetylcysteine modulates the pathogenesis of *P. aeruginosa* isolated from the lungs of cystic fibrosis patient. J. Nanomed. Nanotechnol. 5: 1.

He, J., K. Abdelraouf, K.R. Ledesma, D.S.-L. Chow and V.H. Tam. 2013. Pharmacokinetics and efficacy of liposomal polymyxin B in a murine pneumonia model. Int. J. Antimicrob. Agents. 42: 559–564.

Hoiby, N., T. Bjarnsholt, M. Givskov, S. Molin and O. Ciofu. 2010. Antibiotic resistance of bacterial biofilms. Int. J. Antimicrob. Agents. 35: 322–332.

Hussein, M.A. and K.C. Anderson. 2004. Role of liposomal anthracyclines in the treatment of multiple myeloma. Semin. Oncol. 31: 147–160.

Jones, M.N., S.E. Francis, F.J. Hutchinson, P.S. Handley and I.G. Lyle. 1993. Targeting and delivery of bactericide to adsorbed oral bacteria by use of proteoliposomes. Biochim. Biophys. Acta. 1147: 251–261.

Jones, M.N., M. Kaszuba, M.D. Reboiras, I.G. Lyle, K.J. Hill, Y.-H. Song, et al. 1994. The targeting of phospholipid liposomes to bacteria. Biochim. Biophys. Acta. 1196: 57–64.

Juang, V., H.-P. Lee, A.M.-Y. Lin and Y.-L. Lo. 2016. Cationic PEGylated liposomes incorporating an antimicrobial peptide tilapia hepcidin 2–3: An adjuvant of epirubicin to overcome multidrug resistance in cervical cancer cells. Int. J. Nanomedicine. 11: 6047–6064.

Katayama, K., Y. Kato, H. Onishi, T. Nagai and Y. Machida. 2003. Double liposomes: Hypoglycemic effects of liposomal insulin on normal rats. Drug Dev. Ind. Pharm. 29: 725–731.

Kirby, C. and G. Gregoriadis. 1984. Dehydration-rehydration vesicles: A simple method for high yield drug entrapment in liposomes. Bio/Technology. 2: 979–984.

Labana, S., R. Pandey, S. Sharma and G. Khuller. 2002. Chemotherapeutic activity against murine tuberculosis of once weekly administered drugs (isoniazid and rifampicin) encapsulated in liposomes. Int. J. Antimicrob. Agents. 20: 301–304.

Laouini, A., C. Jaafar-Maalej, I. Limayem-Blouza, S. Sfar, C. Charcosset and H. Fessi. 2012. Preparation, characterization and applications of liposomes: state of the art. J. Colloid. Sci. 1: 147–168.

Li, Y., C. Tang, E. Zhang and L. Yang. 2017. Electrostatically entrapped colistin liposomes for the treatment of *Pseudomonas aeruginosa* infection. Pharm. Dev. Technol. 22: 436–444.

Liu, J., Z. Wang, F. Li, J. Gao, L. Wang and G. Huang. 2015. Liposomes for systematic delivery of vancomycin hydrochloride to decrease nephrotoxicity: Characterization and evaluation. Asian J. Pharm. Sci. 10: 212–222.

López-Sagaseta, J., E. Malito, R. Rappuoli and M.J. Bottomley. 2016. Self-assembling protein nanoparticles in the design of vaccines. Comput. Struct. Biotechnol. J. 14: 58–68.

Ma, Y., Z. Wang, W. Zhao, T. Lu, R. Wang, Q. Mei, et al. 2013. Enhanced bactericidal potency of nanoliposomes by modification of the fusion activity between liposomes and bacterium. Int. J. Nanomedicine. 8: 2351–2360.

Manjappa, A.S., K.R. Chaudhari, M.P. Venkataraju, P. Dantuluri, B. Nanda, C. Sidda, et al. 2011. Antibody derivatization and conjugation strategies: application in preparation of stealth immunoliposome to target chemotherapeutics to tumor. J. Control. Release. 150: 2–22.

Maruyama, K., E. Holmberg, S.J. Kennel, A. Klibanov, V.P. Torchilin and L. Huang. 1990. Characterization of *in vivo* immunoliposome targeting to pulmonary endothelium. J. Pharm. Sci. 79: 978–984.

Moghadas-Sharif, N., B.S. Fazly Bazzaz, B. Khameneh and B. Malaekeh-Nikouei. 2015. The effect of nanoliposomal formulations on *Staphylococcus epidermidis* biofilm. Drug Dev. Ind. Pharm. 41: 445–450.

Moribe, K. and K. Maruyama. 2002. Pharmaceutical design of the liposomal antimicrobial agents for infectious disease. Curr. Pharm. Des. 8: 441–454.

Mozafari, M.R. 2005. Liposomes: An overview of manufacturing techniques. Cell. Mol. Biol. Lett. 10: 711–719.

Mugabe, C., M. Halwani, A.O. Azghani, R.M. Lafrenie and A. Omri. 2006. Mechanism of enhanced activity of liposome-entrapped aminoglycosides against resistant strains of *Pseudomonas aeruginosa*. Antimicrob. Agents Chemother. 50: 2016–2022.

Mui, B., L. Chow and M.J. Hope. 2003. Extrusion technique to generate liposomes of defined size. Methods Enzymol. 367: 3–14.

Nayar, R., M.J. Hope and P.R. Cullis. 1989. Generation of large unilamellar vesicles from long-chain saturated phosphatidylcholines by extrusion technique. Biochim. Biophys. Acta. 986: 200–206.

Pandey, R., S. Sharma and G. Khuller. 2004. Lung specific stealth liposomes as antitubercular drug carriers in guinea pigs. Indian J. Exp. Biol. 42: 562–566.

Robert, N.J., C.L. Vogel, I.C. Henderson, J.A. Sparano, M.R. Moore, P. Silverman, et al. 2004. The role of the liposomal anthracyclines and other systemic therapies in the management of advanced breast cancer. Semin. Oncol. 31(6 Suppl 13): 106–146.

Robinson, A.M., J.E. Creeth and M.N. Jones. 1998. The specificity and affinity of immunoliposome targeting to oral bacteria. Biochim Biophys Acta. 1369: 278–286.

Robinson, A.M., J.E. Creeth and M.N. Jones. 2000. The use of immunoliposomes for specific delivery of antimicrobial agents to oral bacteria immobilized on polystyrene. J. Biomater. Sci. Polym. Ed. 11: 1381–1393.

Robinson, A.M., M. Bannister, J.E. Creeth and M.N. Jones. 2001. The interaction of phospholipid liposomes with mixed bacterial biofilms and their use in the delivery of bactericide. Colloids Surf. A Physicochem. Eng. Asp. 186: 43–53.

Saadat, E., R. Dinarvand and P. Ebrahimnejad. 2016. Encapsulation of nystatin in nanoliposomal formulation: Characterization, stability study and antifungal activity against *Candida albicans*. Pharmaceutical and Biomedical Research. 2: 44–54.

Sabbahy, L. 2017. Infectious diseases in ancient Egypt. Lancet Infect. Dis. 17: 594.

Sachetelli, S., H. Khalil, T. Chen, C. Beaulac, S. Sénéchal and J. Lagacé. 2000. Demonstration of a fusion mechanism between a fluid bactericidal liposomal formulation and bacterial cells. Biochim. Biophys. Acta. 1463: 254–266.

Sanderson, N.M. and M.N. Jones. 1996. Targeting of cationic liposomes to skin-associated bacteria. Pestic. Sci. 46: 255–261.

Schiffelers, R.M., G. Storm, T. Marian, L.E. Stearne-Cullen, J.G. den Hollander, H.A. Verbrugh, et al. 2001. *In vivo* synergistic interaction of liposome-coencapsulated gentamicin and ceftazidime. J. Pharmacol. Exp. Ther. 298: 369–375.

Schinabeck, M.K., L.A. Long, M.A. Hossain, J. Chandra, P.K. Mukherjee, S. Mohamed, et al. 2004. Rabbit model of *Candida albicans* biofilm infection: Liposomal amphotericin B antifungal lock therapy. Antimicrob Agents Chemother. 48: 1727–1732.

Sharma, A. and U.S. Sharma. 1997. Liposomes in drug delivery: Progress and limitations. Int. J. Pharm. 154: 123–140.

Solleti, V.S., M. Alhariri, M. Halwani and A. Omri. 2014. Antimicrobial properties of liposomal azithromycin for *Pseudomonas infections* in cystic fibrosis patients. J. Antimicrob. Chemother. 70: 784–796.

Stano, P., S. Bufali, C. Pisano, F. Bucci, M. Barbarino, M. Santaniello, et al. 2004. Novel camptothecin analogue (gimatecan)-containing liposomes prepared by the ethanol injection method. J. Liposome. Res. 14: 87–109.

Szoka, Jr, F. and D. Papahadjopoulos. 1980. Comparative properties and methods of preparation of lipid vesicles (liposomes). Annu. Rev. Biophys. Bioeng. 9: 467–508.

Tseng, B.S., W. Zhang, J.J. Harrison, T.P. Quach, J.L. Song, J. Penterman, et al. 2013. The extracellular matrix protects *Pseudomonas aeruginosa* biofilms by limiting the penetration of tobramycin. Environ. Microbiol. 15: 2865–2878.

Vitas, A.I., R. Diaz and C. Gamazo. 1997. Protective effect of liposomal gentamicin against systemic acute murine brucellosis. Chemotherapy. 43: 204–210.

Vyas, S., V. Sihorkar and P. Dubey. 2001. Preparation, characterization and *in vitro* antimicrobial activity of metronidazole bearing lectinized liposomes for intra-periodontal pocket delivery. Die Pharmazie. 56: 554–560.

Wallace, S.J., R.L. Nation, J. Li and B.J. Boyd. 2013. Physicochemical aspects of the coformulation of colistin and azithromycin using liposomes for combination antibiotic therapies. J. Pharm. Sci. 102: 1578–1587.

Wingard, J.R., M.H. White, E. Anaissie, J. Raffalli, J. Goodman and A. Arrieta. 2000. A randomized, double-blind comparative trial evaluating the safety of liposomal amphotericin B versus amphotericin B lipid complex in the empirical treatment of febrile neutropenia. L Amph/ABLC Collaborative Study Group. Clin. Infect. Dis. 31: 1155–1163.

Wu, P.-C., Y.-H. Tsai, C.-C. Liao, J.-S. Chang and Y.-B. Huang. 2004. The characterization and biodistribution of cefoxitin-loaded liposomes. Int. J. Pharm. 271:31–39.

Wu, F., S.G. Bhansali, M. Tamhane, R. Kumar, L.A. Vathy, H. Ding, et al. 2012. Noninvasive real-time fluorescence imaging of the lymphatic uptake of BSA–IRDye 680 conjugate administered subcutaneously in mice. J. Pharm. Sci. 101: 1744–1754.

Yang, Z., J. Liu, J. Gao, S. Chen and G. Huang. 2015. Chitosan coated vancomycin hydrochloride liposomes: Characterizations and evaluation. Int. J. Pharm. 495: 508–515.

Yoneyama, H. and R. Katsumata. 2006. Antibiotic resistance in bacteria and its future for novel antibiotic development. Biosci. Biotechnol. Biochem. 70: 1060–1075.

Index